Plant Succession and Vegetation Dynamics

Contents

diversity

MDPI

Editorial

Vegetation Changes in Space and Time—A Special Issue on Plant Succession and Vegetation Dynamics

Thomas Fickert [1,*] and Michael Richter [2]

[1] German Alpine Club, Baden-Wuerttemberg Branch, Nature Conservation Unit, Fritz-Walter-Weg 19, 70372 Stuttgart, Germany
[2] Institute for Geography, Friedrich-Alexander-University, Kochstraße 4/4, 91054 Erlangen, Germany; sairecabur@posteo.de
* Correspondence: thomas.fickert@posteo.de

1. Introduction

Plant cover on Earth is far from static. Rather it is subject to continuous alterations on different spatial and temporal scales. Changes can be very obvious or barely noticeable [1], they can be gradual or abrupt [2,3], they can be locally restricted or extend over large areas [4], they can be running over different periods of time, i.e., from days or weeks to millions of years [5], they might include variation in plant physiological processes or shifts of entire biomes [6], and they can be linked to a variety of drivers [7]. These processes are commonly subsumed under the term vegetation dynamics and cover a wide spatio-temporal range [8–13], from local or seasonal changes (symphenology [14], i.e., at a recurring rhythm linked to climate) to secular developments on a spatially large and temporally very long-term scale (synchronology [15]) controlled by vegetation history, e.g., the post-glacial vegetation development. Between the two extremes, cyclic processes (regeneration) and processes of directional change (succession) are arranged. they are indicative of changes over time in species populations, communities, and whole ecosystems, for example, on new substrates or after disturbances. While succession is one of the most fundamental and longest-studied concepts in plant community ecology [5], a comprehensive successional theory is still lacking [16,17]. The processes and patterns of succession and vegetation dynamics are common subjects in the field of biogeography, aimed at recording and explaining the distribution of species, population communities, and ecosystems, as well as their changes in space and time [18]. In times of substantial global changes, this topic is more relevant than ever, to better predict how ecosystems will respond to both land-use and climate change, and to improve the design of promising ecosystem restoration strategies [13,16,17]. This paper aims to review the existing concepts and models on types, trajectories, and processes recognized within vegetation dynamics [19,20], and to put the papers published in the MDPI *Diversity* Special Issue "Plant Succession and Vegetation Dynamics" into a contextual and conceptual framework. The Special Issue was intended to offer a platform to showcase recent findings in the study of plant succession and vegetation dynamics. with case studies from various plant communities, starting from different origins, covering different spatial and temporal scales, and employing different research methods, this SI provides a nice compilation of recent studies on a classical yet highly relevant object.

2. Views and Hypotheses on Vegetation Dynamics over Time

The early records of (bio)geographical observation date back to Greek scholars like Theoprastus and Aristotle in 300 BC (for a detailed overview on the science history of bio-

https://doi.org/10.3390/d17070482

geography, see [18,21]). Perceptions of nature at that time differed widely, from a virtually static nature (equilibrium) to a dynamic one (non-equilibrium). During the Renaissance, European scientists started studying biogeographical foundations, which basically followed two fundamentally different biogeographical paths, a historical–evolutionary and an ecologically oriented one (Figure 1, [18]).

Two important representatives of the formerwere Charles Darwin (1809–1882) and Alfred Russel Wallace (1823–1913), whose insights through their exceptional journeys challenged the creationist view of the unchangeability of nature [22,23]. The recognition of cyclical climate fluctuations [24,25] and the drift of the lithospheric plates [26] subsequently underpinned their visionary views. While the historical–evolutionary biogeography made long-lasting evolutionary processes responsible for the species composition and distribution, main drivers within an ecologically oriented biogeography at that time were the prevailing environmental conditions. Introducing an ecological approach for explaining the distribution of species and communities, Alexander von Humboldt (1796–1859) is regarded as the founder of modern ecological biogeography ([27], Figure 1).

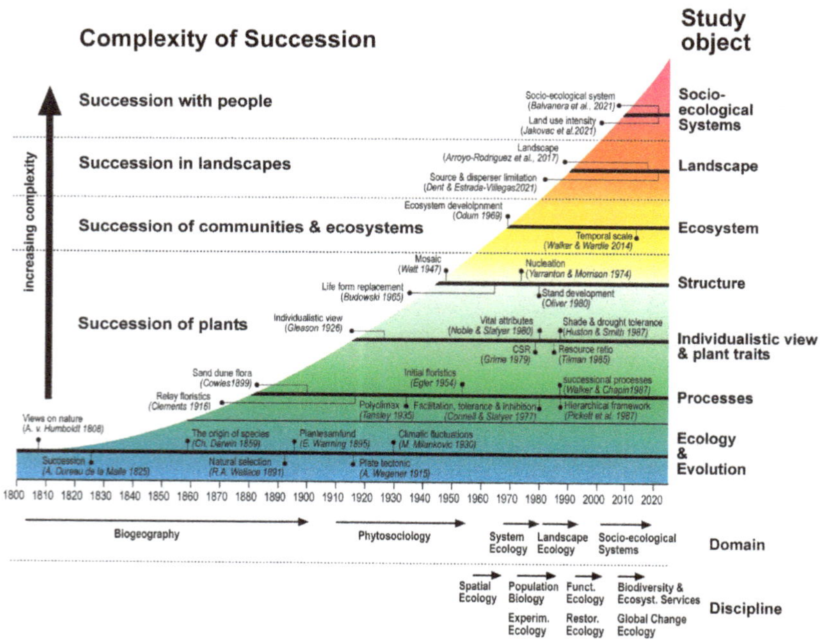

Figure 1. The development of successional models with time. Fundamentals come from early accounts of biogeographical pioneers ([22,23,25–29], in blue); during the last century and a half, many different models evolved (each of these continues to develop) that might be arranged into four groups—succession of plants ([30–45] in green), succession of communities and ecosystems [5,46], (in yellow), succession in landscapes ([47,48], in orange), and succession with people ([49,50], in red), successively increasing in complexity. Date of milestone publications for the different groups of models are indicated by the pin including publication year and authors' name. The arrows below the timeline indicate the domain, discipline, and time of development (adapted and modified from [16]).

Based on these early biogeographical accounts, the concept of succession (from Latin, "succession" meaning progress) of plants was developed during the 20th century, despite the basic idea being introduced earlier by the French naturalist Adolphe Dureau de la Malle [28], who described the vegetation development after a forest clearcut. The term succession, however, was not yet used. Later, the Danish botanist Eugen Warming described

Plant Succession and Vegetation Dynamics

Guest Editors

Thomas Fickert
Michael Richter

MDPI

Basel • Beijing • Wuhan • Barcelona • Belgrade • Novi Sad • Cluj • Manchester

Guest Editors
Thomas Fickert
Faculty of Arts and Humanities
University of Passau
Passau
Germany

Michael Richter
Institute for Geography
Friedrich-Alexander-University
Erlangen
Germany

Editorial Office
MDPI AG
Grosspeteranlage 5
4052 Basel, Switzerland

This is a reprint of the Special Issue, published open access by the journal *Diversity* (ISSN 1424-2818), freely accessible at: https://www.mdpi.com/journal/diversity/special_issues/E6YS80O3G1.

For citation purposes, cite each article independently as indicated on the article page online and as indicated below:

Lastname, A.A.; Lastname, B.B. Article Title. *Journal Name* **Year**, *Volume Number*, Page Range.

ISBN 978-3-7258-5123-2 (Hbk)
ISBN 978-3-7258-5124-9 (PDF)
https://doi.org/10.3390/books978-3-7258-5124-9

Cover image courtesy of Thomas Fickert

About the Editors

Thomas Fickert

Thomas Fickert is currently working in the Nature Conservation Unit of the German Alpine Club, in the Baden-Württemberg branch. He finished a PhD in vegetation geography at the Friedrich Alexander University of Erlangen-Nuremberg with Micheal Richter as his supervisor. After completing the PhD, he worked between 2004 and 2018 as a research assistant at the Chair of Physical Geography at the University of Passau. His research interest is on vegetation dynamics in general (e.g., a project on mangrove regeneration in the Caribbean after a severe hurricane impact) and particularly on plant succession in glacier forelands in the Eastern Alps (ongoing project).

Michael Richter

Michael Richter held a professorship at the Institute of Geography, FAU Erlangen, until he retired in 2011. His research interests include plant geography, climate ecology, and high mountain geography, with a regional focus on the Andes, Mediterranean region, and Central Asia. Finished projects were located amongst others in Chiapas (Mexico), Loja (Ecuador), Atacama (Chile), Southwestern USA, Ticino (Switzerland), and Italy. Two long-term monitoring projects dealt with the long-term effects of Super El Niño 1997/98 and subsequent ones dealt with vegetation development in the Piura region (Northwestern Peru) and with the interrelation of land-use change and effects for natural hazard risk in the Sierra Madre de Chiapas (Southern Mexico).

the vegetation development on sand dunes on the coast of Denmark in his work Plantesamfund [29]. His interpretation considerably influenced two American contemporaries, Henry Cowles and Frederik Clements, who are generally regarded as the forefathers of the scientific concept of succession, although without any doubt Warming deserves that honor. In a broader understanding, all vegetation changes in a certain direction via various stages, and the underlying processes are referred to as successions.

Cowles interpreted the vegetation of differently aged sand dunes as stages of a directional vegetation sequence [30], so he can be credited to be the founder of the chronosequence approach or space for time substitution in the sense of Pickett's [51], where a temporal sequence is derived from a spatial juxtaposition of different locations [52]. Clements instead became famous for his mono-climax concept [31,53], which ever since has been disputed. According to his conception, succession always develops towards a climate-controlled final plant association called climax, and colonization of open ground is taking place through the immigration of species from the surrounding area (primary succession) or due to a diaspore bank already present on the site (secondary succession). Schematically, he divided the controlling processes of succession into the following steps [9,31]:

1. Nudation (disclosure of the substrate by disturbance);
2. Migration (immigration of species);
3. Ecesis (establishment);
4. Competition;
5. Reaction (location modification);
6. Stabilization (establishment of a stable climax society).

Clements regarded the two factors of competition and site modification as central mechanisms for the evolving stages of succession. Clements' holistic view was controversially discussed. In particular, the European School of Vegetation Science at that time doubted the existence of a single final plant association, because variable site conditions or different disturbances should result in different endpoints even under equal climatic conditions [20]. Consequently, the British plant ecologist Sir Arthur George Tansley attempted to address this dilemma by introducing the term polyclimax [32].

In the Clements' "Relay Floristics" model, particular sets of plant species appear and disappear on a site during succession (Figure 2). Each set of species appears at a specific stage, making the site conditions unsuitable for itself but more suitable for the next set of species. While generally considering this to be possible, Egler [33] proposed the "Initial Floristic Composition" model. In this model, diaspores of all future species are already present at the beginning of a successional sequence but certain groups of species and/or lifeforms become dominant at certain stages during succession (Figure 2).

In the 1970s, Connell and Slatyer [34] presented their widely recognized and still influential hypothesis, trying to integrate all concepts of succession developed so far in three trajectories (Figure 3), with early settlers promoting, tolerating, or inhibiting the establishment of later successional species. The three successional pathways presented [34] were not considered exclusive [35,36], rather sub-elements of the three models are complementary and might explain the different phases of the succession or for different species.

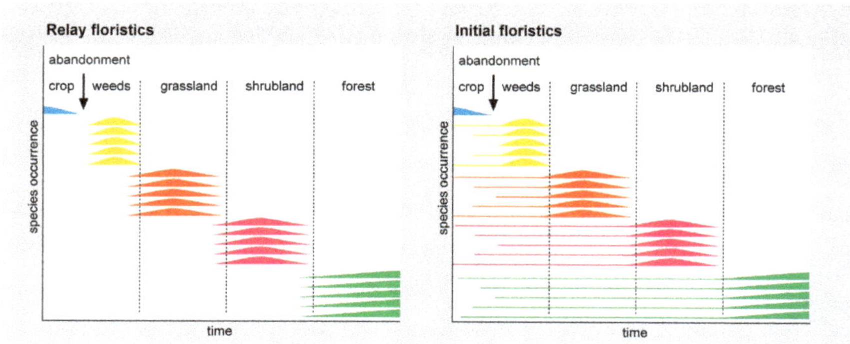

Figure 2. Two contrasting trajectories of vegetation succession: **Left**: Relay floristics of Clements [31]: sets of plant species successively appear and disappear. Species make the site conditions successively unsuitable for themselves and more suitable for another set of species. **Right**: Initial floristics of Egler [33]: all species are present at the beginning of a successional sequence but certain groups of species become dominant or drop out at certain stages during succession (adapted and modified from [19]).

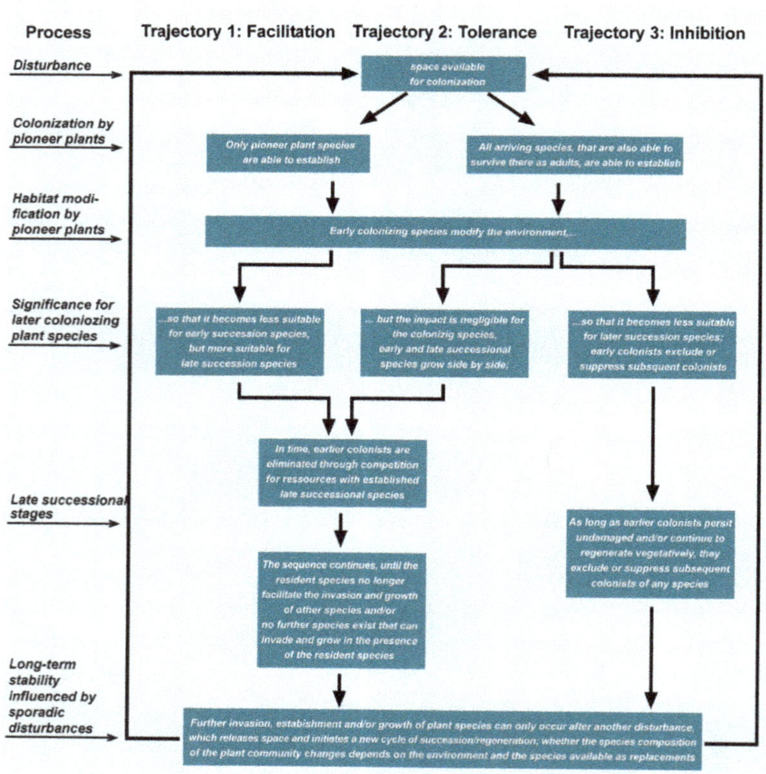

Figure 3. The three successional pathways presented by Connell and Slatyer [34]: in pathway 1, the establishment of later species is facilitated by the early pioneers; in pathway 2, the early pioneers have no effect on the later species; and in pathway 3, the establishment of later species is inhibited by the early pioneers. All three pathways are equally presumable.

Henry Gleason, a contemporary of Frederik Clements, vehemently challenged Clements' perception of a climax. He proposed an individualistic view of the succession of plants [37] (Figure 1), assuming that the biotic and abiotic heterogeneity in space and time (i.e., non-equilibrium) makes biocenoses largely unpredictable. Plant communities would represent random conglomerates of plant species controlled by environmental conditions, arranged along ecological gradients. In addition, interspecific competition for resources as well as disturbances and the species pool of the surrounding play an important role in Gleason's succession model.

The two early concepts of Clements and Gleason—though far apart in their basic assumptions—had a very stimulating effect on the development of succession theories. Initially treated as alternatives, they are considered complementary today [20]. In essence, the current doctrines on succession and vegetation dynamics are founded on opinions and concepts that have been developed in the aftermath of Clements and Gleason.

Although initially not much supported, Gleason's ideas became more popular in the middle of the 20th century with an increasing focus on population ecology. Data-driven mathematical–statistical models replaced the "philosophical" views on succession and underpinned Gleason's individualistic view of the arrangement of species along ecological gradients. Grime's CSR model [38] is well known, which distinguishes three groups of plant strategies dominating in particular successional stages (Figure 4):

- Ruderals (R, mainly annual and perennial herbs) immediately colonize after disturbances, grow fast, reproduce rapidly, and have a short lifespan.
- Competitors (C, including various lifeforms from perennial herbs to trees) obtain limiting resources through fast horizontal or vertical expansion and growth, and are adapted to productive, little disturbed habitats.
- Stress tolerators (S, lichens, mosses, perennial plants) persist through slow growth and conservative resource use, and are adapted to unproductive, stressful habitats with little disturbance only.

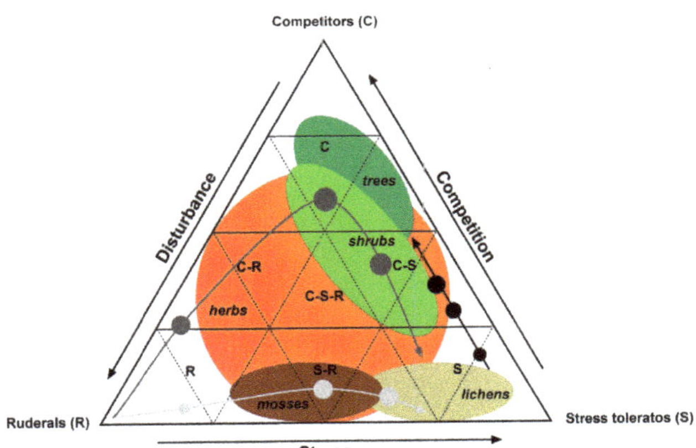

Figure 4. C-S-R model of Grime [38]. Plants are differently adapted to disturbance and stress. Ruderals (R) occur on productive but disturbed sites, competitors (C) on productive but undisturbed sites, and stress tolerators (S) on unproductive and undisturbed sites. Along the axes of the triangle combinations (S-R, C-R, C-S), and different lifeforms tend to be located in different parts of the triangle (colored ovals). Successional trajectories for primary succession on bedrock (black circles) and secondary succession in productive environments (dark-gray circles) and unproductive environments (light-gray circles). The size of the circles along the successional trajectories indicates the vegetation biomass (adapted and modified from [16]).

This group of successional theories belong to the models of Noble and Slatyer [39], Tilman [40], and Huston and Smith [41], making "vital" life history attributes, plant traits, and below- (e.g., nitrogen, water) or aboveground resources (e.g., light) responsible for the (dis)appearance of species during succession, thus explaining the large variety of successional pathways observed in nature.

Still other theories on the succession of plants focus on structural changes during succession (Figure 1). In many vegetation types, a mosaic of patches in different successional stages is present, for example, after a disturbance or when the dominant plants become senescent and die [42]. Such cyclic successions are a characteristic feature of forests where small-scale disturbances like treefall gaps initiate patch development. Vegetation dynamics following large-scale disturbances such as clearcutting for agriculture and subsequent abandonment were described by Budowski [43] and Oliver [44] who recognized different successional stages dominated by particular functional groups and life history traits. Yarranton and Morrison [45] proposed the idea that shrubs and trees might serve as nuclei, as safe sites for regeneration, enabling other species to establish in a favorable environment (less solar radiation, cooler temperature, and more humidity, entrapped organic material, improved soil conditions) below their crowns. Due to parallels to the nucleation in physics where the formation of a new organized structure is initiated by interactions of pre-existing disordered components as in a self-assembling process, Yarranton and Morrison [45] called this process nucleation.

In the 1960s, the idea evolved that succession can be observed not only from a species or lifeform point of view, but also with regard to ecosystems (Figure 1). Maybe the first heading in this direction was Odum [46]. Changes in ecosystem characteristics (energetics, nutrient cycling, community structure, life history, selection pressure, and homeostasis) during succession were made responsible for ecosystem development over time. Temporal scales are also key to the concept of Walker and Wardle [5], where plant succession is explained by processes operating at different timescales. Short-term processes such as soil nutrient fluxes and plant physiology are believed to drive plant succession while long-term geological (parent material), evolutionary (species pool), and soil-forming processes (soil properties) operating on a timescale of thousands to millions of years limit succession, having a modifying effect.

More recently, different spatial scales are considered to influence succession ([47,48], Figure 1). A larger regional scale determines climate, topography, disturbance regime, and species pool, while a smaller landscape scale refers to spatially heterogeneous areas consisting of a mosaic of different land cover types (such as forests, agricultural areas, meadows, etc.), determining mesoclimate, seed availability, dispersal agents, etc. Spatial scales explain the large variation in successional pathways observed across and within landscapes.

The latest add-on in the study of succession and vegetation dynamics is the integration of humans as relevant factor, understanding the land cover on Earth as a socio-ecological system (Figure 1). Land-use intensity (duration, extent, type, use of chemicals, etc.) affects the local species pool, which is the seed source for colonizers and, in turn, affects the direction of successional pathways [49,50]. A high land-use intensity commonly results in a high number of plant species adapted to disturbance such as weeds, lianas, or invasive species. As natural regeneration would lead to an undesirable outcome, planting or at least a human-assisted regeneration should be carried out. Low land-use, in contrast, preserve more intact landscapes, allowing natural regeneration from more diverse and pristine seed sources.

3. Current Perspectives on Vegetation Dynamics

Vegetation dynamics cover a wide spatio-temporal range, from small-scale or annually recurring symphenological processes to large-scale or secular synchronological developments. Between the two extremes, short-to-long-term vegetation changes are found as a result of gradual site modifications and/or sudden disturbances that initiate succession or regeneration, but the transitions are usually gradual, and their different types can occur together, next to each other or one after the other (Figure 5).

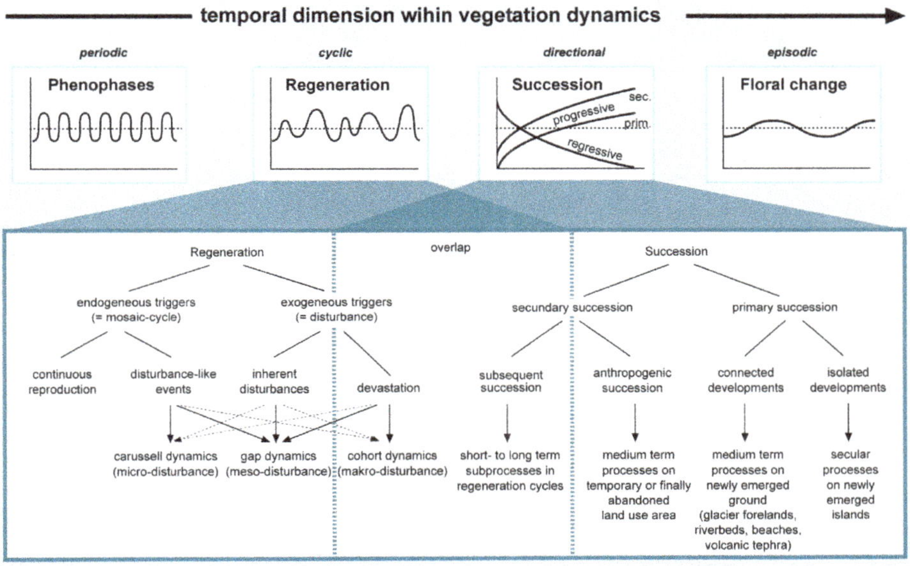

Figure 5. Types, arrangement and overlap of cyclic regeneration and directional succession within vegetation dynamics (adapted and modified from [54,55]). Arrows indicate processes initiated by particular events or developments.

3.1. Directional Succession

Succession as a directional sequence of different vegetation types or plant communities at a site [17,56] can be subdivided into progressive and regressive successions with regard to the direction of development (Figure 5). In both cases, different stages can be distinguished by species composition, vegetation structure, and/or stand physiognomy. Progressive successions are usually characterized by an increase in the number of species, the number of individuals, ground cover, increased structural complexity (e.g., lifeform diversity), and a replacement of earlier colonizers by succeeding species [57]. In regressive succession, changes occur backwards from a more advanced to an earlier stage of development [58]. Reverse developments are often caused by anthropogenic factors (e.g., overgrazing, lowering of the groundwater table, etc.), but also might be a result of natural events (drought, fire, etc.) or the gradual deterioration of the site (e.g., soil acidification caused by the vegetation itself).

Vegetation development on new terrain without any diaspores and soil formation (e.g., fresh lava, freshly piled sandbanks surrounded by the sea, new islands, recently deglaciated glacier forelands, etc.) are called primary succession ([59], Figure 5), characterized by the following main processes [60]:

I. Getting there: diaspores must reach the new ground, which requires suitable plants to be nearby for dispersal.
II. Establish: seeds must germinate and establish themselves; success might also depend on the ability of species to dormancy, the stress tolerance of the species, and the presence of "safe sites".
III. Growth: largely controlled by the site resources (nutrients, water, light, etc.).

Early colonizers are commonly short-lived, often annual taxa, which are fast-growing ("R-strategists" according to Grime ([38,61,62], Figure 4) and often wind- or self-pollinated species.

Directional vegetation developments on sites already covered by plants and/or possessing a seed bank are called secondary successions ([62], Figure 5) and proceed considerably faster than primary successions. It is a continuously ongoing process in undisturbed plant communities, affecting the dominant species composition, environmental characteristics, and plant community dynamics. This is shown by Duan et al. [63] in this Special Issue for *Picea jezoensis* forests in Northeast China. The quantification of changes in *Picea jezoensis* dominance status provides a scientific basis for evaluating the status and for the protection and restoration of *Picea jezoensis* forests. More obvious and somewhat transitional to regeneration dynamics are secondary successions after natural or anthropogenic disturbances and in (semi-)natural systems.

Primary succession is commonly slower than secondary succession, but proceeds more quickly, if locations adjoin existing vegetation from where diaspores can access the new ground via plants' natural diaspore dispersal pathways. Differences between primary succession and secondary succession also exist in terms of species diversity. While in the former, the species diversity commonly increases with ongoing development, in the latter, the early stages are often significantly more species-rich than later shrub and forest formations [54]. While secondary succession is usually swift, there are exceptions as De Giuli et al. [64] report in this Special Issue. Even 80 years after the subalpine forest fire in the Northern European Alps, the fire affected slopes are still not wooded and show higher species numbers and clearly distinct floristic communities compared to unburned sites. This pattern is likely driven by disturbance-related environmental changes, such as increased light availability and highly eroded soils, preventing a fast return to the subalpine forest under this so-called arrested succession.

3.2. Cyclic Regeneration

Directional (acyclical) succession is contrasted with cyclical changes due to regularly occurring disturbances and site-internal modifications of environmental factors. Vegetation changes of endogenous or exogenous origin are summarized under the term regeneration (Figure 5). Endogenous causes are those that arise from the ecosystem itself (e.g., "mosaic cycles", [65], including continuous self-reproduction in plant communities with a more or less constant species stock, or small-to-medium-sized "disturbance-like" events such as gaps by species' die-off (Remmert (1991 [65], Figure 6). Sometimes large areas are affected by the so-called "cohort die-offs" if individuals in even-aged populations reach the end of their lifespan at the same time, as described for *Metrosideros* forests in Hawaii [66] or *Nothofagus* populations in Patagonia [67].

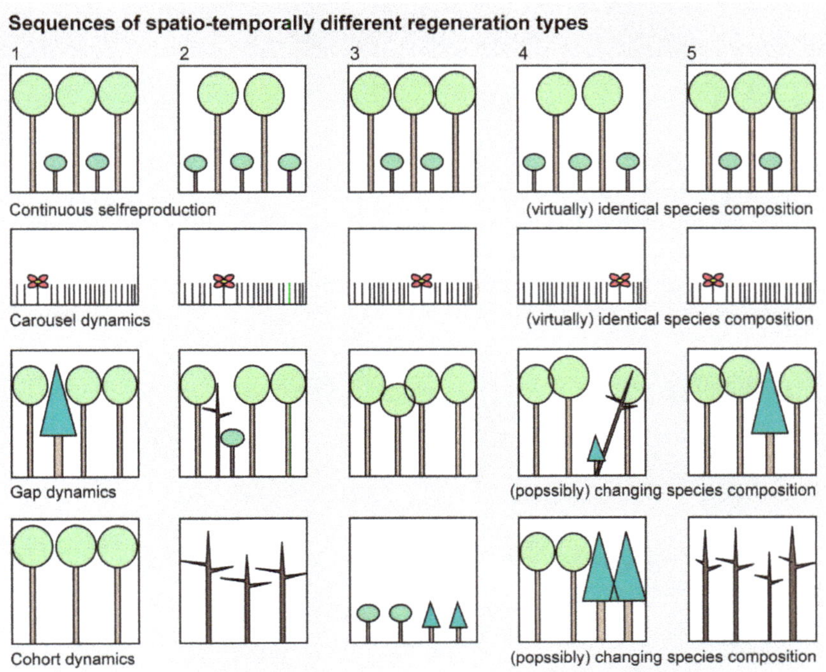

Figure 6. Sequence of development steps for different types of disturbances and regeneration (adapted and modified from [54,55]).

Equally important for the dynamics in plant communities are exogenous disturbances that affect plant communities from outside (see also Figure 5). As defined by White and Pickett [68], disturbances are discrete events in time that disrupt ecosystem, community, or population structure and change resources, substrate availability, or the physical environment. Due to their patchy nature, dynamic processes caused by the disturbances are called "patch dynamics", independent of their dimension [9,68]. Garate-Quispe et al. [69] in this Special Issue describe the factors influencing the natural regeneration in abandoned small-scale gold mining areas in the Peruvian Amazon. The focus is on the influence of the distance to the forest edge and the time since abandonment for spontaneous natural succession. Basal area and species diversity here are related to time after abandonment and inversely related to the distance to forest edges. Here, variation in species composition is mainly explained by the distance to the forest edge rather than the abandonment time. In a subsequent contribution to this Special Issue [70] by the same team of researchers, aboveground biomass, forest structure, and species diversity during secondary succession were analyzed in the same study area, where two different types of goldmining (heavy machinery and suction pumping) were formerly used. The study shows that heavy machinery more negatively affects natural regeneration than suction pumping. Tropical secondary forests give important insights on how to support restoration and enhance conservation as Rodríguez-León et al. [71] show in this Special Issue for Colombian Amazon. The authors evaluated aboveground biomass, species diversity, forest structure, and soil properties in secondary forests along a chronosequence to identify key indicators for effective restoration management.

In order to structure the confusing variety of disturbances, Böhmer and Richter [55] propose a rough classification by three size classes (Figure 7), which are responsible for different regeneration models:

- Macro-disturbances such as wildfires or strong storms trigger mass extinctions comparable to endogenously triggered cohort extinctions.
- Meso-disturbances with individual trees or groups of trees dying/falling due to exogenous factors (gap dynamics sensu stricto).
- Micro-disturbances allowing an interim invasion of certain species due to the small-scale gaps caused by the death of individual, mostly non-tree species. While these species disappear in one location, they easily colonize newly created gaps ("carousel dynamics" [72]).

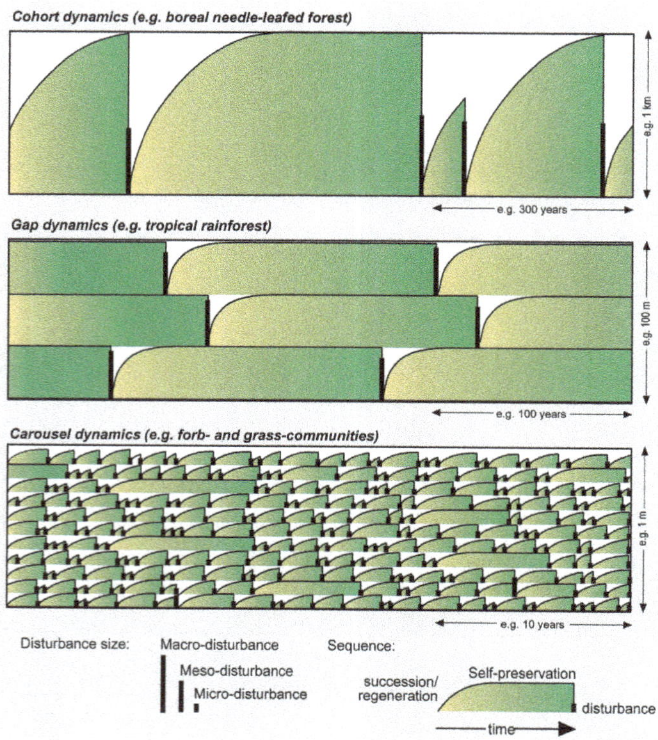

Figure 7. Spatio-temporal dimensions of disturbances (adapted and modified from [55]).

Moreno-Gonzalez et al. [73], in this Special Issue, describe the early vegetation recovery after a macro-disturbance by the 2008–2009 explosive eruption of the Chaitén Volcano in Chile. There was a gradient of decreasing disturbance with distance from the crater. High levels of devastation without surviving species, scarcely standing-dead trees and logs, and no tree regeneration were found close to the crater. Farther away from the crater, trees were resprouting, understory plants were regrowing, and seedlings were present, creating completely different starting points for regeneration/succession. Large-scale disturbances often also result from anthropogenic activities such as forest clearcutting. This is highly problematic if climate-resistant tree species such as *Quercus robur* are involved. Nasibullina et al. [74] report, in this Special Issue, that the natural regeneration of the oaks in the clearcut oak–lime forests of European Russia is successful, so planting oak seedlings or sowing acorns, i.e., active restoration, in combination with the natural restoration of lime trees are recommended to regain climate-resistant mixed forests here. Vegetation dynamics (and associated vegetation patterns such as species diversity) are sometimes controlled by co-varying disturbance and environmental gradients as shown by Abella et al. [75] in this Special Issue. In their case study, fire history (i.e., number of fires) and parent material (i.e., soil coarseness and chemistry) in combination were responsible for diversity patterns and plant community features including fuel, tree structure, and understory vegetation within the pinyon–juniper woodland of the Southwestern USA.

Last but not least, phytoglobalization [76] has become an important topic in the current ecological science. The highly dynamic spread of invasive alien plant species such as *Impatiens glandulifera* has become a serious threat to native, less-competitive taxa worldwide. Allelopathy has been suggested to be one of the possible mechanisms supporting the invasion of exotic plants. The study of Kupcinskiene et al. [77], in this Special Issue, about the allelopathic potential and phenolic content of *I. glandulifera* confirms the supreme competitive ability of this species over *I. parviflora* and native co-occurring species.

4. A Broad Range of Questions Needs a Broad Range of Methods

The processes of cyclic regeneration and directional succession play a major role for species composition and vegetation structure in plant communities. The controlling factors, however, are controversially discussed since the early accounts by pioneers of biological and biogeographical research in the 19th and 20th century. A better understanding of successional pathways, the processes involved, and vegetation dynamics, in general, is crucial in times of rapid environmental change, increasing the frequency and intensity of natural and man-made disturbances, as it has important implications for ecosystem restoration and overall vegetation management. Due to the broad array of dynamic processes and patterns of vegetation change in space and time involved, several different investigation methods are available [59]. The appropriate choice is eventually depending on the temporal and spatial scales of the study (see Figure 8), each with its respective pros and cons. Generally, a distinction is made between direct and indirect recording methods. Indirect methods include comparisons between the current state and the conditions documented in historical documents (records, photographs, paintings, etc.). Diaspore banks, datable macrofossil remains, and/or pollen analyses have been proven useful for a reconstruction over long periods of secular development. This method is used by O'Connell and Wolters in this Special Issue for the reconstruction of vegetation history and land use during the last 4000 years in Kilmore townland, Dingle peninsula, south-western Ireland in this Special Issue [78]. They found the main factor determining the direction and magnitude of change of woody vegetation and biodiversity to be farming activity.

The use of chronosequences (space for time substitution according to [51]), assuming a temporal sequence derived from spatially different sites, is another well-established indirect method. It is useful when the site age can be determined with sufficient accuracy, for example, in the case of lava flows, exogenous disturbance events (fire, storm, flooding, insect calamities, etc.), or in the case of retreating glaciers. The comparisons of current vegetation recordings with historical surveys of the same area (without precisely fixed long-term sample sites) allow for an accurate assessment of the degree and direction of the vegetation change. Such comparisons are transitional from the indirect to the direct methods. Direct methods include the comparisons of vegetation maps of the same areas but from different dates or repeated recordings of precisely localized and dated sample sites set up in the past ("Permanent Plots"), which allow a very precise evaluation of the course of succession.

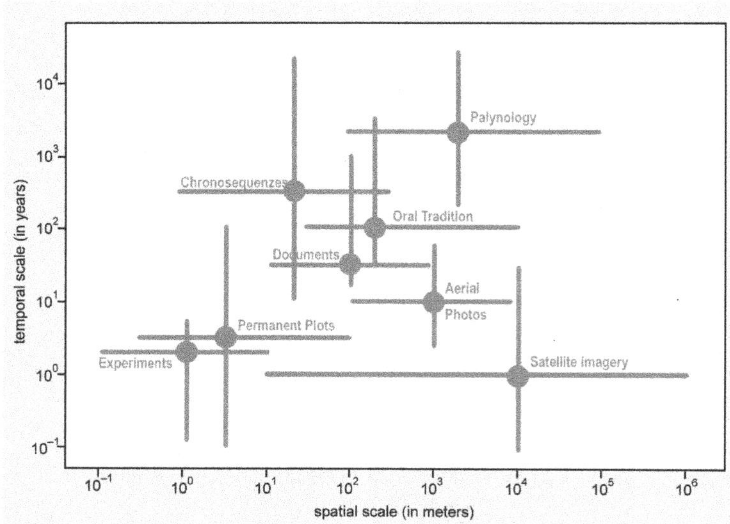

Figure 8. Spatial and temporal scales at which certain methods for studying vegetation changes come into play. The lines show the span of the spatio-temporal resolution, the crossing points, and the main areas of application of the methods (adapted from [59]).

The papers presented in this Special Issue give evidence of the broad range of spatial and temporal scales on which vegetation dynamics might be studied, regarding type, plant communities, geographical location, and investigation methods employed. We wish all readers an insightful reading and many new findings.

Acknowledgments: We would like to thank all the authors for their contribution to this SI; our thanks go to the staff members at the MDPI editorial office for their support.

Conflicts of Interest: The authors declare no conflicts of interest.

References

1. Kneitel, J. Successional Changes in Communities. *Nat. Educ. Knowl.* **2010**, *3*, 41.
2. Cooper, W.S. The fundamentals of vegetation change. *Ecology* **1926**, *7*, 391–413. [CrossRef]
3. De Jong, R.; Verbesselt, J.; Zeileis, A.; Schaepman, M.E. Shifts in Global Vegetation Activity Trends. *Remote Sens.* **2013**, *5*, 1117–1133. [CrossRef]
4. Chang, C.C.; Turner, B.L. Ecological succession in a changing world. *J. Ecol.* **2019**, *107*, 503–550. [CrossRef]
5. Walker, L.R.; Wardle, D. A Plant succession as an integrator of contrasting ecological time scales. *Trends Ecol. Evol. Sept.* **2014**, *29*, 504–510. [CrossRef]
6. Franklin, J.; Serra-Diaz, J.M.; Syphard, A.D.; Regan, H.M. Global change and terrestrial plant community dynamics. *Proc. Natl. Acad. Sci. USA* **2016**, *113*, 3725–3734. [CrossRef]
7. Janečková, P.; Tichý, L.; Walker, L. Global drivers influencing vegetation during succession: Factors and implications. *J. Veg. Sci.* **2024**, *35*, 13297. [CrossRef]
8. Knapp, R. Vegetation Dynamics, Vegetation Dynamics. In *Handbook of Vegetation Science*; Springer: Dordrecht, The Netherlands, 1974; Volume 8.
9. Glenn-Lewin, D.C.; van der Maarel, E. Patterns and processes of vegetation dynamics. In *Plant Succession—Theory and Prediction*; Glenn-Lewin, D.C., Peet, R.K., Veblen, T.T., Eds.; Population and Community Biology Series 11; Chapman and Hall: London, UK, 1992; pp. 11–59.
10. Miles, J. *Vegetation Dynamics*; Springer Science & Business Media: Berlin/Heidelberg, Germany, 2012.
11. Pickett, S.T.A.; Cadenasso, M.L.; Meiners, S.J. Ever since Clements: From Succession to Vegetation Dynamics and Understanding to Intervention. *Appl. Veg. Sci.* **2009**, *12*, 9–21. Available online: http://www.jstor.org/stable/27735041 (accessed on 19 June 2025). [CrossRef]

12. Pickett, S.T.A.; Cadenasso, M.L.; Meiners, S.J. Vegetation Dynamics. In *Vegetation Ecology*; van der Maarel, E., Franklin, J., Eds.; John Wiley & Sons, Ltd.: Hoboken, NJ, USA, 2013.
13. Franklin, O.; Harrison, S.P.; Dewar, R.; Farrior, C.E.; Brännström, Å.; Dieckmann, U.; Pietsch, S.; Falster, D.; Cramer, W.; Loreau, M.; et al. Organizing principles for vegetation dynamics. *Nat. Plants* **2020**, *6*, 444–453. [CrossRef]
14. Dierschke, H.; Goedecke, F. Forty years of symphenological research in a submontane calcareous beech forest under the influence of climate change. *Flora Mediterr.* **2022**, *31*, 257–270. [CrossRef]
15. Braun-Blanquet, J. Plant Sociology. In *The Study of Plant Communities*; McGraw-Hill Book Company: New York, NY, USA, 1932.
16. Poorter, L.; Amissah, L.; Bongers, F.; Hordijk, I.; Kok, J.; Laurance, S.G.W.; Lohbeck, M.; Martínez-Ramos, M.; Matsuo, T.; Meave, J.A.; et al. Successional theories. *Biol. Rev. Camb. Philos. Soc.* **2023**, *98*, 2049–2077. [CrossRef] [PubMed]
17. Poorter, L.; van der Sande, M.; Amissah, L.; Bongers, F.; Hordijk, I.; Kok, J.; Laurance, S.; Martinez-Ramos, M.; Matsuo, T.; Meave, J.; et al. A comprehensive framework for vegetation succession. *Ecosphere* **2024**, *15*, e4794. [CrossRef]
18. Blumler, M.A.; Cole, A.; Flenley, J.; Schickhoff, U. History of biogeographical thought. In *The SAGE-Handbook of Biogeography*; Millington, A., Blumler, M., Schickhoff, U., Eds.; SAGE Publications Ltd.: New York, NY, USA, 2011; pp. 23–42.
19. Pulsford, S.; Lindenmayer, D.; Driscoll, D. A succession of theories: Purging redundancy from disturbance theory. *Biol. Rev. Camb. Philos. Soc.* **2014**, *91*, 12163. [CrossRef]
20. Hufnagl, L.; Mics, F. Introductory Chapter: Vegetation Dynamics, Basic Phenomena, and Processes. In *Vegetation Dynamics, Changing Ecosystems and Human Responsibility*; InTech Open: Rijeka, Croatia, 2022. [CrossRef]
21. Trudgill, S. The interactions between all neighbouring organisms: The roles of Charles Darwin, Ernst Haeckel and Eugenius Warming in the evolution of ideas on plant dynamics. *Prog. Phys. Geogr.* **2012**, *36*, 853–861. [CrossRef]
22. Darwin, C. *On the Origin of Species by Means of Natural Selection, or the Preservation of Favoured Races in the Struggle for Life*; John Murray: London, UK, 1895.
23. Wallace, A.R. *Natural Selection and Tropical Nature*; MacMillan and Co.: London, UK; New York, NY, USA, 1895.
24. Köppen, W.; Wegener, A. *Die Klimate der Geologischen Vorzeit*; Borntraeger: Berlin, Germany, 1924.
25. Milanković, M. Mathematische Klimalehre und astronomische Theorie der Klimaschwankungen. In *Handbuch der Klimatologie, Bd. 1 (A): Allgemeine Klimalehre*; Köppen, W., Geiger, R., Eds.; Borntraeger: Berlin, Germany, 1930.
26. Wegener, A. *Die Entstehung der Kontinente und Ozeane*; Vieweg & Sohn: Braunschweig, Germany, 1915.
27. von Humboldt, A.S. *Ansichten der Natur Mit Wissenschaftlichen Erläuterungen*; Cotta: Tübingen, Germany, 1808.
28. Dureau de la Malle, A.J.C.A. Memoire sur l'alternance ou sur ce problème: La succession alternative dans la reproduction des especes végétales vivant en société, est-elle une loi générale de la nature? *Ann. Des Sci. Nat.* **1825**, *15*, 353–381.
29. Warming, E. *Plantesamfund—Grundtræk af den Økologiske Plantegeografi*; Philipsens: Copenhagen, Denmark, 1895.
30. Cowles, H. Ecological relation of the vegetation on sand dunes of Lake Michigan. *Bot. Gaz.* **1899**, *27*, 95–117+167–202+281–308+361–388. [CrossRef]
31. Clements, F.E. *Plant Succession: An Analysis of the Development of Vegetation*; Publication No. 242; Carnegie Institution of Washington: Washington, WA, USA, 1916.
32. Tansley, A.G. The use and abuse of vegetational concepts and terms. *Ecology* **1935**, *16*, 284–307. [CrossRef]
33. Egler, F.E. Vegetation science concepts. 1. Initial floristic composition, a factor in old-field vegetation development. *Vegetatio* **1954**, *4*, 412–417. [CrossRef]
34. Connell, J.H.; Slatyer, R.O. Mechanisms of succession in natural communities and their role in community stability and organization. *Am. Nat.* **1977**, *111*, 1119–1144. [CrossRef]
35. Pickett, S.T.A.; Collins, S.L.; Armesto, J.J. Models, mechanisms and pathways of succession. *Bot. Rev.* **1987**, *53*, 335–371. [CrossRef]
36. Walker, L.; Chapin, F.S., III. Interactions among Processes Controlling Successional Change. *Oikos* **1987**, *50*, 131–135. [CrossRef]
37. Gleason, H.A. The individualistic concept of the plant association. *Bull. Torrey Bot. Club* **1926**, *53*, 7–26. [CrossRef]
38. Grime, J.P. *Plant Strategies and Vegetation Processes*; John Wiley & Sons Ltd.: Chichester, UK, 1979.
39. Noble, I.; Slatyer, R. The use of vital attributes to predict successional changes in plant communities subject to recurrent disturbances. *Vegetatio* **1980**, *43*, 5–21. [CrossRef]
40. Tilman, D. The resource-ratio hypothesis of plant succession. *Am. Nat.* **1985**, *125*, 827–852. [CrossRef]
41. Huston, M.; Smith, T. Plant succession– life history and competition. *Am. Nat.* **1987**, *130*, 168–198. [CrossRef]
42. Watt, A.S. Pattern and process in the plant community. *J. Ecol.* **1947**, *35*, 1–22. [CrossRef]
43. Budowski, C.G. Distribution of tropical American rain-forest species in the light of successional processes. *Turrialba* **1965**, *15*, 40–42.
44. Oliver, C.D. Forest development in North America following major disturbances. *For. Ecol. Manag.* **1980**, *3*, 153–168. [CrossRef]
45. Yarranton, G.; Morrison, R. 0Spatial dynamics of a primary succession: Nucleation. *J. Ecol.* **1974**, *62*, 417–428. [CrossRef]
46. Odum, E.P. The strategy of ecosystem development: An understanding of ecological succession provides a basis for resolving man's conflict with nature. *Science* **1969**, *164*, 262–270. [CrossRef]

47. Arroyo-Rodriguez, V.; Melo, F.P.; Martinez-Ramos, M.; Bongers, F.; Chazdon, R.L.; Meave, J.A.; Norden, N.; Santos, B.A.; Leal, I.R.; Tabarelli, M. Multiple successional pathways in human-modified tropical landscapes: New insights from forest succession, forest fragmentation and landscape ecology research. *Biol. Rev.* **2017**, *92*, 326–340. [CrossRef] [PubMed]
48. Dent, D.H.; Estrada-Villegas, S. Uniting niche differentiation and dispersal limitation predicts tropical forest succession. *Trends Ecol. Evol.* **2021**, *36*, 700–708. [CrossRef] [PubMed]
49. Balvanera, P.; Paz, H.; Arreola-Villa, F.; Bhaskar, R.; Bongers, F.; Cortés, S.; del Val, E.; García-Frapolli, E.; Gavito, M.E.; Gonzalez Esquivel, C.E. Social ecological dynamics of tropical secondary forests. *For. Ecol. Manag.* **2021**, *496*, 119369. [CrossRef]
50. Jakovac, C.C.; Junqueira, A.B.; Crouzeilles, R.; Peña-Claros, M.; Mesquita, R.C.; Bongers, F. The role of land-use history in driving successional pathways and its implications for the restoration of tropical forests. *Biol. Rev.* **2021**, *96*, 1114–1134. [CrossRef]
51. Pickett, S.T.A. Space-for-time substitution as an alternative to long-term studies. In *Long-Term Studies in Ecology: Approaches and Alternatives*; Likens, G.E., Ed.; Springer: New York, NY, USA; pp. 110–135.
52. Walker, L.R.; Wardle, D.A.; Bardgett, R.D.; Clarkson, B.D. The use of chronosequences in studies of ecological succession and soil development. *J. Ecol.* **2010**, *98*, 725–736. [CrossRef]
53. Clements, F.E. Nature and structure of the climax. *J. Ecol.* **1936**, *24*, 252–284. [CrossRef]
54. Richter, M. *Allgemeine Pflanzengeographie*; Teubner Studienbücher Geographie: Stuttgart, Germany, 1997; p. 256.
55. Böhmer, H.J.; Richter, M. Regeneration of Plant Communities—An Attempt to Establish a Typology and Zonal System. *Plant Res. Dev.* **1997**, *45*, 74–88.
56. Krebs, C. *Ecology—The Experimental Analysis of Distribution and Abundance*; Pearson International: Mississauga, ON, Canada, 2013.
57. Müller-Dombois, D.; Ellenberg, H. *Aims and Methods of Vegetation Ecology*; Wiley & Sons: New York, NY, USA, 1974.
58. Peltzer, D.; Wardle, D.; Allison, V.; Baisden, T.; Bardgett, R.; Chadwick, O.; Condron, L.M.; Parfitt, R.; Porder, S.; Richardson, S.; et al. Understanding ecosystem retrogression. *Ecol. Monogr.* **2010**, *80*, 509–529. [CrossRef]
59. Walker, L.R.; del Moral, R. *Primary Succession and Ecosystem Rehabilitation*; Cambridge University Press: Cambridge, UK, 2003; p. 442.
60. Bradshaw, A.D. Introduction: Understanding the fundamentals of succession. In *Primary Succession on Land*; Miles, J., Walton, D.W.H., Eds.; Blackwell Scientific Publications: Oxford, UK, 1993; pp. 1–3.
61. MacArthur, R.H.; Wilson, E.O. *The Theory of Island Biogeography*; Princeton University Press: Princeton, NJ, USA, 1967.
62. Chapin, F.S., III. Physiological control over plant establishment in primary succession. In *Primary Succession on Land*; Miles, J., Walton, D.W.H., Eds.; Blackwell Scientific Publications: Oxford, UK, 1993; pp. 161–178.
63. Duan, J.; Jia, Z.; Ge, S.; Li, Y.; Kang, D.; Li, J. Dominant Species Composition, Environmental Characteristics and Dynamics of Forests with Picea jezoensis Trees in Northeast China. *Diversity* **2024**, *16*, 731. [CrossRef]
64. De Giuli, M.; Winkler, M.; Deola, T.; Henschel, J.; Sass, O.; Wolff, P.; Jentsch, A. Arrested Succession on Fire-Affected Slopes in the Krummholz Zone and Subalpine Forest of the Northern Limestone Alps. *Diversity* **2024**, *16*, 366. [CrossRef]
65. Remmert, H. The mosaic-cycle concept of ecosystems—An overview. In *Ecological Studies Bd. 85*; Remmert, H., Ed.; Springer: Berlin/Heidelberg, Germany, 1991; pp. 1–21.
66. Müller-Dombois, D. Natural Dieback in Forests. *BioScience* **1987**, *37*, 575–583. [CrossRef]
67. Rebertus, A.J.; Veblen, T.T.; Kitzberger, T. Gap formation and dieback in Fuego-Patagonian Nothofagus forests. *Phytocoenologia* **1993**, *23*, 581–599. [CrossRef]
68. White, P.S.; Pickett, S.T.A. (Eds.) Natural disturbance and patch dynamics: An introduction. In *Natural Disturbance and Patch Dynamics*; Acadamic Press: San Diego, CA, USA, 1985; pp. 3–13.
69. Garate-Quispe, J.; Velásquez Ramírez, M.; Becerra-Lira, E.; Baez-Quispe, S.; Abril-Surichaqui, M.; Rodriguez-Achata, L.; Muñoz-Ushñahua, A.; Nascimento Herbay, P.; Fernandez-Mamani, Y.; Alarcon-Aguirre, G.; et al. Influence of Distance from Forest Edges on Spontaneous Vegetation Succession Following Small-Scale Gold Mining in the Southeast Peruvian Amazon. *Diversity* **2023**, *15*, 793. [CrossRef]
70. Garate-Quispe, J.; Herrera-Machaca, M.; Pareja Auquipata, V.; Alarcón Aguirre, G.; Baez Quispe, S.; Carpio-Vargas, E.E. Resilience of Aboveground Biomass of Secondary Forests Following the Abandonment of Gold Mining Activity in the Southeastern Peruvian Amazon. *Diversity* **2024**, *16*, 233. [CrossRef]
71. Rodríguez-León, C.H.; Sterling, A.; Trujillo-Briñez, A.; Suárez-Córdoba, Y.D.; Roa-Fuentes, L.L. Forest Attribute Dynamics in Secondary Forests: Insights for Advancing Ecological Restoration and Transformative Territorial Management in the Amazon. *Diversity* **2025**, *17*, 39. [CrossRef]
72. van der Maarel, E.; Sykes, M.T. Small scale species turnover in a limestone grassland: The carousel model and some comments on the niche concept. *J. Veg. Sci.* **1993**, *4*, 179–188. [CrossRef]
73. Moreno-Gonzalez, R.; Díaz, I.A.; Christie, D.A.; Lara, A. Early Vegetation Recovery After the 2008–2009 Explosive Eruption of the Chaitén Volcano, Chile. *Diversity* **2025**, *17*, 14. [CrossRef]
74. Nasibullina, A.; Tiebel, K.; Wagner, S. Succession as a Natural Tool for Restoration of Oak—Lime Forests on Aspen-Covered Clearcuts. *Diversity* **2024**, *16*, 376. [CrossRef]

75. Abella, S.R.; Chiquoine, L.P.; Bailey, E.C.; Porter, S.L.; Morrison, C.D.; Farris, C.A.; Fox, J.E. Diversity in Burned Pinyon–Juniper Woodlands Across Fire and Soil Parent Material Gradients. *Diversity* **2025**, *17*, 88. [CrossRef]
76. Richter, M. Vegetation development before, during, and after El Niño 1997/98 in Northwestern Perú. *Lyonia* **2005**, *8*, 19–27. Available online: http://www.lyonia.org/viewIssue.php?volumeID=18 (accessed on 19 June 2025).
77. Kupcinskiene, E.; Budreviciute, R.; Jasionyte, V.; Simanaviciute, L.; Jociene, L.; Krokaite-Kudakiene, E.; Rekasius, T.; Marozas, V. Allelopathic Properties of the Species Comprising Communities of Invasive *Impatiens* spp. and Antioxidant System of Invaders' Populations. *Diversity* **2025**, *17*, 20. [CrossRef]
78. O'Connell, M.; Wolters, S. Holocene flora, vegetation and land-use changes on Dingle peninsula, south-western Ireland, as reflected in pollen analytical, archaeological and historical records. *Diversity* **2025**, *17*, 456. [CrossRef]

diversity

MDPI

Article

Influence of Distance from Forest Edges on Spontaneous Vegetation Succession Following Small-Scale Gold Mining in the Southeast Peruvian Amazon

Jorge Garate-Quispe [1,2,3,*], Manuel Velásquez Ramírez [4], Edwin Becerra-Lira [5], Sufer Baez-Quispe [1], Milagro Abril-Surichaqui [6,7], Liset Rodriguez-Achata [8], Adenka Muñoz-Ushñahua [5], Pedro Nascimento Herbay [5], Yoni Fernandez-Mamani [2], Gabriel Alarcon-Aguirre [1], Marx Herrera-Machaca [2], Litcely Hilares Vargas [5], Ronald Corvera Gomringer [5] and Dennis del Castillo Torres [5]

1 Departamento Académico de Ingeniería Forestal y Medio Ambiente, Facultad de Ingeniería, Universidad Nacional Amazónica de Madre de Dios, Puerto Maldonado 17001, Peru
2 Ecology & Restoration of Tropical Ecosystems Research Group (ECORET), Universidad Nacional Amazónica de Madre de Dios, Puerto Maldonado 17001, Peru
3 Department of Evolutionary Biology, Ecology and Environmental Sciences, Universitat de Barcelona, 08028 Barcelona, Spain
4 Proyecto Recuperación de Áreas Degradadas, Instituto de Investigaciones de la Amazonía Peruana (IIAP), Puerto Maldonado 17001, Peru
5 Instituto de Investigaciones de la Amazonía Peruana (IIAP), Puerto Maldonado 17001, Peru
6 Science Support Peru S.A.C., Lima 15026, Peru
7 Universidad Peruana Cayetano Heredia (UPCH), Lima 15102, Peru
8 Departamento Académico de Ciencias Básicas, Universidad Nacional Amazónica de Madre de Dios, Puerto Maldonado 17001, Peru
* Correspondence: jgarate@unamad.edu.pe; Tel.: +51-974257923

Abstract: Few studies describe the factors that influence the natural regeneration in abandoned gold mining areas in the Amazon. Here we focus on the influence of the distance to the forest edge and abandonment time in a spontaneous succession of degraded areas by gold mining in the southeastern Peruvian Amazon. We assessed woody species composition (DBH $\geq$ 1 cm) and forest stand structure across a chronosequence (2–23 years). A total of 79 species belonging to 30 families were identified. The natural regeneration was dominated by Fabaceae, Malvaceae, and Urticaceae. Together, they represented 60% of the importance index. *Cecropia membranacea* and *Ochroma pyramidale* were the dominant pioneer species at the initial successional stage. The basal area and species diversity were directly related to time after abandonment and inversely related to the distance to forest edges. The distance-based redundancy analysis showed that more of the variation in species composition was explained by distance to the forest edge than the abandonment time. Our study revealed that regeneration was relatively slow and provided evidence that the distance to the forest edge is important for natural regeneration in areas degraded by gold mining.

Keywords: degraded ecosystems; Madre de Dios; natural regeneration; secondary succession; species richness

1. Introduction

Mining has been a crucial activity for the world economy and the development of human societies [1,2]. However, mining-related activities cause large-scale disturbance and fragmentation when original habitats are removed, excavated, or buried by the deposition of sterile material [3,4]. Even though mining-related activities take part only in less than 1% of the earth's land surface [1], they are a major cause of environmental degradation [5,6], serious health issues in local communities [7], and a threat to tropical ecosystems [8].

Amazon rainforests harbor a big part of worldwide terrestrial biodiversity [8]. However, they are being heavily affected by anthropogenic activities and changing environmental conditions [9]. The National Peruvian Forestry and Wild Fauna Entity (also known as SERFOR) confirmed that between 2017 and 2018, deforestation in Peru (128,069 ha/year) increased [10]. Artisanal and small-scale gold mining (ASGM) in Madre de Dios is the main cause of vegetation and soil degradation [2,4]. A common method to restore post-mining areas, in general, is reforestation (restoration or rehabilitation), which includes complex landscape preparation, planting, and maintenance techniques [3]. However, spontaneous succession can, in many cases, lead to the establishment of a functional forest ecosystem in disturbed areas, including post-mining sites [11,12], but it can be limited by the level of degradation [6] in soil quality, both in terms of structure and chemical properties [13,14]. On the other hand, ASGM is still an important part of the local economy in many developing countries [1].

Previous studies have found that naturally recovered areas are commonly more diverse, provide more ecosystem services, and are more robust against disturbances. At the same time, restored forests often are more similar to plantations with few fast-growing species [3,11,12]. However, such cover will depend on the impact level of the perturbation, the type of mining carried out, the available seed sources, and the distance to seed sources. Natural regeneration (passive restoration) is the process in which trees establish from seeds that fall and germinate in situ without anthropic intervention [15,16]. This restoration method does not include any complementary activities to improve restoration [17]. Under these conditions, the establishment and development of natural regeneration are affected by several factors: poor seed dispersal, site morphology, poor soil nutrient availability, drought, and competition between herbaceous or invasive species [4,18].

Natural regeneration has been favored for lands degraded by mining due to its low cost and importance for successional trajectory management [19]. There are several studies that suggest that the distance to forest edges influences the natural regeneration of abandoned pastures. Alarcon-Aguirre et al. [4] reported that nearby forests and fragments that were not affected by gold mining in the Madre de Dios Amazon provided seeds that favored vegetation development (distances between 22 m to 152 m). However, distance to the forest showed an inverse relationship with species abundance, recovery time after abandonment was directly and significantly correlated with the basal area, species richness, and diversity. Günter et al. [20] found a significant influence of distance on the floristic composition of natural regeneration in an Ecuadorian forest, where unique species were replaced by more common ones, with higher prevalence in places that were further away from reference forests [20]. Cubiña and Aide [21] reported that distance from forest edges had an influence on seed rain and soil seed bank in a tropical pasture in Puerto Rico.

In addition, the natural regeneration of degraded areas by mining in tropical forests in Brazil revealed that the number of individuals, species, families, and diversity decreases with greater distance from the remaining forest fragment [3]. Studies have also reported a positive relationship between abandonment time and the increase in abundance and diversity of species [1,6,15,22].

The Amazon forest facilitates the study of plant community formation due to the rapid recovery of diversity and structure of successional communities, even in case of anthropogenic disruption [23]. While successions occur under similar climatic, environmental, and soil conditions, they might differ in abandonment time [24]. Chronosequence studies represent a useful approach to understanding successional dynamics in tropical forests [25] by sampling the natural regeneration of areas of different stages of succession [26]. Chronosequence studies are widely used because they provide results with minimal effort and extend over long periods of time [27,28]. There are several articles related to the secondary succession of tropical forests in areas abandoned by agriculture, livestock, pastures, etc. [18,28–30] and on the recovery of vegetation cover after mining in the Brazilian Amazon [6,30,31]. However, studies on natural colonization in areas abandoned by gold mining and the influence of distance from the forest edge on floristic diversity, structure, and composition are particularly rare in the Peruvian Amazon, especially in Madre de Dios, a department known as the "Capital of the Biodiversity of Peru".

The present study analyzes the influence of the distance to the forest edge and abandonment time on spontaneous succession (natural regeneration) in abandoned gold-mining lands in the southeast of the Peruvian Amazon (Madre de Dios). In this study, we address the speed of regeneration in the early years after mining abandonment and how the rate of natural regeneration in abandoned mining is influenced by abandonment time and distance to the forest edge. This is one of the first studies in the Peruvian Amazon on the influence of the distance to the forest edge, abandonment time, and influence of ecological factors and their interactions on the establishment of species in the spontaneous succession of areas abandoned by the ASGM. The study highlights the importance of understanding the regular plant community in the early stage of secondary forest, and it could be used as a reference point for restoration practices in the Peruvian Amazon, Department of Madre de Dios.

2. Materials and Methods

2.1. Study Area

The study took place in abandoned gold mining areas (12°31′34.22″ S and 69°23′45.77″ W) in San Jacinto native community, Madre de Dios department, located 35 km northeast of Puerto Maldonado city, southeastern Peruvian Amazon. Average annual temperature and precipitation (between the years 1980–2022) were 25.4 °C and 2120 mm, respectively. The dry season lasts from June to September, in which precipitation is lower than 100 mm per month (Figure 1).

The study area is a patchwork of abandoned mining areas with different times of abandonment after ASGM (Figure 2). Twenty areas were selected with covered soil of natural regeneration [32] of different abandoned times to establish a 23-year chronosequence which includes initial (<5 years), intermediate (5 to <10 years), and advanced successional stage (≥10 years) [33]. Covered soil, as defined by Velásquez-Ramírez [34], is an impacted area by gold mining and covered ground by natural regeneration with predominant sand particles.

Time elapsed from the abandonment of gold extraction, distance to the forest edge, plot locations, and discard areas with re-entries of gold mining activity was determined by using multi-temporal Landsat (1975–2010), RapidEye, and Planet imagery (2010–2021).

The gold deposits in Madre de Dios are classified as secondary deposits or placer deposits because they were formed by the weathering and erosion of hard rock ores in the Andes, which produced fine grains of gold that were then carried down rivers to the lowlands of Madre de Dios. These fine gold particles settle in deposits along the inside of river meanders and have formed near-surface deposits in the floodplains of these rivers over time due to changes in river geomorphology and seasonal inundation [7].

In the study area, to obtain gold, the miners used an open-pit mining process. In such operations, all aboveground vegetation must be cut, and the soil must be completely removed. Then, during dredging operations, gold-containing soils are pumped from the deep pits (up to 10 m deep) to a sluice box, which collects gold particles. In the San Jacinto Native Community, the small area mines have an aggregated spatial pattern (15 to 20 ha of area). This high density of small mines causes a vast amount of deforestation and forest fragmentation [35]. Mining impacts the natural dynamics of vegetation and soil degradation, transforming the landscape into disturbed material without well-developed soil horizons and stony substrates that limit the rate and form of natural recolonization of abandoned sites by plants and soil biota [2,4]. The mines in the San Jacinto Native Community are generally abandoned after exploitation, and no restoration or rehabilitation actions are applied.

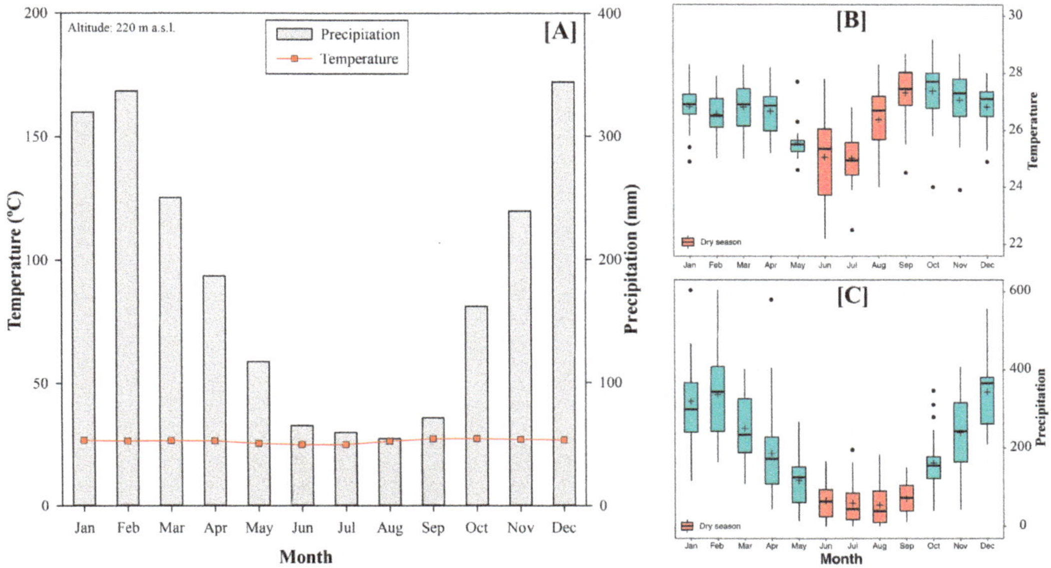

Figure 1. The climatic conditions at the study area (1980–2022), San Jacinto Native Community in the Peruvian Amazon. (**A**) Vertical bars show the mean monthly sums of precipitation, and the red line shows the monthly average air temperature. (**B**,**C**) Box plots of monthly mean temperature and precipitation for the study area (1980–2022).

Prior to mining activities in the study area, the San Jacinto Native Community area consisted of low-terrace rainforest. The canopy reached a height of 30–35 m above the forest floor, with some emergent trees reaching up to 45 m. These forests represented common low-terrace forest types in the Amazonia of Madre de Dios, which are dominated by *Bertholletia excelsa*, *Pseudolmedia laevis*, *Iriartea deltoidea*, *Poulsenia armata*, *Quararibea wittii*, *Astrocaryum murumuru* and *Euterpe precatoria*.

2.2. Sampling and Vegetation Structure

Vegetation was sampled in 20 plots, each 10 m × 25 m. All individuals with a diameter at breast height (DBH) bigger than 1 cm were inventoried. Total height (TH) and DBH were measured for each individual. Botanical material was collected and stored for identification in the Alwyn Gentry herbarium (HAG) of Amazon National University of Madre de Dios.

Woody vegetation was classified according to the state of development of the individuals [36], considering the diameter, seedling (1 cm to <5 cm DBH), sapling (5 cm to <10 cm DBH), and trees (>10 cm DBH). Scientific names were revised and corrected using the TNRS application (https://tnrs.biendata.org).

The structural and vertical characteristics of vegetation were categorized according to diameter classes (5 cm range), and then average height was obtained for each diameter class. This characterization was carried out for each of the four categories of abandonment time.

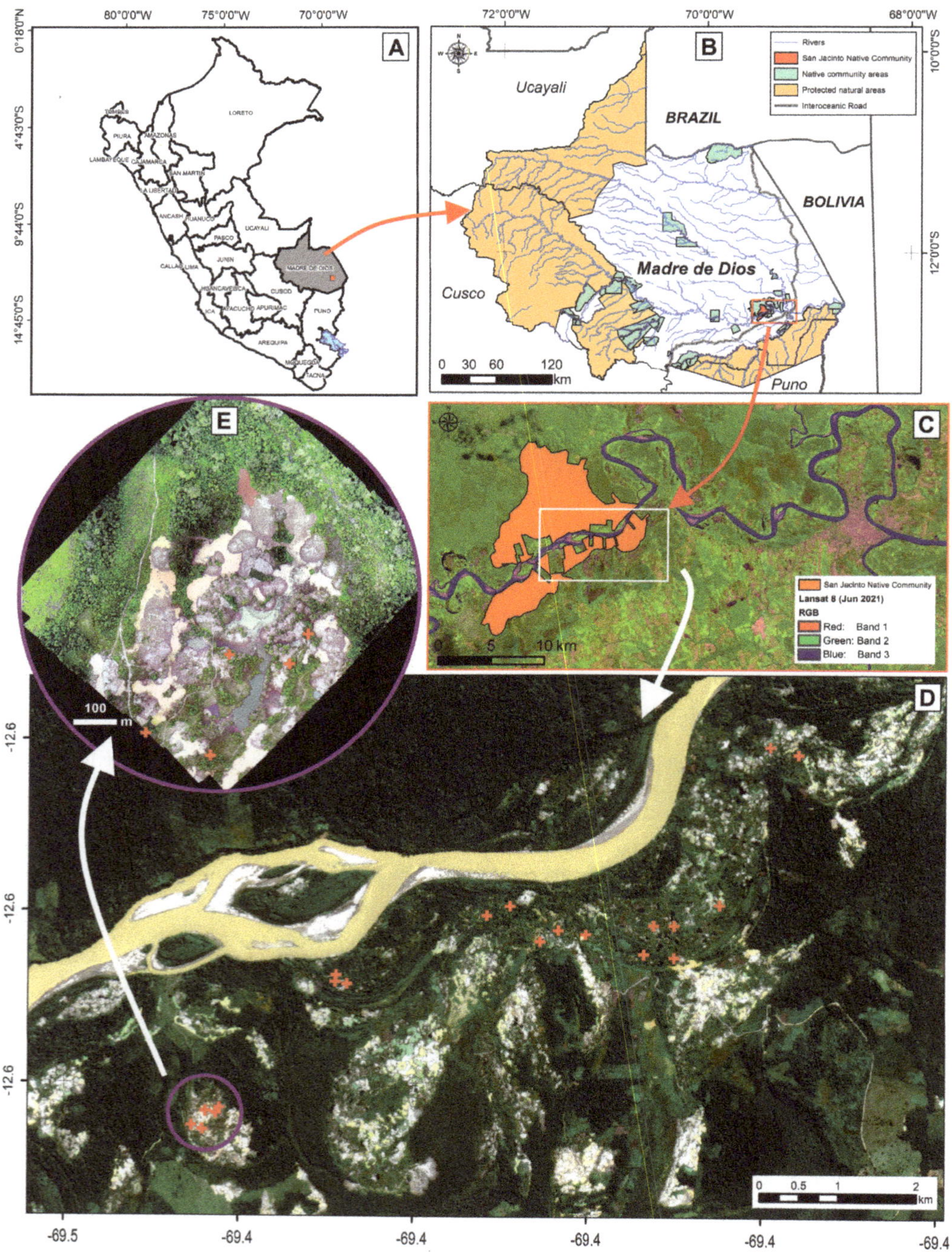

Figure 2. The geographic location of the study area at San Jacinto Native Community, Madre de Dios (**B**), Perú (**A**). (**C**) Landsat-8 image (June 2021) over San Jacinto Native Community (**D**) Spatial distribution of sampling plots. (**E**) Sampling plots in images from drone flight integrated into an orthomosaic photo (July 2021).

2.3. Species Importance and Diversity Index

The Importance Index (IIE), a simplified version of the Importance Value Index, was calculated as the sum of relative dominance and relative abundance [37]. The importance value index is a measure of how dominant a species is at a growth level [38].

$$\text{Abun}(\%) = \frac{\text{Number of individuals belong to one species}}{\text{Total number of individuals}} \times 100 \quad (1)$$

$$\text{Dom}(\%) = \frac{\text{Basal area belong to one species}}{\text{Basal area of all species}} \times 100 \quad (2)$$

$$\text{IIE} = \text{Abun}(\%) + \text{Dom}(\%) \quad (3)$$

where IIE is the Importance Index, Abun (%) is the relative abundance, and Dom (%) is the relative dominance.

Floristic attributes for each plot were analyzed considering the richness of species and richness of genera and families. In order to characterize and compare the biodiversity of our samples, Shannon-Weaver (Equation (4)), Margalef (Equation (5)), and α-Fisher (Equation (6)) diversity indices were calculated. Calculations were performed using the Paleontological Statistics v4.11 Statistical package.

$$\text{Shannon index} = -\sum_i \frac{n_i}{n} \ln \frac{n_i}{n} \quad (4)$$

$$\text{Margalef's richness index} = \frac{S-1}{\ln(n)} \quad (5)$$

$$S = \alpha \ln\left(1 + \frac{n}{\alpha}\right) \quad (6)$$

where n is the total number of individuals, ni is the number of individuals of species i, S is the number of species, and α is Fisher's alpha.

The species accumulation curve was generated considering the species richness values in the evaluated plots. Likewise, tree species richness established after natural regeneration was extrapolated for up to 100 plots using the program EstimateS 9.1 [39].

2.4. Response of the Species to the Time of Abandonment and Distance to the Forest Edge

Huisman-Olff-Fresco (HOF) models were used to analyze the influence of abandonment time and distance to the forest edge in species establishment [40]. These models provided a hierarchical set of five models based on the increasing complexity of biological information [41,42]. According to Huisman et al. [41], models are classified as: Model I, no significant trend in space or time; Model II, an increasing or decreasing trend where the maximum is equal to the upper bound M; Model III: an increasing or decreasing trend where the maximum is below the upper bound M; Model IV: increase and decrease by the same rate: symmetrical response curve; Model V: increase and decrease by different rates: skewed data. Only the HOF response curves of the species that were found in more than 50% of the evaluated plots were modeled.

2.5. Data Analysis

Average basal area, abundance, species richness, and vegetation diversity were evaluated according to abandonment time category with the Student–Newman–Keuls post-hoc test (SNK).

To analyze the relationship between abandonment time and distance to the forest edge on the floristic attributes of natural regeneration, correlation analysis, and multiple linear regression were performed. The Spearman's rank correlation coefficient was used because the parametric assumptions were not met. Regarding the multiple linear regression analysis, the floristic diversity attributes were considered as response variables, and the distance to the forest edge and the abandonment time of the plots as predictor variables.

Prior to the analysis, the square root transformation was performed for each of the variables to comply with the parametric assumptions. The hierarchical partition of the variance was performed using the hier.part package in R [43] to analyze the importance of each predictor in the response variable of the multiple regression analysis.

To evaluate and represent the floristic affinity between the evaluated plots, the non-metric multidimensional scaling (nMDS) analysis was used, and stress was used as a quantitative measure of the nMDS ordination fit [44]. To identify the grouping of plots based on their floristic composition, a Similarity Profile Analysis (SIMPROF) was carried out. These analyzes were performed in Primer v7 statistical package.

To analyze the influence of the distance to the forest edge and the abandonment time on the floristic composition, the Distance-based Redundancy Analysis (dbRDA), based on the Bray–Curtis distance, was performed. Species abundance transformed matrix was used considering the square root transformation and excluding the species that were only present in three plots. To determine the individual effect of the predictor variables (distance to the forest edge and abandonment time), the variance was partitioned using the "rdacca.hp" package [45].

Finally, the analyses mentioned before were all performed at a significance level of 5% for type 1 errors. The figures were generated using the R statistical packages in the R-Studio environment [43], SigmaPlot 15 (Inpixon Inc., Palo Alto, CA, USA), and Primer Permanova+ v7 (PRIMER-E Ltd., Auckland, New Zealand).

3. Results

3.1. Structure, Diversity, and Floristic Composition

The abandonment time of the evaluated plots varied between 2 and 23 years, and the distance to the forest edge ranged between 35 and 560 m (Table 1). In the 20 evaluated plots, 931 individuals were recorded, distributed in 79 species and 30 families. Species richness per plot ranged from 4 to 23, while abundance per plot ranged from 12 to 73. Species richness and diversity indices (Shannon, Margalef, and α-Fisher) were higher in plots with longer abandonment time and less distance to the forest edge.

Families with the highest species richness were Fabaceae (20), Malvaceae (8), and Moraceae (6). Fabaceae, Malvaceae, and Urticaceae presented the highest values of the importance index (>18%), and they represented 49% of the abundance and 70% of the total relative dominance of the individuals found. On the other hand, Asteraceae had the highest number of individuals (17.72% of the total) but represented a very low dominance (2.6%) (Table 2). Regarding the relation to the time of abandonment and distance to the forest edge, plots showed variable results. The majority (75%) of the plots belonged to 10 years of abandonment time (Table 1).

Table 1. Characteristics of the plots and abundance, species richness, and diversity indices of natural regeneration found in the study area and reference values for primary forests of Madre de Dios.

N	Abandonment Time	Distance to the Forest Edge	Species Richness	Abundance	Shannon-H	Margalef	α-Fisher
1	10	60	13	73	1.76	2.80	4.60
2	3	110	10	65	1.77	2.16	3.30
3	4	125	11	68	1.99	2.37	3.72
4	20	40	15	61	2.20	3.41	6.35
5	23	35	20	66	2.41	4.54	9.76
6	11	110	21	66	2.73	4.77	10.63
7	14	70	23	55	2.69	5.49	14.86
8	5	150	10	28	2.01	2.70	5.57
9	2	260	8	43	1.41	1.86	2.90
10	6	250	5	12	1.42	1.61	3.22
11	7	139	16	50	2.16	3.83	8.14
12	7	227	9	41	1.60	2.15	3.56
13	3	560	4	21	1.12	0.99	1.47
14	6	120	15	73	1.74	3.26	5.72

Table 1. *Cont.*

N	Abandonment Time	Distance to the Forest Edge	Species Richness	Abundance	Shannon-H	Margalef	α-Fisher
15	4	318	8	19	1.78	2.38	5.21
16	4	195	12	48	1.74	2.84	5.14
17	5	175	11	45	1.69	2.63	4.64
18	5	171	11	42	1.59	2.68	4.85
19	8	250	13	20	2.32	4.01	16.10
20	6	150	12	35	2.11	3.09	6.45
Reference values for the old-growth forest in Madre de Dios [46]			82.7 ± 13 (61–125)	215 ± 51 (129–323)	3.9 ± 0.4 (2.4–4.5)	-	53.3 ± 16 (24.5–90.8)
Reference values for early secondary forests (5 years after abandonment) in the Amazon [47]			41.7 ± 10.4 (26–70)	-	2.8 ± 0.4 (1.8–3.7	-	-

Table 2. Abundance, species richness, and family importance index of tree vegetation for the study area.

Family	Species Richness	Abundance	Relative Abundance (%)	Basal Area (m²)	Relative Dominance (%)	Importance Index (%)
Fabaceae	20	141	15.15	0.7369	28.76	21.95
Malvaceae	8	155	16.65	0.5498	21.46	19.05
Urticaceae	3	161	17.29	0.5139	20.05	18.67
Asteraceae	4	165	17.72	0.0658	2.57	10.15
Piperaceae	1	78	8.38	0.0644	2.51	5.45
Moraceae	6	15	1.61	0.2121	8.28	4.94
Euphorbiaceae	5	22	2.36	0.1316	5.14	3.75
Verbenaceae	2	60	6.44	0.0225	0.88	3.66
Cannabaceae	4	17	1.83	0.0530	2.07	1.95
Rubiaceae	2	16	1.72	0.0552	2.16	1.94
Myrtaceae	1	13	1.40	0.0134	0.52	0.96
Hypericaceae	2	14	1.50	0.0086	0.33	0.92
Muntingiaceae	1	10	1.07	0.0119	0.46	0.77
Bixaceae	1	8	0.86	0.0121	0.47	0.67
Primulaceae	1	3	0.32	0.0240	0.93	0.63
Salicaceae	3	8	0.86	0.0075	0.29	0.58
Caryophyllaceae	1	4	0.43	0.0179	0.70	0.56
Meliaceae	1	1	0.11	0.0219	0.85	0.48
Poaceae	1	8	0.86	0.0025	0.10	0.48
Solanaceae	2	6	0.64	0.0037	0.14	0.39
Melastomataceae	4	6	0.64	0.0034	0.13	0.39
Polygonaceae	1	4	0.43	0.0082	0.32	0.38
Peraceae	1	4	0.43	0.0046	0.18	0.30
Rutaceae	1	2	0.21	0.0099	0.39	0.30
Malpighiaceae	1	4	0.43	0.0034	0.13	0.28
Boraginaceae	4	2	0.21	0.0007	0.03	0.12
Araliaceae	1	1	0.11	0.0021	0.08	0.10
Phyllanthaceae	1	1	0.11	0.0007	0.03	0.07
Simaroubaceae	1	1	0.11	0.0006	0.02	0.06
Gentianaceae	1	1	0.11	0.0002	0.01	0.06

Figure 3 shows the comparison of the species accumulation curve produced by the analytical formula and extrapolation by randomizing the samples with estimates. At least 90% of naturally regenerating species had been sampled in the study area, as shown in the accumulation curve (Figure 3).

A higher density of individuals was found in the lower diameter category (DBH < 5 cm), independent of the abandonment time. Towards higher diameter categories, a decreasing trend appeared in relation to the density, showing an inverted-J distribution pattern in the chronosequence (Figure 4). On the other hand, the proportion of individuals in the <5 cm DBH category decreased throughout the chronosequence, from 79.6% in plots with 1–5 years of abandonment to 64.8% in those with more than 10 years of abandonment. This decrease is paired with an increase in the proportion of individuals (between 4.5% to 5%) in the diameter categories between 5–20 cm. Regarding the total height, a gradual increase was found according to diameter categories (Figure 4), and the tallest individuals were found in the plots with the longest abandonment time (Figure 4D).

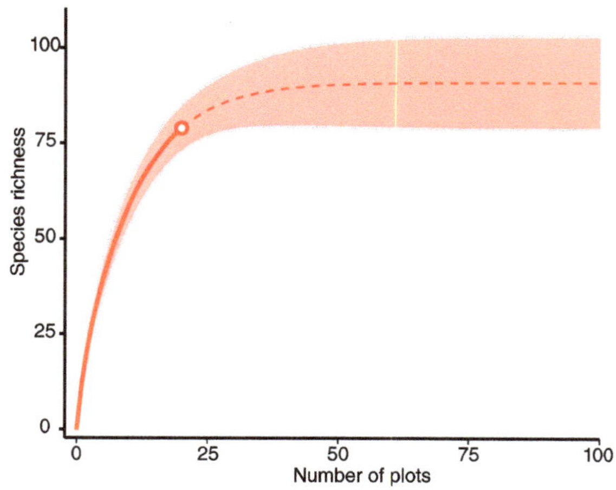

Figure 3. Species accumulation curve of natural regeneration in study areas. Reference samples are indicated by a solid white dot. Dashed lines: extrapolation curves. Shaded area for each solid line: 95% confidence interval for the expected richness of rarefied classes. Shaded area for each dashed line: 95% confidence interval for the expected extrapolated class richness up to a sample size of 80 based on a bootstrap method with 9999 replicates.

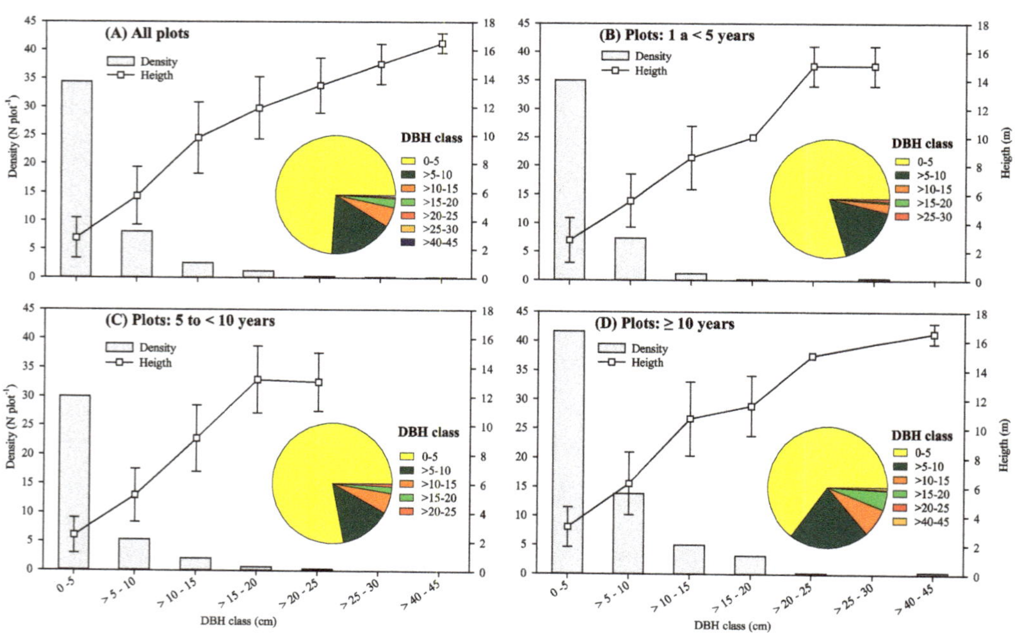

Figure 4. The density and average height of individuals according to diameter categories of the vegetation in study areas. DBH = diameter at breast height.

3.2. Species according to Abandonment Time Categories

In areas with shorter abandonment time (1–5 years), a greater abundance and dominance of pioneer species were observed, such as *Ochroma pyramidale* and *Cecropia membranacea*. The longer the abandonment time, the greater presence of early succession species was observed. In the period of 5–10 years of abandonment, species of *Ficus* and *Alchornea* were found. In the plots with more than 10 years of abandonment, several species of the genera *Inga*, *Croton*, and *Ficus* were found (Figure 5).

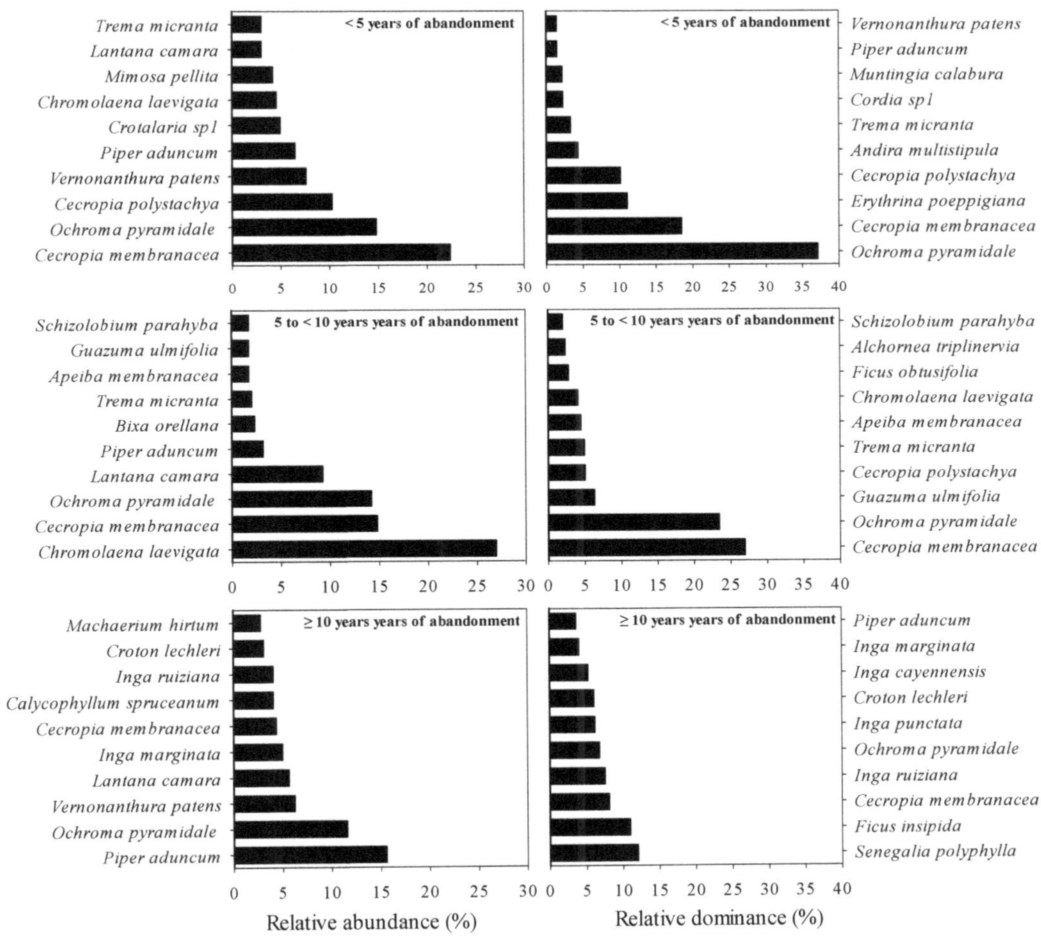

Figure 5. Relative abundance and relative dominance of the first ten species according to different abandonment time categories.

3.3. Species Occurrence according to the Gradient of Distance to the Forest Edge and Abandonment Time

Figures 6 and 7 show the HOF response curves of the 12 most frequent species found in the study area, where it is apparent that the other remaining species do not intensively colonize degraded areas. In most species modeled (*Cecropia membranacea, Ochroma pyramidale, Piper aduncum, Lantana camara, Psidium guajava, Vernonanthura patens, Apeiba tibourbou* y *Vismia gracilis*), HOF response curves were obtained. Different trends were found depending on the variable analyzed (abandonment time and distance to the forest edge) (Figures 6 and 7). Thus, *C. membranacea* and *O. pyramidale* showed a biased response along the gradient of distance to the forest edge. Both species increased in abundance up to distances between 100–200 m and then decreased along the gradient. However, the response of these two species was different in relation to abandonment time. A decreasing trend in abundance was found for *C. membranacea. Meanwhile*, for *O. pyramidale*, it was relatively abundant in most of the plots in the abandonment time gradient. Moreover, *P. aduncum*, *L. camara*, and *A. tibourbou* abundance showed a tendency to decrease with greater distance from the forest edge but increased with the longer abandonment time gradient.

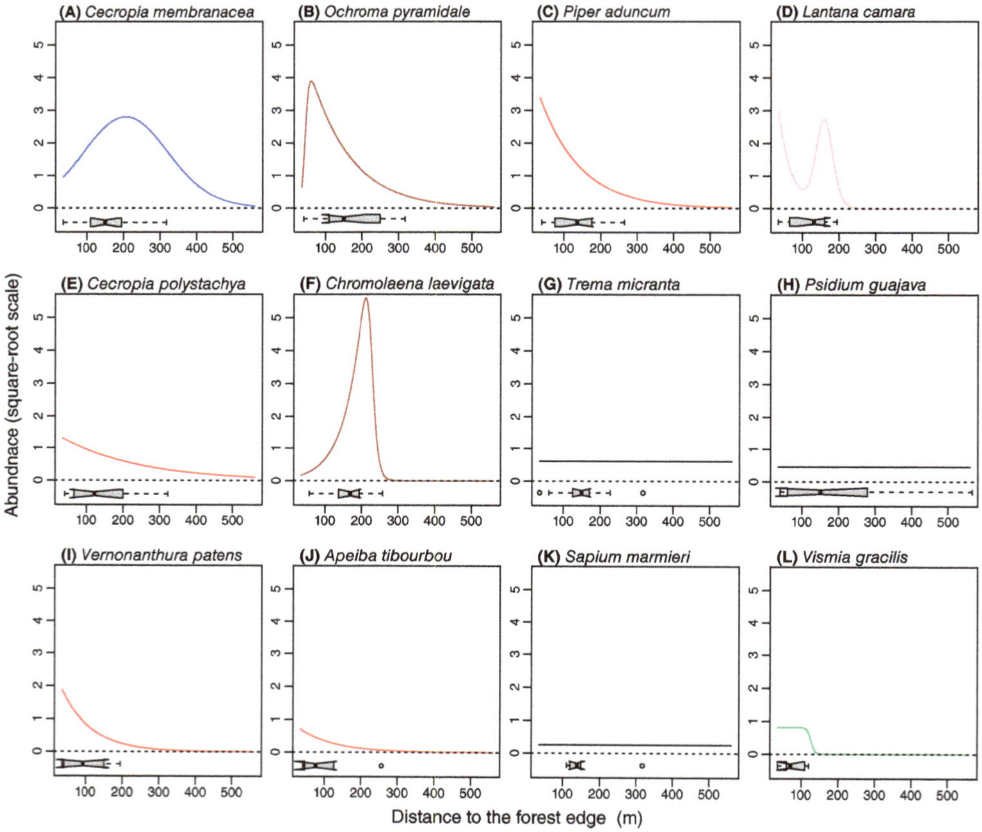

Figure 6. Response curves derived from HOF models of the most common species in relation to the gradient of distance to the forest edge of plots in the study area.

3.4. Time and Distance Relationship

Variability was found in the plots according to abandonment time and distance to the forest edge. According to categories of abandonment time, a significant increase (ANOVA, p-value < 0.05) was found in the species richness, diversity, and basal area of the vegetation in the degraded areas evaluated throughout the chronosequence; however, this pattern varied according to floristic attribute (Figure 8).

Abandonment time was found to be directly related to the basal area, richness, and species diversity (rho > 0.3; p-value < 0.05). On the other hand, distance to the forest edge was inversely related to the same vegetation attributes (rho < −0.4; p-value < 0.05). This suggests that plots with a longer period of abandonment have greater development of natural regeneration and that the plots closest to the reference forest have a greater recovery of the vegetation cover (Table 3A).

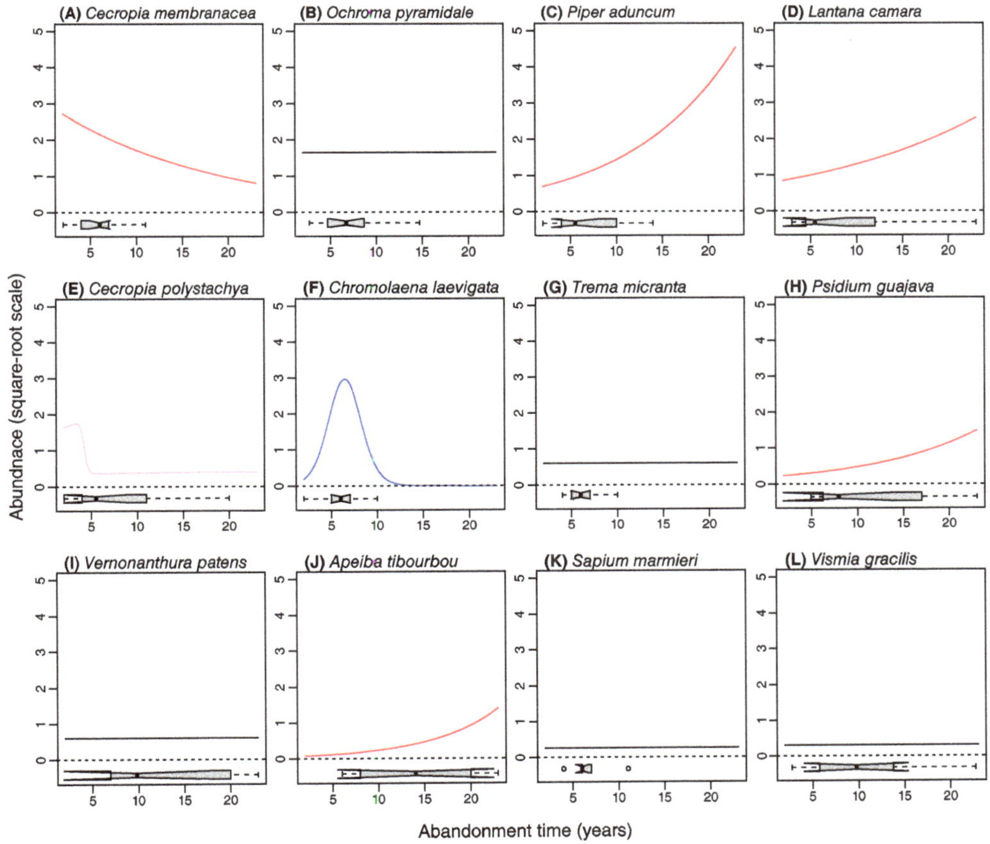

Figure 7. Response curves derived from HOF models of the most common species in relation to the abandonment time gradient in the study area.

However, the observed pattern changes slightly when we analyze the relationships between the floristic attributes with the distance of the forest edge and abandonment time according to size categories (seedling, sapling, and tree) of the evaluated individuals. The diversity of seedling species was inversely and significantly related to the distance to the forest edge. At the same time, the lowest and non-significant correlation was found between the time of abandonment and the abundance of the same category (Table 3B). Regarding the tree category, a significant relationship between abandonment time and the number of individuals was found (rho = 0.55 p-value < 0.05) (Table 3D).

Significant multiple time-distance regressions were fitted for basal area, species richness, number of individuals, and diversity indices (Shannon, Margalef, and α-Fisher). Most of the floristic attributes showed a direct relationship (partial correlation) with the abandonment time and an inverse relationship with the distance to the forest edge (Table 4). Models for species richness and abundance were the ones with higher adjusted R-squared values ($R^2 > 0.6$). That is, more than 60% of the variability in species richness and abundance could be explained by the abandonment time and the distance to the forest edge. In variance partitioning of the independent effects, we found that distance to the forest edge contributes more to the variance of species richness and abundance than the abandonment time. Nevertheless, the difference in the individual effect was the highest in abundance (Table 4).

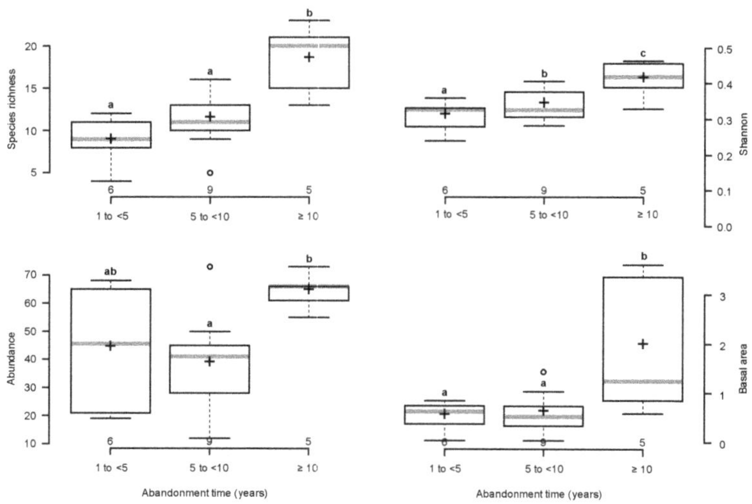

Figure 8. Boxplots of the abundance, species richness, Shannon diversity, and basal area of woody vegetation according to categories of abandonment time. The gray line represents the mean. Different letters indicate that the difference between the groups is significant by Student–Newman–Keuls post hoc testing (p-value < 0.05).

Table 3. Correlation between abandonment time and distance to the forest edge with floristic attributes in the study area. Spearman's correlation coefficient, * p-value < 0.05; ** p-value < 0.01; *** p-value < 0.001; ns = not significant. The letter "a" after the correlation coefficient means that the Pearson correlation was used.

	Basal Area	Species Richness	Individuals	Diversity Index		
				Shannon	Margalef	α-Fisher
(A) All individuals						
Abandonment time	0.48 *	0.76 ***	0.31 ns	0.67 **	0.75 ***	0.71 ***
Distance to the forest edge	−0.58 **	−0.77 ***	−0.81 ***	−0.63 **	−0.63 **	−0.45 *
(B) Seedling category						
Abandonment time	0.48 *	0.52 *	0.09 ns	0.42 ns	0.56 *	0.43 ns
Distance to the forest edge	−0.58 **	−0.86 ***	−0.63 ***	−0.80 ***	−0.77 ***	−0.69 ***
(C) Sapling category						
Abandonment time	0.49 *	0.61 **	0.30 ns	0.58 *	0.61 **	0.41 ns
Distance to the forest edge	−0.46 ns	−0.54 *	−0.51a *	−0.33a (ns)	−0.30a (ns)	−0.20 ns
(D) Tree category						
Abandonment time	0.45 ns	0.59 *	0.55 *	0.62 **	0.59 *	0.52 *
Distance to the forest edge	−0.50 *	−0.36 ns	−0.48 *	−0.33a (ns)	−0.35 ns	−0.37 ns

3.5. Ordination Analysis of Floristic Composition

The non-metric multidimensional scaling (nMDS) analysis diagram showed reduced-dimensional representations of the plots evaluated according to their floristic composition (Figure 9). These representations vary according to the matrix of the abundance used, either considering all registered individuals (Figure 9A) or according to size categories: seedlings (Figure 9B), saplings (Figure 9C), and trees (Figure 9D). Stress found in all the representations was acceptable (<0.20) [44]. Regarding the nMDS of all the individuals and in the seedlings category, a distribution gradient of the plots was found according to the distance to the forest edge. Taking into consideration all the individuals (Figure 9A), low values in the nMDS2 axis are related to plots with a greater distance from the forest edge. On the other hand, in the seedlings category, high values of the nMDS2 axis were found for the plots with a greater distance from the forest edge (Figure 9B).

3.6. Abandonment Time and Distance to the Forest Edge in the Structuring of Plant Communities

The distance-based redundancy analysis (dbRDA) showed that the adjusted model, with abandonment time and distance to the forest edge, significantly explained variation in the floristic composition of the communities (adj-R^2 = 0.15, p-value = 0.002). By partitioning the variance, it was found that distance to the forest edge (p-value = 0.001) and abandonment time (p-value = 0.013) significantly influence the floristic composition. However, distance to the forest edge had a greater individual effect than the abandonment time, 11.9% compared to 5.6% of the total variation of the communities (Figure 10).

Table 4. Multiple regression between abandonment time and distance to the forest edge, with the basal area, abundance, species richness, and species diversity in study areas.

Variable	Predictors	Partial Correlation	Hierarchical Partitioning			p-Value	Model Adjusted R^2
			Independent	Joint	Independent Effect (%)		
Basal Area	Abandonment time Distance to the forest edge	0.175 −0.329	0.134 0.195	0.113	40.81 59.19	0.034	0.25
Species richness	Abandonment time Distance to the forest edge	0.341 −0.578	0.274 0.396	0.231	40.91 59.09	<0.001	0.631
Abundance	Abandonment time Distance to the forest edge	−0.378 −0.764	0.099 0.543	0.039	15.47 84.53	<0.001	0.642
Shannon diversity index	Abandonment time Distance to the forest edge	0.38 −0.382	0.27 0.272	0.193	49.89 50.11	0.001	0.488
Margalef diversity index	Abandonment time Distance to the forest edge	0.421 −0.391	0.297 0.282	0.206	51.29 48.71	<0.001	0.53
α-Fisher diversity index	Abandonment time Distance to the forest edge	0.469 −0.027	0.277 0.104	0.103	72.79 27.21	0.017	0.308

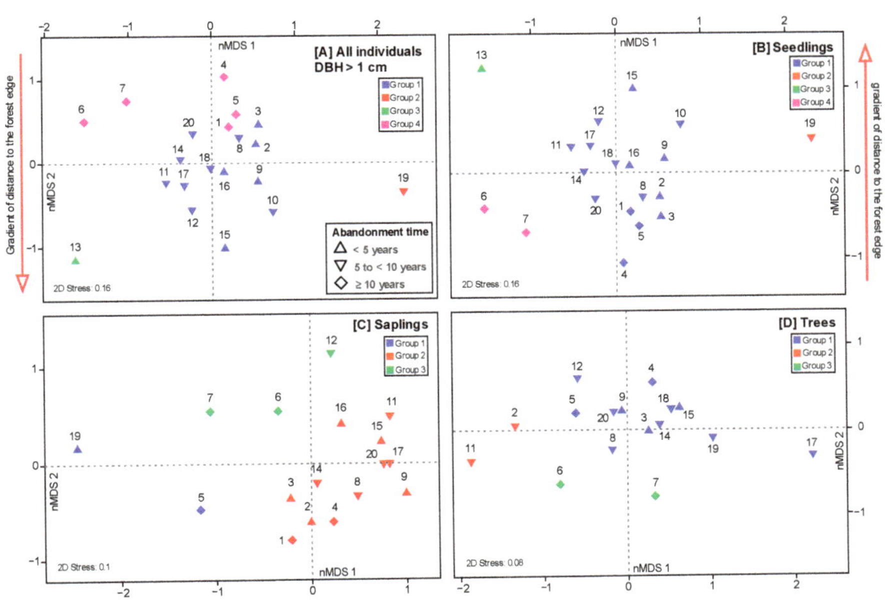

Figure 9. Non-metric multidimensional scaling (nMDS) analysis diagram of floristic similarity between the 20 naturally regenerated plots evaluated. All individuals (**A**) and according to size categories: (**B**) seedlings, (**C**) saplings, and (**D**) trees. Groups (different colors) represent statistically significant clusters of plots (p-value < 0.05) according to the SIMPROF test, and the shape of the symbols represents the category of abandonment time.

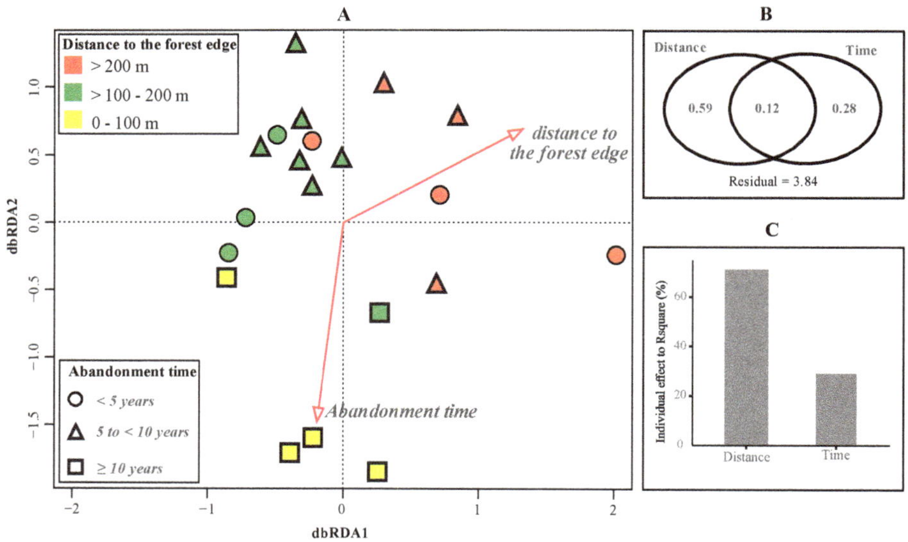

Figure 10. Graphic representation of the dbRDA (**A**) floristic similarity between the plots evaluated in the study area. (**B**) Venn diagram of the variance partition according to distance to the forest edge and abandonment time and (**C**) individual effects of distance to the forest edge and abandonment time in the floristic composition.

4. Discussion

Recovery of tropical forests after the removal of vegetation cover varies between regions and depends on local conditions [30,48]. Variability can also be influenced by a series of possible environmental and ecological factors [2,49,50]. It is known that in areas severely degraded by mining, the heterogeneity of the habitats generated and a long distance from a propagule source are the main factors affecting the dispersal and establishment of vegetation [27,51]. In this study, we found differences in structure, abundance, species diversity, and floristic composition of spontaneous vegetation according to abandonment time and distance to the reference forest edge in areas abandoned by mining activity in the San Jacinto native community, Peruvian Amazonia. Our analysis showed that the influence of distance to the forest edge was significantly higher (p-value = 0.001) than abandonment time (p-value = 0.013) on the variation of floristic attributes and composition of the vegetation.

4.1. Species according to Abandonment Time Category

In terms of species richness, tropical forests are highly resilient to disturbances, p.e. Rozendaal et al. [52] found that tree species richness increased very rapidly during secondary succession, with 80% recovery of old-growth values after only 20 years. The values of richness and diversity of species found from natural regeneration in the study areas are much lower than the averages reported in neotropical secondary forests [52] and in the forests of Madre de Dios [46]. For example, on average, the richness and diversity of α-Fisher were found to represent only 15% and 12%, respectively, of the average values in the natural forests of Madre de Dios (Table 1). In comparison with early secondary forests in the Amazon (5 years after abandonment) [47], our study revealed that the natural regeneration following the abandonment of gold mining activities was rather slow. On average, the species richness found in the present study represents only 30% of the average values in early secondary forests in the Amazon [47] (Table 1). This low species richness in the early secondary succession is due to the fact that the first colonizers of abandoned areas are pioneer and herbaceous plants [18]. However, our values of richness and diversity are higher than those reported in a study of natural regeneration in reforested areas of 20 years of abandonment due to mining activity in Madre de Dios [53].

According to the results of the HOF response curves, *C. membranacea* and *O. pyramidale* clearly dominate these successional forests in plots severely degraded by mining activity. The greater abundance and dominance of pioneer tree species (for example, species of the genus *Cecropia*, *Ochroma*, and *Vernonanthura*) in plots with shorter abandonment times (1–5 years) would facilitate the recruitment of old-growth specialists that become dominant in late stages of succession [4,30]. Since they can accumulate organic matter in the soil and improve its structure, creating microsites, reducing runoff, attracting dispersers, and promoting the colonization of species, they can function as nurse plants [49,54].

The diameter distribution pattern showed changes throughout the different abandonment times. The highest density was found in the first diameter class (<5 cm), with a strong decrease in the following classes. This inverted J pattern suggests a good recruitment potential [55], and it is typical in primary forests [56], secondary succession forests [33], and reforested areas in the Amazon [53,57]. The average height values according to diameter class followed the pattern expected for a successional chronosequence, that is, an increase in the average height of the largest diameter classes in areas with a longer abandonment time [33].

4.2. Abandonment Time and Distance to the Forest Edge

The abundance, richness, and diversity of natural regeneration were inversely correlated with distance to the forest edge, as reported by previous studies [50,53,58]. This would suggest that closer proximity to the remaining reference forest could accelerate the recovery of the vegetation [15]. Because the recovery of biodiversity in tropical forests depends on the recruitment of native species and their accumulation over time [59], the fragmentation of Amazonian forests significantly changes the abundance and richness of species, mainly at distances of less than 500 m [60]. However, our results showed different responses between floristic attributes and distance to the forest edge in a recent study of lands degraded by gold mining in Madre de Dios. Alarcón-Aguirre et al. [4] found no significant inverse relationship between distance to the forest edge with abundance, species richness, and basal area. This difference in the results could be explained by the high degree of landscape fragmentation in our study, compared to Alarcón-Aguirre et al. [4].

The structure and diversity of secondary forests can be recovered over time [22,29]. In this study, we found that species richness increased throughout chronosequence, a typical pattern in successional forests [59,61]. The increase of diversity in natural regeneration would be associated with the probability of the coexistence of fast-growing pioneer species that would favor the establishment of shade-tolerant species throughout the succession [30]. Abundance, species richness, and diversity of tree vegetation are directly related to abandonment time and inversely related to distance to the forest edge. A slower recovery of natural regeneration was evidenced for plots distant from the forest edges. This can be explained by the remoteness of the remaining forest due to the high degree of fragmentation and landscape degradation, reduced seed input, and the potential for natural seed dispersal from nearby remaining habitats [20,21,27]. Mi et al. [23] and Mesquita et al. [9] found significant interactions between disturbance regimes and distance from habitat conditions on floristic composition for most of the sites studied. According to Mi et al. [23] and Mesquita et al. [9], the explanation is that stochastic processes dominate forest succession in post-disturbance restoration areas in tropical forests.

Regarding the natural regeneration of forests, although it increases biomass and biodiversity, it has an uncertain path for recovery because it depends on the specific impact generated in the past on these areas [62]. Therefore, the passive restoration of the evaluated areas would be ineffective due to the forest fragmentation [63] and the landscape and ecological conditions of the study area [15], as they could limit the supply of seeds. Slower natural regeneration hinders the process of ecological succession and the recovery of the forests [2]. However, when the distance to the remaining forest edges is not too great, natural regeneration, which is a passive restoration, could be considered a valuable option for biodiversity restoration [20]. Moreover, in late succession, seed sources and their dispersal are the most important biotic factors as they are more limited by distance [49]. In addition, passive restoration is considered a cost-effective restoration strategy in comparison with active restoration [22]. However, passive restoration is a slow process, and it may take decades or longer to achieve desired ecosystem functions or species compositions without active restoration [64]. Previous studies in areas degraded by mining in Madre de Dios have reported that the initial cost of reforestation ranged between 3464 and 9250 US$ ha^{-1} [53,65]. Because of the high degree of fragmentation and landscape degradation, active restoration should be considered for the recovery of degraded areas [63] by gold mining in the study area.

Even though various studies show, that abandonment time explains the greatest variability in the recovery of vegetation communities in a secondary succession [9,22,26,29,30]. Our study shows that the distance to the forest edge has more influence on the abundance, species diversity, and composition of tree communities after abandonment by mining activities. A low variance explanation of abandonment time in the recovery of the vegetation was demonstrated by previous studies [29,47]. An explanation would be that the recovery rate of the diversity and composition of secondary forest species are determined by a complex interaction of specific local and historical factors and landscape structure [22,29].

4.3. Abandonment Time and Distance to the Forest Edge in the Structuring of Forest Communities

In the nMDS, we found a gradient in the floristic similarity between the evaluated plots with respect to abandonment time, mainly in all the individuals (DBH > 1 cm). In the dbRDA, we found that the distance to the forest edge had a greater influence on floristic similarity than the time of abandonment. In addition, we found a greater Bray–Curtis similarity between plots near the edge of the forest (47% $\pm$ 0.1) than between the plots furthest away (13.3% $\pm$ 14.9). A heterogeneity was revealed in the floristic composition of the vegetation that was established in these areas abandoned by mining activity. The reason is that distance to the forest edge causes a significant effect on forest communities that are established under these conditions [3,22], and it could take centuries for their species composition to reach the levels of a reference forest [30]. Moreover, in most cases where the recovery of vegetation cover depends on the structure and spatial distribution of primary forests, it would directly affect the diversity of tree species [64]. On the other hand, Goosem et al. [48] found an influence of fragmentation on species richness and composition, where a low level of dispersion is related to limited reference forest seed dispersers [50], mainly birds in the case of large fruits [31]. This was demonstrated by an experimental study in Costa Rica, in which fragmentation strongly affected the presence of forest-dependent birds, the abundance of late-successional seedling recruitment, and species richness in forest restoration plantations [50].

Several studies found a strong impact of the distance to the forest edges on seed rain, which would justify the patterns of diversity and floristic composition found in our study [3,15]. In addition, Nolasco et al. [58] found that abandonment time in successional forests in the northern Brazilian Amazon is one of the variables that best explain the variation in floristic composition, contrary to what was reported in the present study. These differences might be explained due to higher fragmentation in our study area. In contrast, in the study of Nolasco et al. [58], dispersal by distance was less significant because successional forests are dispersed within a continuous forest matrix. Therefore, proximity to forest fragments is crucial for the success of restoration because it can increase the success of restoration and reduce costs [59].

4.4. Natural Regeneration Outlook for Forest Recovery

Our study provides evidence of natural regeneration in areas degraded by ASGM, as well as proof that passive restoration is not an adequate tool for the recovery of vegetation cover in highly fragmented landscapes. Thus, 79 species and 30 botanical families were identified with abandonment time after mining activity that varies between 2 and 23 years. Since the natural regeneration of woody vegetation is very slow, passive restoration should not be considered as an alternative to the recovery of degraded areas in Madre de Dios [2]. Among the main reasons for the slowness of natural recolonization would be the remoteness of existing forest fragments combined with reduced seed input caused by a high level of fragmentation. In general, 95% of some studies about biodiversity stated a direct relationship between diversity and ecosystem functioning, that only 20–50% of species are needed to maintain most biogeochemical ecosystem processes [66]. In this way, we found that a greater floristic composition (greater biodiversity) would allow directing the environmental recovery of natural regeneration towards the restoration of ecological processes. We also found that there are key species that would allow the maintenance of ecological processes, as would be the case of an assisted restoration with *P. aduncum*, *L. camara*, and *A. tibourbou*. This would improve the ecological integrity (complexity of biological assemblages, including species composition and representation of all functional groups) needed for restoration [67]. It is important to acknowledge the processes and factors that influence natural regeneration since an inadequate restoration plan can cause negative impacts on the process, such as the establishment of invasive species or alterations in successional trajectories [15,68].

Studies of recovery of areas degraded by gold mining, whether by anthropogenic action or spontaneously, have practical implications for subsequent management depending on the type of mining (small and semi-mechanized), the level of fragmentation of the landscape, soil quality, rivers, and residual water disposal from mining (seasonal flooding). These implications were confirmed in the study where unassisted succession processes can lead to the establishment of woody vegetation with greater or lesser demand of time. These results could be further supported by soil quality data through the physical-chemical analysis of the study. Finally, the results lay the groundwork for further studies that should consider the evaluation of natural vegetation dynamics in areas degraded by gold mining and be compared with processes assisted by the level of disturbance from gold mining.

4.5. Limitations and Future Research Directions

This study has certain limitations. First, the study focused solely on degraded areas in a native community of Madre de Dios, which affects the generalizability of the results. The sector or type of mining activity was not considered. Therefore, more studies are needed at a regional level to enhance the findings of the study. Second, the current study used a chronosequence approach (cross-sectional data), and future studies need to be longitudinal to confirm the recovery of the vegetation over time. In addition to these limitations, future studies can incorporate the other variables that may affect the establishment of natural regeneration in areas degraded by mining activity in Madre de Dios, i.e., functional traits, level of landscape fragmentation, soil invertebrates, soil quality, measuring wood density and rivers (seasonal flooding) should be considered. Despite these limitations, this study provides new insights for a deeper understanding of natural regeneration in areas degraded by ASGM in the Peruvian Amazon.

5. Conclusions

A strong dominance of pioneer species was found in plots with less abandonment time in the areas affected by the mining of the native community of San Jacinto in comparison to plots with longer abandonment times. Our study revealed that the natural regeneration following the abandonment of gold mining activities was rather slow. The recovery of natural regeneration is significantly correlated with the time of abandonment and the distance to the reference forest. Based on our results and due to the high degree of fragmentation of the study area by mining activity, we suggest that passive restoration is not an alternative for the recovery of these degraded areas. Tree species such as *O. pyramidale* and species of the genera *Ficus*, *Cecropia*, and *Inga* are key. All of them could allow an active process of integral ecological restoration in case an assisted restoration takes place in more than 90,000 ha degraded by ASMG in Madre de Dios [35].

Author Contributions: Conceptualization, J.G.-Q., M.V.R. and D.d.C.T.; methodology, J.G.-Q., M.V.R. and S.B.-Q.; validation, E.B.-L. and L.H.V.; formal analysis, J.G.-Q. and Y.F.-M.; investigation, J.G.-Q., E.B.-L., S.B.-Q., Y.F.-M. and M.H.-M.; writing—original draft preparation, J.G.-Q.; writing—review and editing, M.V.R., M.A.-S., L.R.-A., P.N.H., L.H.V., R.C.G., G.A.-A. and D.d.C.T.; visualization, M.V.R., G.A.-A. and M.A.-S.; resources, S.B.-Q., P.N.H. and R.C.G.; data curation, A.M.-U. and M.H.-M.; funding acquisition, M.V.R. and. D.d.C.T.; project administration, R.C.G. All authors have read and agreed to the published version of the manuscript.

Funding: This research was funded by Consejo Nacional de Ciencia y Tecnología (CONCYTEC) and Fondo Nacional de Desarrollo Científico y Tecnológico (FONDECYT), grant number CONTRATO-LORETO-IIAP-122-2018-FONDECYT-BM-E041.

Institutional Review Board Statement: Not applicable.

Data Availability Statement: The data presented in this study are available on request from the corresponding author.

Acknowledgments: This study is made possible by funding provided by the Consejo Nacional de Ciencia y Tecnología (CONCYTEC) and Fondo Nacional de Desarrollo Científico y Tecnológico (FONDECYT) (Proyecto CONTRATO—LORETO-IIAP-122-2018-FONDECYT-BM- E041), the Peruvian National Protected Areas Service (Servicio Nacional de Áreas Naturales Protegidas—SERNANP), KU Leuven University in Belgium, under the terms of cooperative scientific agreement with the Peruvian Amazon Research Institute (Instituto de Investigaciones de la Amazonía Peruana—IIAP). Additional funding was provided by the NGO Solidaridad Network South America through its "Responsible Gold in Madre de Dios" project, Franco Arista, the project leader, and his collaborators William Moreno, Jhonatan Jaramillo, and Diana Wu; and the Vice-Rectorate for Research of the Universidad Nacional Amazónica de Madre de Dios. This research is dedicated to the Community of Fortuna and the Native Community of San Jacinto. The community in Peru might use this research as a tool to manage their land restoration. We are grateful for the support of specialists from the Research Institute of the Peruvian Amazon (IIAP), Biodiversity International, Centro de Innovación Científica Amazónica (CINCIA), Universidad Nacional Amazónica de Madre de Dios (UNAMAD), Universidad Nacional de Piura (UNP) and the National Agrarian University La Molina (UNALM). We would also like to thank Foster Brown for checking the English version of the manuscript.

Conflicts of Interest: The authors declare no conflict of interest.

References

1. Cherlet, M.; Hutchinson, C.; Reynolds, J.; Hill, J.; Sommer, S.; von Maltitz, G. *World Atlas of Desertification, Mapping Land Degradation and Sustainable Land Management Opportunities*, 3rd ed.; Publication Office of the European Union: Luxembourg, 2018.
2. Timsina, S.; Hardy, N.G.; Woodbury, D.J.; Ashton, M.S.; Cook-Patton, S.C.; Pasternack, R.; Martin, M.P. Tropical surface gold mining: A review of ecological impacts and restoration strategies. *Land Degrad. Dev.* **2022**, *33*, 3661–3674. [CrossRef]
3. Reitschmiedová, E.; Dvorščík, P.; Mudrák, O.; Šimáňová, D.; Frouz, J. Differences in colonization strategies of three common pioneer woody species in post mining heaps. *J. Environ. Manag.* **2022**, *319*, 115668. [CrossRef]
4. Alarcón-Aguirre, G.; Sajami Quispe, E.; Vásquez Zavaleta, T.; Ponce Tejada, L.V.; Ramos Enciso, D.; Rodríguez Achata, L.; Garate-Quispe, J. Vegetation dynamics in lands degraded by gold mining in the southeastern Peruvian Amazon. *Trees For. People* **2023**, *11*, 100369. [CrossRef]
5. Prach, K.; Walker, L.R. Differences between primary and secondary plant succession among biomes of the world. *J. Ecol.* **2019**, *107*, 510–516. [CrossRef]
6. Rocha-Nicoleite, E.; Campos, M.L.; Colombo, G.T.; Overbeck, G.E.; Müller, S.C. Forest restoration after severe degradation by coal mining: Lessons from the first years of monitoring. *Braz. J. Bot.* **2018**, *41*, 653–664. [CrossRef]
7. Leiva, J.D. Appropriate technologies and the geosocial evolution of informal, small-scale gold mining in Madre de Dios, Peru. *Extr. Ind. Soc.* **2022**, *12*, 101165. [CrossRef]
8. Stoll, E.; Roopsind, A.; Maharaj, G.; Velazco, S.; Caughlin, T.T. Detecting gold mining impacts on insect biodiversity in a tropical mining frontier with SmallSat imagery. *Remote Sens. Ecol. Conserv.* **2022**, *8*, 379–390. [CrossRef]
9. Mesquita, R.; Massoca, P.; Jakovac, C.C.; Bentos, T.; Williamson, G. Amazon Rain Forest Succession: Stochasticity or Land-Use Legacy? *Bioscience* **2015**, *65*, 849–861. [CrossRef]

10. Tito, R.; Salinas, N.; Cosio, E.G.; Boza Espinoza, T.E.; Muñiz, J.G.; Aragón, S.; Nina, A.; Roman-Cuesta, R.M. Secondary forests in Peru: Differential provision of ecosystem services compared to other post-deforestation forest transitions. *Ecol. Soc.* **2022**, *27*, 12. [CrossRef]
11. Zapata-Carbonell, J.; Bégeot, C.; Carry, N.; Choulet, F.; Delhautal, P.; Gillet, F.; Girardclos, O.; Mouly, A.; Chalot, M. Spontaneous ecological recovery of vegetation in a red gypsum landfill: Betula pendula dominates after 10 years of inactivity. *Ecol. Eng.* **2019**, *132*, 31–40. [CrossRef]
12. Lenart, J.; Schuchová, K.; Kašing, M.; Falteisek, L.; Cimalová, Š.; Bílá, J.; Ličbinská, M.; Kupka, J. The abandoned underground mine as a semi-natural ecosystem: The story of Flaschar's Mine (Czechia). *CATENA* **2022**, *213*, 106178. [CrossRef]
13. Velásquez-Ramírez, M.; del Castillo Torres, D.; Guerrero Barrantes, J.; Vásquez Bardales, J.; Thomas, E.; Cusi Auca, E.; Chinen Gushiken, M.; Muñoz Diaz, B.; Russo, R.; Corvera Gomringer, R. Soil recovery of alluvial gold mine spoils in the Peruvian Amazon using *Stylosanthes guianensis*, a promising cover crop. *Land Degrad. Dev.* **2021**, *32*, 5143–5153. [CrossRef]
14. Soto-Benavente, M.; Rodriguez-Achata, L.; Olivera, M.; Arostegui, V.; Colina, C.; Garate, J. Health risks due to the presence of heavy metals in agricultural products cultivated in areas abandoned by gold mining in the Peruvian Amazon. *Sci. Agropecu.* **2020**, *11*, 49–59. [CrossRef]
15. Conrado, D.; Rey-Benayas, J.; Costa, G.; Santos, S. Tree Communities in Three-Year-Old Post-Mining Sites Under Different Forest Restoration Techniques in the Brazilian Amazon. *Forests* **2020**, *11*, 527. [CrossRef]
16. Ribeiro, S.; Cerveira, A.; Soares, P.; Fonseca, T. Natural Regeneration of Maritime Pine: A Review of the Influencing Factors and Proposals for Management. *Forests* **2022**, *13*, 386. [CrossRef]
17. Hariharan, P.; Raman, T.R.S. Active restoration fosters better recovery of tropical rainforest birds than natural regeneration in degraded forest fragments. *J. Appl. Ecol.* **2022**, *59*, 274–285. [CrossRef]
18. Moyo, B.; Ravhuhali, K.E. Abandoned Croplands: Drivers and Secondary Succession Trajectories under Livestock Grazing in Communal Areas of South Africa. *Sustainability* **2022**, *14*, 6168. [CrossRef]
19. Brasil, A.B.; Schwartz, G.; Noronha, N.C.; Gama, M.A.P.; Ferreira, G.C. Natural regeneration for restoration of degraded areas after bauxite mining: A case study in the Eastern Amazon. *Ecol. Eng.* **2021**, *171*, 106392. [CrossRef]
20. Günter, S.; Weber, M.; Erreis, R.; Aguirre, N. Influence of distance to forest edges on natural regeneration of abandoned pastures: A case study in the tropical mountain rain forest of Southern Ecuador. *Eur. J. For. Res.* **2007**, *126*, 67–75. [CrossRef]
21. Cubiña, A.; Aide, T.M. The Effect of Distance from Forest Edge on Seed Rain and Soil Seed Bank in a Tropical Pasture. *Biotropica* **2001**, *33*, 260–267. [CrossRef]
22. Barros, T.C.; Elias, F.; Romano, L.L.; Ferreira, J. Natural recovery of plant species diversity in secondary forests in Eastern Amazonia: Contributions to passive forest restoration. *Rev. Bras. Bot.* **2020**, *43*, 165–175. [CrossRef]
23. Mi, X.; Swenson, N.G.; Jia, Q.; Rao, M.; Feng, G.; Ren, H.; Bebber, D.P.; Ma, K. Stochastic assembly in a subtropical forest chronosequence: Evidence from contrasting changes of species, phylogenetic and functional dissimilarity over succession. *Sci. Rep.* **2016**, *6*, 32596. [CrossRef]
24. Quintana, J.R.; Martín-Sanz, J.P.; Valverde-Asenjo, I.; Molina, J.A. Drought differently destabilizes soil structure in a chronosequence of abandoned agricultural lands. *CATENA* **2023**, *222*, 106871. [CrossRef]
25. O'Brien, S.A.; Dehling, D.M.; Tylianakis, J.M. The recovery of functional diversity with restoration. *Ecology* **2022**, *103*, e3618. [CrossRef] [PubMed]
26. Rocha, G.P.E.; Vieira, D.L.M.; Simon, M.F. Fast natural regeneration in abandoned pastures in southern Amazonia. *For. Ecol. Manag.* **2016**, *370*, 93–101. [CrossRef]
27. Li, T.; Yang, H.; Yang, X.; Guo, Z.; Fu, D.; Liu, C.; Li, S.; Pan, Y.; Zhao, Y.; Xu, F.; et al. Community assembly during vegetation succession after metal mining is driven by multiple processes with temporal variation. *Ecol. Evol.* **2022**, *12*, e8882. [CrossRef] [PubMed]
28. van der Sande, M.T.; Powers, J.S.; Kuyper, T.W.; Norden, N.; Salgado-Negret, B.; Silva de Almeida, J.; Bongers, F.; Delgado, D.; Dent, D.H.; Derroire, G.; et al. Soil resistance and recovery during neotropical forest succession. *Philos. Trans. R. Soc. B Biol. Sci.* **2023**, *378*, 20210074. [CrossRef]
29. Matsuo, T.; Martínez-Ramos, M.; Bongers, F.; van der Sande, M.T.; Poorter, L. Forest structure drives changes in light heterogeneity during tropical secondary forest succession. *J. Ecol.* **2021**, *109*, 2871–2884. [CrossRef]
30. Villa, P.M.; Martins, S.V.; de Oliveira Neto, S.N.; Rodrigues, A.C.; Safar, N.V.H.; Monsanto, L.D.; Cancio, N.M.; Ali, A. Woody species diversity as an indicator of the forest recovery after shifting cultivation disturbance in the northern Amazon. *Ecol. Indic.* **2018**, *95*, 687–694. [CrossRef]
31. Martins, W.B.R.; de Matos Rodrigues, J.I.; de Oliveira, V.P.; Ribeiro, S.S.; dos Santos Barros, W.; Schwartz, G. Mining in the Amazon: Importance, impacts, and challenges to restore degraded ecosystems. Are we on the right way? *Ecol. Eng.* **2022**, *174*, 106468. [CrossRef]
32. Velasquez-Ramirez, M.; Corvera, R.; Guerrero, J.; Thomas, E.; Russo, R.; Cusi, E.; Becerra, E.; Nascimiento, P.; Muñoz, A.; Vasquez, J.; et al. *Guía Práctica: Recuperación de la Salud del Suelo para la Restauración de Áreas Degradadas por la Minería Aurífera Aluvial en la Amazonia Peruana*, 1st ed.; Instituto de Investigaciones de la Amazonia Peruana (IIAP): Iquitos, Peru, 2022; ISBN 9786120065365.
33. de Jesus Silva, C.V.; dos Santos, J.R.; Galvao, L.S.; da Silva, R.D.; Moura, Y.M. Floristic and structure of an Amazonian primary forest and a chronosequence of secondary succession. *Acta Amaz.* **2016**, *46*, 133–150. [CrossRef]

34. Velásquez-Ramírez, M.; Vega Ruiz, C.; Gomringer, R.; Pillaca, M.; Thomas, E.; Stewart, P.M.; Gamarra Miranda, L.A.; Dañobeytia, F.R.; Guerrero Barrantes, J.A.; Gushiken, M.C.; et al. Mercury in soils impacted by alluvial gold mining in the Peruvian Amazon. *J. Environ. Manag.* **2021**, *288*, 112364. [CrossRef] [PubMed]
35. Caballero, J.; Messinger, M.; Román-Dañobeytia, F.; Ascorra, C.; Fernandez, L.; Silman, M. Deforestation and Forest Degradation Due to Gold Mining in the Peruvian Amazon: A 34-Year Perspective. *Remote Sens.* **2018**, *10*, 1903. [CrossRef]
36. Sáenz, G.; Finegan, B. *Monitoreo de la Regeneración Natural con Fines de Manejo Forestal. Manejo Forestal Tropical N° 15*; CATIE: Turrialba, Costa Rica, 2000.
37. Lozada, J. Consideraciones metodológicas sobre los estudios de comunidades forestales. *Rev. For. Venez.* **2010**, *54*, 77–88.
38. Sugianto, S.; Muslih, A.M.; Ar-Rasyid, U.H.; Anhar, A. Vegetation analysis of rawa singkil wildlife reserve in Rantau Gedang Village, Singkil Distric, Aceh Singkil Regency, Aceh Province. *IOP Conf. Ser. Earth Environ. Sci.* **2021**, *667*, 012068. [CrossRef]
39. Zhao, J.; Chen, J.; Chen, C.; Lu, S.; Song, C.; Liu, S.; Li, J.; Zhuang, C. Is it sufficient? Assessment of two sampling methods for urban plant species richness investigations. *Urban For. Urban Green.* **2023**, *79*, 127824. [CrossRef]
40. Oksanen, J.; Minchin, P.R. Continuum theory revisited: What shape are species responses along ecological gradients? *Ecol. Model.* **2002**, *157*, 119–129. [CrossRef]
41. Huisman, J.; Olff, H.; Fresco, L.F.M. A hierarchical set of models for species response analysis. *J. Veg. Sci.* **1993**, *4*, 37–46. [CrossRef]
42. Ahmadi, K.; Alavi, S.J.; Hosseini, S.M. Modeling response curves of European yew (*Taxus baccata* L.) using HOF models along the environmental gradient in north of Iran. *Acta Ecol. Sin.* **2022**, *42*, 383–391. [CrossRef]
43. R Core Team. *R: A Language and Environment for Statistical Computing*; R Foundation for Statistical Computing: Vienna, Austria, 2022.
44. Dexter, E.; Rollwagen-Bollens, G.; Bollens, S.M. The trouble with stress: A flexible method for the evaluation of nonmetric multidimensional scaling. *Limnol. Oceanogr. Methods* **2018**, *16*, 434–443. [CrossRef]
45. Lai, J.; Zou, Y.; Zhang, J.; Peres-Neto, P.R. Generalizing hierarchical and variation partitioning in multiple regression and canonical analyses using the rdacca.hp R package. *Methods Ecol. Evol.* **2022**, *13*, 782–788. [CrossRef]
46. Dueñas, H.; Peña, J.; Báez, S.; Bejar, N.; Berrocal, S.; Garate, J. Influencia de la conservación de bosques en la diversidad y composición florística arbórea en el sureste de la Amazonía peruana. *Cienc. Amaz.* **2019**, *7*, 111–125. [CrossRef]
47. Jakovac, C.C.; Peña-Claros, M.; Kuyper, T.W.; Bongers, F. Loss of secondary-forest resilience by land-use intensification in the Amazon. *J. Ecol.* **2015**, *103*, 67–77. [CrossRef]
48. Goosem, M.; Paz, C.; Fensham, R.; Preece, N.; Goosem, S.; Laurance, S.G.W. Forest age and isolation affect the rate of recovery of plant species diversity and community composition in secondary rain forests in tropical Australia. *J. Veg. Sci.* **2016**, *27*, 504–514. [CrossRef]
49. Banin, L.F.; Raine, E.H.; Rowland, L.M.; Chazdon, R.L.; Smith, S.W.; Rahman, N.E.B.; Butler, A.; Philipson, C.; Applegate, G.G.; Axelsson, E.P.; et al. The road to recovery: A synthesis of outcomes from ecosystem restoration in tropical and sub-tropical Asian forests. *Philos. Trans. R. Soc. B Biol. Sci.* **2023**, *378*, 20210090. [CrossRef]
50. San-José, M.; Werden, L.; Joyce, F.H.; Reid, J.L.; Holl, K.D.; Zahawi, R.A. Effects of landscape structure on restoration success in tropical premontane forest. *Sci. Rep.* **2022**, *12*, 13452. [CrossRef] [PubMed]
51. van Breugel, M.; Craven, D.; Lai, H.R.; Baillon, M.; Turner, B.L.; Hall, J.S. Soil nutrients and dispersal limitation shape compositional variation in secondary tropical forests across multiple scales. *J. Ecol.* **2019**, *107*, 566–581. [CrossRef]
52. Rozendaal, D.M.A.; Bongers, F.; Aide, T.M.; Alvarez-Dávila, E.; Ascarrunz, N.; Balvanera, P.; Becknell, J.M.; Bentos, T.V.; Brancalion, P.H.S.; Cabral, G.A.L.; et al. Biodiversity recovery of Neotropical secondary forests. *Sci. Adv.* **2019**, *5*, eaau3114. [CrossRef]
53. Garate-Quispe, J.S.; Canahuire-Robles, R.; Surco-Huacachi, O.; Alarcón-Aguirre, G. Desarrollo estructural y composición florística arbórea en áreas afectadas por minería aurífera en la Amazonía peruana: A 20 años de su reforestación. *Rev. Mex. Biodivers.* **2021**, *92*, e923437. [CrossRef]
54. Fagundes, M.V.; Oliveira, R.S.; Fonseca, C.R.; Ganade, G. Nurse-target functional match explains plant facilitation strength. *Flora Morphol. Distrib. Funct. Ecol. Plants* **2022**, *292*, 152061. [CrossRef]
55. Astudillo-Sánchez, E.; Pérez, J.; Troccoli, L.; Aponte, H. Composición, estructura y diversidad vegetal de la Reserva Ecológica Comunal Loma Alta, Santa Elena, Ecuador. *Rev. Mex. Biodivers.* **2019**, *90*, e902871. [CrossRef]
56. Bonilla-Mata, R.; Acosta-Vargas, L.G. Dynamic and growth of the forests of the Isla del Coco National Park, Costa Rica. *Rev. Biol. Trop.* **2020**, *68*, S89–S102. [CrossRef]
57. Vacek, Z.; Cukor, J.; Vacek, S.; Podrázský, V.; Linda, R.; Kovařík, J. Forest biodiversity and production potential of post-mining landscape: Opting for afforestation or leaving it to spontaneous development? *For. J.* **2018**, *64*, 116–126. [CrossRef]
58. Villa, P.M.; Martins, S.V.; Diniz, É.S.; Nolasco de Oliveira Neto, S.; Neri, A.V.; Pinto-Junior, H.; Nunes, J.A.; Bueno, M.L.; Ali, A. Taxonomic and functional beta diversity of woody communities along Amazon forest succession: The relative importance of stand age, soil properties and spatial factor. *For. Ecol. Manag.* **2021**, *482*, 118885. [CrossRef]
59. König, L.A.; Medina-Vega, J.A.; Longo, R.M.; Zuidema, P.A.; Jakovac, C.C. Restoration success in former Amazonian mines is driven by soil amendment and forest proximity. *Philos. Trans. R. Soc. B Biol. Sci.* **2023**, *378*, 20210086. [CrossRef]
60. Hendy, A.; Hernandez-Acosta, E.; Valério, D.; Ferreira, N.; Reis, C.; Rodrigues, E.; Silva, E.; Andes, J.; Prado, F.; Margarete, V.; et al. Where boundaries become bridges: Mosquito community composition, key vectors, and environmental associations at forest edges in the central Brazilian Amazon. *PLoS Negl. Trop. Dis.* **2023**, *17*, e0011296. [CrossRef] [PubMed]

61. Liu, W.; Fu, S.; Yan, S.; Ren, C.; Wu, S.; Deng, J.; Li, B.; Han, X.; Yang, G. Responses of plant community to the linkages in plant-soil C:N:P stoichiometry during secondary succession of abandoned farmlands, China. *J. Arid Land* **2020**, *12*, 215–226. [CrossRef]
62. de Paula, F.R.; Ruschel, A.R.; Felizzola, J.F.; Frauendorf, T.C.; de Barros Ferraz, S.F.; Richardson, J.S. Seizing resilience windows to foster passive recovery in the forest-water interface in Amazonian lands. *Sci. Total Environ.* **2022**, *828*, 154425. [CrossRef] [PubMed]
63. Longworth, J.B.; Williamson, G.B. Composition and Diversity of Woody Plants in Tree Plantations Versus Secondary Forests in Costa Rican Lowlands. *Trop. Conserv. Sci.* **2018**, *11*, 194008291877329. [CrossRef]
64. Rocchini, D.; Bacaro, G.; Chirici, G.; Da Re, D.; Feilhauer, H.; Foody, G.M.; Galluzzi, M.; Garzon-Lopez, C.X.; Gillespie, T.W.; He, K.S.; et al. Remotely sensed spatial heterogeneity as an exploratory tool for taxonomic and functional diversity study. *Ecol. Indic.* **2018**, *85*, 983–990. [CrossRef]
65. Román-Dañobeytia, F.; Huayllani, M.; Michi, A.; Ibarra, F.; Loayza-Muro, R.; Vázquez, T.; Rodríguez, L.; García, M. Reforestation with four native tree species after abandoned gold mining in the Peruvian Amazon. *Ecol. Eng.* **2015**, *85*, 39–46. [CrossRef]
66. Purvis, A.; Hector, A. Getting the measure of biodiversity. *Nature* **2000**, *405*, 212–219. [CrossRef] [PubMed]
67. Suding, K.; Higgs, E.; Palmer, M.; Callicott, J.B.; Anderson, C.B.; Baker, M.; Gutrich, J.J.; Hondula, K.L.; LaFevor, M.C.; Larson, B.M.H.; et al. Committing to ecological restoration. *Science* **2015**, *348*, 638–640. [CrossRef] [PubMed]
68. Yunanto, T.; Amanah, F.; Siregar, I.Z. The influence of fast-growing species composition on natural regeneration of coal mine reclamation. *IOP Conf. Ser. Earth Environ. Sci.* **2021**, *918*, 012048. [CrossRef]

diversity

MDPI

Article

Resilience of Aboveground Biomass of Secondary Forests Following the Abandonment of Gold Mining Activity in the Southeastern Peruvian Amazon

Jorge Garate-Quispe [1,2,*], Marx Herrera-Machaca [2,3], Victor Pareja Auquipata [2], Gabriel Alarcón Aguirre [2], Sufer Baez Quispe [4] and Edgar Eloy Carpio-Vargas [5]

1. Postgraduate Unit of Statistical and Computer Engineering, Faculty of Statistical and Computer Engineering, Universidad Nacional del Altiplano de Puno, Puno P.O. Box 291, Peru
2. Departamento Académico de Ingeniería Forestal y Medio Ambiente, Facultad de Ingeniería, Universidad Nacional Amazónica de Madre de Dios, Puerto Maldonado 17001, Peru; mherrera@unamad.edu.pe or herrermm@wfu.edu (M.H.-M.); vpareja@unamad.edu.pe (V.P.A.); galarcon@unamad.edu.pe (G.A.A.)
3. Programa Restauración de Ecosistemas, Centro de Innovación Científica Amazónica (CINCIA), Av. Ucayali Mz 4-Z Lt. 9-A, Puerto Maldonado 17001, Peru
4. Herbario Alwyn Gentry (HAG), Vicerrectorado de Investigación, Universidad Nacional Amazónica de Madre de Dios, Puerto Maldonado 17001, Peru; sbaez@unamad.edu.pe
5. Academic Department of Statistical and Computer Engineering, Faculty of Statistical and Computer Engineering, Universidad Nacional del Altiplano de Puno, Puno P.O. Box 291, Peru; ecarpio@unap.edu.pe

* Correspondence: garate.qjs@gmail.com

Abstract: Amazon rainforests are critical for providing a wide range of ecosystem services. In the Southeastern Peruvian Amazon; however, goldmining activities are causing severe soil degradation and forest loss. We analyzed aboveground biomass (AGB), forest structure, and species diversity recovery during secondary succession in 179 forest plots. Our study provides the first field-based quantification of AGB recovery following the abandonment by two types of goldmining (heavy machinery and suction pumping) in Madre de Dios (Peru). We found that successional secondary forests in areas subjected to suction pumping were more resilient than those in areas subjected to heavy machinery. After 20 years, mean AGB in suction pumping mining areas had reached 56% of reference forest AGB, while in areas of heavy machinery mining it was only 18%. Mining type, stand age, and distance from the forest edge had a significant effect on AGB. The influence of the distance from the forest edge on AGB varies according to mining type because the effects of species diversity on AGB are mediated by the distance from the forest edge. Our results clearly showed the dynamics of AGB recovery across a secondary succession after goldmining, and the contrasting responses of AGB between the two mining types. Our study disentangles the importance of key factors in forest recovery after mining and improves understanding of the resilience of biomass accumulation in these highly degraded ecosystems.

Keywords: chronosequence; ecological succession; forest recovery; forest structure; Madre de Dios; natural regeneration; recovery of impacted areas; species composition; species diversity

1. Introduction

Tropical forests are a crucial component of the Earth system [1], because they harbor great biodiversity (63% of tree species globally) [1] and provide carbon storage, which is relevant to the regulation of global climate change [2,3]. These forests are being transformed rapidly as a result of human activities [4]. Such deforestation and forest degradation generate between 10 and 15% of global greenhouse gas emissions [5]. Therefore, its loss would have a significant impact on climate change [6]. In addition, the Amazon rainforests are very important for providing hydrological services and harboring a high species diversity [7].

The mining industry is essential for the development of the world economy and human societies [8,9]. Although mining activities occur on a small fraction of the Earth's land surface (less than 1%) [8], they can generate severe environmental impacts [10,11] and have potentially adverse effects on the health of the local population [12], as well as representing one of the most serious threats to tropical ecosystems. [13,14].

More than half of the global forests are disturbed and knowing the rate of forest recovery is important for estimating the ecosystem services that these forests can provide [15]. Despite the importance of tropical forests and their high levels of degradation, they remain poorly monitored [7]. The Amazon forests of Madre de Dios are being affected by rapid economic and population growth, and among all human activities, gold mining generates the most severe impacts [16]. Goldmining causes severe environmental degradation through deforestation, mercury pollution, and soil degradation and eliminates the capacity of ecosystems to accumulate biomass [16]. However, areas degraded by mining are not usually evaluated because they occupy relatively small land areas [17].

Land use changes in tropical forests, however, are occurring at a faster rate than in other regions of the planet, and this imbalance between deforestation and forest restoration has serious effects on the global carbon cycle [18]. Agriculture and goldmining are the primary drivers of deforestation in the southern Peruvian Amazon [19], although mining activity causes the most severe impacts [20,21]. The resilience of these ecosystems after abandonment by mining activities remains to be studied in detail [21]. Forest biomass is key in the carbon cycle process and is being studied in the context of international climate change mitigation initiatives. [22]. Information on how these areas are recovering after abandonment will allow the generation of more sustainable management models and identify potential sites for future carbon sequestration projects [18]. In this context, aboveground biomass (AGB) is a key indicator to determine the dynamics of carbon recovery in degraded ecosystems [15]. Therefore, biomass estimation is important in quantifying the carbon sink flux, measuring international carbon stocks, and mitigating the increasing CO_2 pollution [23].

Recent studies in the Amazon have reported the ecological mechanisms of human activity influence on the recovery of species diversity, species composition, and AGB [24–27]. Many studies have shown that the stand forest age [28,29], species number [30], species diversity [31], species composition [32], and forest height [33] are positively correlated with AGB levels in secondary forests. In addition, a recent study showed that forest structure (e.g., forest height) conditions the effects of species richness on AGB accumulation [24]. Other factors that may limit the AGB recovery are the degree of land use intensity and soil degradation [34]. Studies of AGB recovery after goldmining in the Amazon are particularly rare. Nevertheless, Kalamandeen et al. [21], in a recent study in the Amazon region of Guyana, found that areas severely degraded by gold mining have a slow biomass recovery compared to other secondary forests (e.g., from agriculture or pasture).

AGB accumulation and changes in species diversity and species composition in primary forests and secondary forests have been monitored in several studies [35,36]. These have shown that secondary forests are an important carbon sink, and in some cases, they can be higher than primary forests [7]. This highlights the importance of secondary forests in mitigating climate change and promoting biodiversity recovery [37]. For this purpose, the chronosequence approach has been one of the most widely used [38]. However, there are few studies that analyze the recovery of biomass in secondary succession after abandonment by mining activities [39], compared to studies in areas degraded by other activities such as agriculture or livestock [21].

In the Southeastern Peruvian Amazon (Madre de Dios), studies on biomass estimates in natural regeneration after abandonment by mining activities are scarce. Most studies have focused on assessing the species diversity, floristic attributes, species composition [20,40,41], reforestation/rehabilitation experiments of these degraded ecosystems [42–46], and monitoring forest loss due to goldmining [19,26]. These studies showed that the degree of resilience of tree communities after abandonment by mining activities depends on the level of landscape fragmentation [41], stand age [20], and distance from the forest edge [41]. Meanwhile, the experiments in these areas demonstrated the rapid acclimatization of native tree species [44,46,47] and cover plants [43] to these severely degraded environments.

This study addresses the recovery of aboveground biomass (AGB) from secondary forests according to two types of mining (heavy machinery and suction pumping) following the abandonment of gold mining activity in the Southeastern Peruvian Amazon. We posed the following research questions: (1) What are the AGB, forest structure, species diversity, and floristic composition of a secondary succession across a chronosequence after 23 years of abandonment? (2) What is the influence of mining type, stand age, and the distance from the forest edge on AGB, forest structure, and tree species diversity of the regenerating community? (3) How do our results in secondary forests after mining compare with reference forests? (4) How is the AGB accumulation affected by stand age, distance from the forest edge, and species diversity according to mining type?

2. Materials and Methods

2.1. Study Area

The study was conducted at 11 mining sites in the Madre de Dios region, Southeastern Peruvian Amazon (Figure 1). The study area is classified as a tropical Amazon forest [46] with an altitude ranging from 220–450 m above sea level. The study area is characterized by warm, humid, and seasonal weather conditions. The annual average rainfall ranges between 1500 mm and 2860 mm [48]. The average annual temperature is around 25 °C, ranging from 16 °C to 39 °C. During the dry season, from June to September, precipitation is lower than 100 mm per month [39,41] (Figure S1).

The forest types that covered the study area were low-terrace forest and low-hill forest [49]. The canopy reaches a height of up to 40 m, with some emergent trees that can reach up to 60 m. Low-terrace forests are located in the alluvial plain of the lowland rainforest, occupying both recent low terraces (floodable) and ancient terraces (non-floodable). Recent low terraces are usually dominated by *Ficus insipida, Calycophyllum spruceanum, Euterpe precatoria, Garcinia macrophylla, Astrocaryum murumuru,* and *Sloanea guianensis.* While ancient terraces are usually dominated by *Bertholletia excelsa* (that reach heights up to 60 m and DBH up to more than 2 m), *Myroxylum balsamun, Clarisia racemosa, Spondias mombin, Hevea guianensis,* and *Iriartea deltoidea.* Low-hill forests develop in lands originated by very ancient fluvial deposits, and they are dominated by *Pseudolmedia laevis, Astrocaryum murumuru, Oenocarpus bataua, Brosimum alicastrum,* and *Cedrelinga cateniformis* [49]. In tropical Amazon forest the proportion of zoochorous species is usually higher in old-growth forests and at the later stage of secondary forests [11,50], and the proportion of anemochorous species is usually higher in the initial successional stage [51].The weathering and erosion of hard rock ores in the Andes produced fine grains of gold. These were carried down the rivers to the lowlands of Madre de Dios, forming placer deposits. These fine gold particles settle in river meanders and form near-surface deposits in floodplains due to changes in river geomorphology and seasonal flooding [52].

At selected sites, two types of mining (heavy machinery and suction pumping) are applied for gold extraction. In suction pumping mining, miners used the open-pit method to extract gold. Such operations require cutting all aboveground vegetation and removing the soil. When dredging, gold-containing soils are pumped from deep pits (up to 10 m) to a sluice box to collect the gold particles [41] (Figure 1C). In highly mechanized mining the gold particles are extracted by excavation using heavy machinery such as front loaders, excavators, and dump trucks for sediment transportation to a sluice box [53,54] (Figure 1C). For this reason, heavy machinery mining causes more deforestation and environmental damage than suction pumping [52]. After mining activities, these areas are abandoned without any strategy for soil or vegetation recovery. The study was conducted in areas degraded and abandoned by goldmining activity, where the vegetation had regenerated naturally.

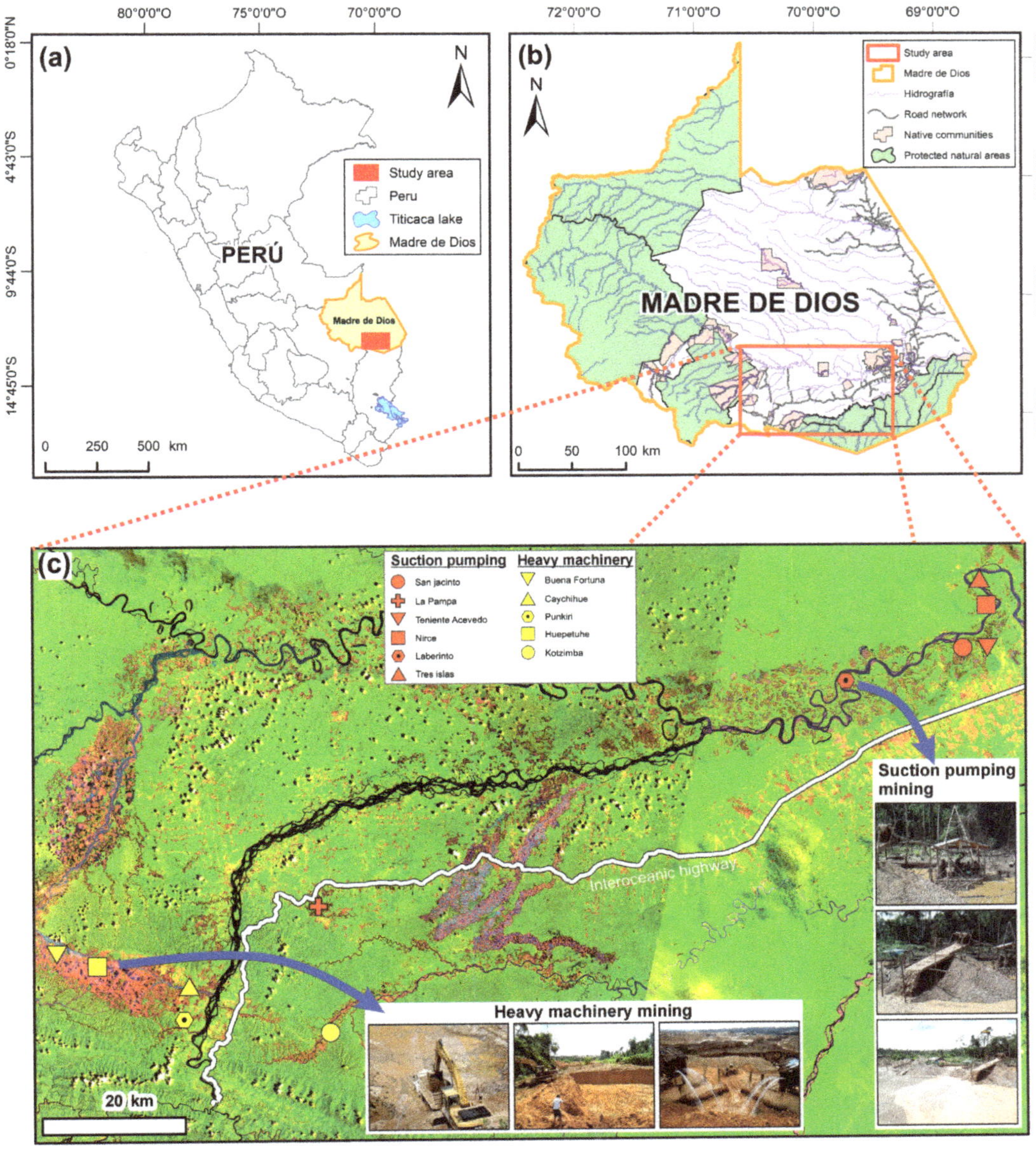

Figure 1. Map of the study area in Peru (**a**) and in Madre de Dios (**b**), Southeastern Peruvian Amazon. (**c**) Landsat-8 images (July 2023) over the study. It shows the location of field sites in the study, and representative photos of the type of mining.

2.2. Study Design and Vegetation Sampling

We compiled data from 11 chronosequence studies (sites) about natural regeneration post goldmine abandonment, including 179 plots, containing 4699 tree stems. For 4 out of the 11 sites, reference forest plots were included (Table 1). We included 17 reference forest plots near the areas where natural regeneration was evaluated, which contained 1139 tree stems. The reference forest has never been affected by goldmining activities for at least 40 years.

Table 1. Number of forest plots sampled in secondary forest after mining abandonment and in the reference forest in the Peruvian Amazon.

N	Site	Mining Type	Number of Plots in Secondary Forest	Distance from the Forest Edge (m)			Stand Age (Years)			Number of Plots in Reference Forest
				Mean ± SD	Min	Max	Mean	Min	Max	
1	Buena Fortuna		6	733	550	900	18	17	20	
2	Kotzimba		3	333	230	500	8	4	9	
3	Caychihue	Heavy machinery	19	607	340	943	10	4	16	
4	Huepetuhe		8	675	600	900	24	24	16	4
5	Punkiri		16	91	25	280	10	7	15	8
6	Laberinto		14	64	22	152	9	2	19	
7	Nirce		27	272	95	517	10	4	22	
8	Teniente Acevedo	Suction pumping	50	617	180	940	7	2	14	
9	Tres islas		12	35	15	70	15	5	23	3
10	La Pampa		6	82	10	176	8	6	12	2
11	San Jacinto		18	175	29	520	8	2	23	
		Total	179	-	-	-				17

Plot size varied from 0.01 to 0.25 ha, with a length varying from 10 to 60 m and width ranging from 10 to 60 m. The size of the plot area varies according to that used in the studies compiled. In addition, the use of small plots is due to the reduced size of areas with natural regeneration after abandonment by mining (varying from 0.05 to 1 ha). In each sample plot, the height and diameter at breast height (DBH; 1.3 m aboveground) of all woody trees with stems $\geq$ 5 cm were measured [55,56]. We excluded trees with DBH < 5 cm because they represent a small proportion of AGB in forests [56].Stand age and the distance from the forest edge for each plot were determined by using multi-temporal Landsat (1975–2010), RapidEye, and Planet imagery (2010–2023). In the case of stand age, this was complemented by interviewing landowners. In this study, the distance from the forest edge was considered as a proxy for fragmentation.

All trees were identified and their AGB estimates made using allometric equations. Species were classified according to the APG IV [57]. Scientific names were revised and corrected using the TNRS application (https://tnrs.biendata.org, accessed on 15 January 2024).

2.3. *Aboveground Biomass (AGB)*

Aboveground biomass (AGB) was calculated, for all trees present in the plots, using the generic pan-tropical allometric biomass equation developed by Chave et al. [58] (Equation (1)):

$$\text{AGB} = 0.0673 \times \left(\rho \times \text{DBH}^2 \times \text{H}\right) \quad (1)$$

where AGB is aboveground biomass in kg, H is the tree height in m, D is the diameter at breast height in cm, and ρ is the wood density in gr cm^{-3}. The wood density of trees was obtained from the global wood density database [59].

Individual tree biomass values within the plot were summed and converted to AGB stock (Mg ha^{-1}) [60]. In order to assess how fast the AGB of the secondary forest is recovering towards the values of the reference forest, we calculated the relative AGB recovery as the AGB of the secondary forest plot compared to the mean AGB of the reference forest plots in the study area [18].

2.4. *Stand Structure and Alpha and Beta-Diversity Calculations*

Forest structure was analyzed by structural indices [58]; we calculated in each plot: (a) the mean of the DBH of all woody species, and (2) the mean of the height of all woody species. In addition, to examine the horizontal structure of the communities [61] we calculated the importance value index (IVI) of each species according to five recovery age classes of natural regeneration and reference forest. The IVI was computed using the method of Curtis and Macintosh [62], using R package BiodiversityR [63]. IVI was computed as follows (Equation (2)):

$$IVI(\%) = \frac{Abun(\%) + Dom(\%) + Frec(\%)}{3} \quad (2)$$

where IVI (%) is the importance value index, Abun (%) is the relative abundance, Dom (%) is the relative dominance, and Frec (%) is the relative frequency.

We used effective numbers of species (Hill numbers) to quantify the species diversity of each plot. Hill numbers are a mathematically unified family of diversity indices that incorporate relative abundance and species richness [64]. The calculations were performed using the R package hillR [65] for the three components of species diversity represented by the Hill series (^{0}D, species richness weighting all species equally; ^{1}D, exponential of Shannon's entropy index weighting typical species; ^{2}D, inverse of Simpson's concentration index weighting dominant species).

For taxonomic β-diversity, to quantify differences in tree community composition among plots, we used the abundance-based Chao–Jaccard dissimilarity index. Chao–Jaccard index was calculated using the R package CommEcol [66].

2.5. Data Analysis

We used both structural equation models (SEM) and generalized linear models (GLM) to disentangle the relative effects of stand age, distance from the forest edge, and species diversity on AGB accumulation according to mining type.

GLMs were used to test for an effect of mining type, stand age, and distance from the forest edge on AGB, species diversity (Hill numbers), and forest structure. For AGB data the Tweedie distributed responses were assumed because data were overdispersed [67]. Whereas Poisson distribution function was used for species diversity (^{0}D, ^{1}D, and ^{2}D) and forest structure (mean DBH and mean height).

To assess the direct and indirect effects of stand age, distance from forest edge, and species diversity (Hill ^{0}D) on AGB, three structural equation models (SEMs) were carried out for the following: (1) each type of mining and (2) all data without differentiating by type of mining. Structural equation modeling was conducted using the "lavaan" package in R [68,69]. To determine path coefficients we used the maximum likelihood method because it is robust to multivariate non-normality [70].

We used a one-way ANOVA (Bonferroni's post hoc tests) or Kruskal–Wallis (Dunn post hoc test) test to detect significant differences ($p < 0.05$) in AGB, species diversity, and stand structure among five recovery age classes (binned into five, five-year age classes) of natural regeneration and reference forests. To check whether the data conforms to normality of residuals and homogeneity of variance (homoscedasticity) assumptions to perform ANOVA, we utilized the Shapiro–Wilk test and Levene's test, respectively. If the normal distribution assumption or homoscedasticity was not satisfied, we used the Kruskal–Wallis test instead of ANOVA. When the difference among groups was significant, we used Bonferroni's post hoc tests in ANOVA and the post hoc Dunn's test in the Kruskall–Wallis test.

We used non-linear regression (Gompertz sigmoidal function) to model and compare the changes in AGB and AGB throughout the chronosequence. SigmaPlot 15 was used for curve fitting Gompertz sigmoidal function. Regression parameter estimates and the respective 95% confidence interval values are presented in scatterplots.

Non-metric multi-dimensional scaling (nMDS) plots were used to visualize the tree community composition at natural regeneration stands and reference forests across stand age, sites, and by type of mining. The NMDS was performed in Primer-E 7 using a Type II Kruskal fit scheme with 9999 restarts and based on a similarity matrix of Chao–Jaccard indices. We considered a stress cut-of 0.2, because nMDS plots with stress > 0.2 are unreliable [71].

We used a permutational multivariate analysis of variance (PERMANOVA) to test for differences in species composition between types of mining and across five recovery age classes of natural regeneration. We conducted PERMANOVA analysis using Primer-E 7, where mining type was treated as a fixed effect and stand age and distance from the forest edge were used as covariates. In addition, we used PERMANOVAs to test for interacting effects of mining type, stand age, and distance from the forest edge on species composition after goldmining abandonment.

All statistical analyses were performed using the R statistical software version 4.3.2 [69], SigmaPlot version 15, InfoStat 2020p, and Primer 7 Permanova+.

3. Results

3.1. *Effects of Mining Type, Stand Age, and Distance from the Forest on AGB*

We found that mining type had a significant effect on AGB (GLM, $p < 0.05$) and species diversity (^{0}D, ^{1}D, and ^{2}D; GLM, $p < 0.001$), but no significant effect was found for forest structure (Table 2), mean DBH (GLM, $p > 0.05$) and mean height (GLM, $p > 0.05$). However, results of the GLM showed that stand age and distance from the forest edge had a significant effect on AGB (GLM, $p < 0.001$; Table 2a), species diversity (GLM, $p < 0.001$; Table 2b–d), mean DBH (GLM, $p < 0.001$; Table 2e), and mean height (GLM, $p < 0.001$; Table 2f). AGB, species diversity, mean DBH, and mean height were positively correlated with stand age and negatively with the distance from the forest edge. Mining type, stand age, and distance from the forest edge explained 73.8% of the variation in AGB (Table 2a).

Table 2. Results of the generalized linear models (GLM) indicating the effects of mining type, stand age and distance from the forest edge on above-ground biomass (AGB), effective number of species (^{0}D, ^{1}D, and ^{2}D), and forest structure (mean DBH and mean height of woody species).

(a) AGB

Source of variation	Estimate	Std. Error	*p*	D^2
Intercept (heavy machinery)	2.0086	0.1986	<0.001	
Mining type (Suction pumping)	0.2972	0.1291	0.023	73.87
Age	0.1592	0.0100	<0.001	
Distance from forest edge	−0.0021	0.0002	<0.001	

(b) Hill ^{0}D

Source of variation	Estimate	Std. Error	*p*	D^2
Intercept (heavy machinery)	2.3214	0.0848	<0.001	
Mining type (Suction pumping)	−0.2468	0.0598	<0.001	68.52
Age	0.0550	0.0043	<0.001	
Distance from forest edge	−0.0024	0.0001	<0.001	

(c) Hill ^{1}D

Source of variation	Estimate	Std. Error	*p*	D^2
Intercept (heavy machinery)	2.0786	0.0970	<0.001	
Mining type (Suction pumping)	−0.3930	0.0680	<0.001	71.65
Age	0.0537	0.0050	<0.001	
Distance from forest edge	−0.0022	0.0001	<0.001	

(d) Hill ^{2}D

Source of variation	Estimate	Std. Error	*p*	D^2
Intercept (heavy machinery)	2.0650	0.1040	<0.001	
Mining type (Suction pumping)	−0.5382	0.0731	<0.001	71.57
Age	0.0441	0.0056	<0.001	
Distance from forest edge	−0.0020	0.0001	<0.001	

(e) Mean DBH

Source of variation	Estimate	Std. Error	*p*	D^2
Intercept (heavy machinery)	2.2480	0.0823	<0.001	
Mining type (Suction pumping)	0.0820	0.0540	0.129	57.38
Age	0.0265	0.0039	<0.001	
Distance from forest edge	−0.0005	0.0001	<0.001	

(f) Mean height

Source of variation	Estimate	Std. Error	*p*	D^2
Intercept (heavy machinery)	2.0645	0.0977	<0.001	
Mining type (Suction pumping)	−0.0531	0.0647	0.412	58.03
Age	0.0187	0.0048	<0.001	
Distance from forest edge	−0.0006	0.0001	<0.001	

Distance from the Forest Edge Strongly Mediates the Effects of Species Diversity on AGB According to Mining Type

The structural equation models (SEM) showed good fits of the model to the data (CFI close to one and lower AIC values; Figure 2). SEMs showed that AGB was affected by the stand age, distance from the forest edge, and species diversity (Hill ^{0}D), but the significance and direction (+/−) of the relationship were dependent on the mining type (Figure 2).

Global SEM results, without differentiating the mining types, indicated that stand age ($\beta = 0.47$, $p < 0.001$) and species diversity had a positive direct effect ($\beta = 0.47$, $p < 0.001$) on AGB. Meanwhile, distance from the forest edge had no significant direct influence ($\beta = -0.10$, $p = 0.059$) on AGB but it showed a negative effect on species diversity (covariance = − 0.46, $p < 0.001$), which ultimately had a direct effect on AGB. The covariance between stand age and species diversity was also significant (covariance = 0.22, $p < 0.001$) (Figure 2A).

In both types of mining, SEMs indicated that AGB increased significantly with increasing stand age, but this direct effect was a little stronger in heavy machinery mining ($\beta = 0.74$, $p < 0.001$) than in suction pumping mining ($\beta = 0.54$, $p < 0.001$). The effect of distance from the forest edge on AGB was significant and negative ($\beta = -0.711$ and $p < 0.001$) in heavy machinery forest stands and non-significant and negative ($\beta = -0.02$, $p = 0.670$) in suction pumping forest stands. Meanwhile, species diversity promoted AGB, but this direct effect was significantly stronger in suction pumping forest stands ($\beta = 0.43$, $p < 0.001$), whereas species diversity non-significantly declined AGB in heavy machinery forest stands ($\beta = -0.08$, $p = 0.496$). The covariance between distance from the forest edge and species diversity was also significant, but this effect was stronger in heavy machinery forest stands ($\beta = -0.74$, $p < 0.001$) than in suction pumping forest stands ($\beta = -0.32$, $p < 0.001$) (Figure 2B,C). Thus, the SEM results are consistent with the results of the analysis of variance and generalized linear models.

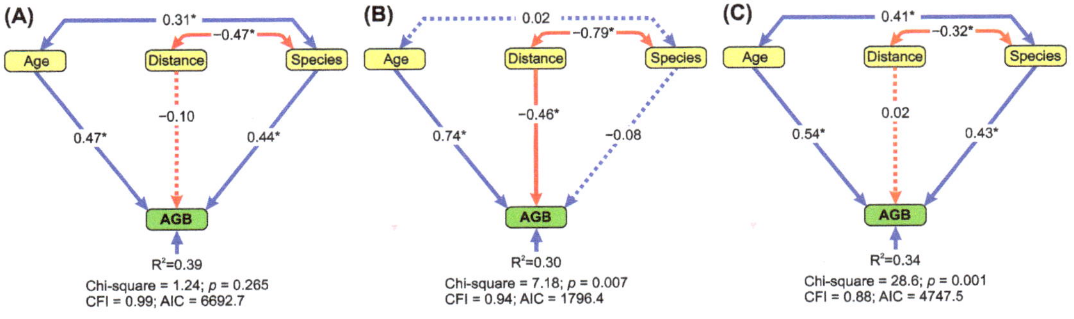

Figure 2. Results of structural equation models (SEM) showing the relationship between aboveground biomass (AGB), stand age (Age), distance from the forest edge (Distance), and number of species (Species) according to type of mining (heavy machinery and suction pumping). The blue arrows indicate positive effects, while the red arrows indicate negative effects. The dash and plain line types represent the non-significant ($p > 0.05$) and significant ($p < 0.05$) paths, respectively. For each path, a standardized regression coefficient value is shown. * Indicates the significant paths at 0.001. (**A**) = global SEM results, without differentiating by type of mining activities (heavy machinery and suction pumping); (**B**) = heavy machinery; and (**C**) = suction pumping.

3.2. Recovery of AGB, Tree Diversity and Forest Structure

Our results revealed that the AGB, tree height, DBH and effective number of species (Hill ^{0}D) increased with age of the stands. However, the pattern of recovery of these variables varied according to mining type. The pattern of AGB accumulation, recovery of DBH, height, and species diversity (Hill ^{0}D) showed a rapid increase in suction pumping mining forest stands in comparison to those of heavy machinery (Figures 3 and 4). High values of AGB and species were found in 20–25-year-old suction pumping stands; they were similar to reference forest stand values.

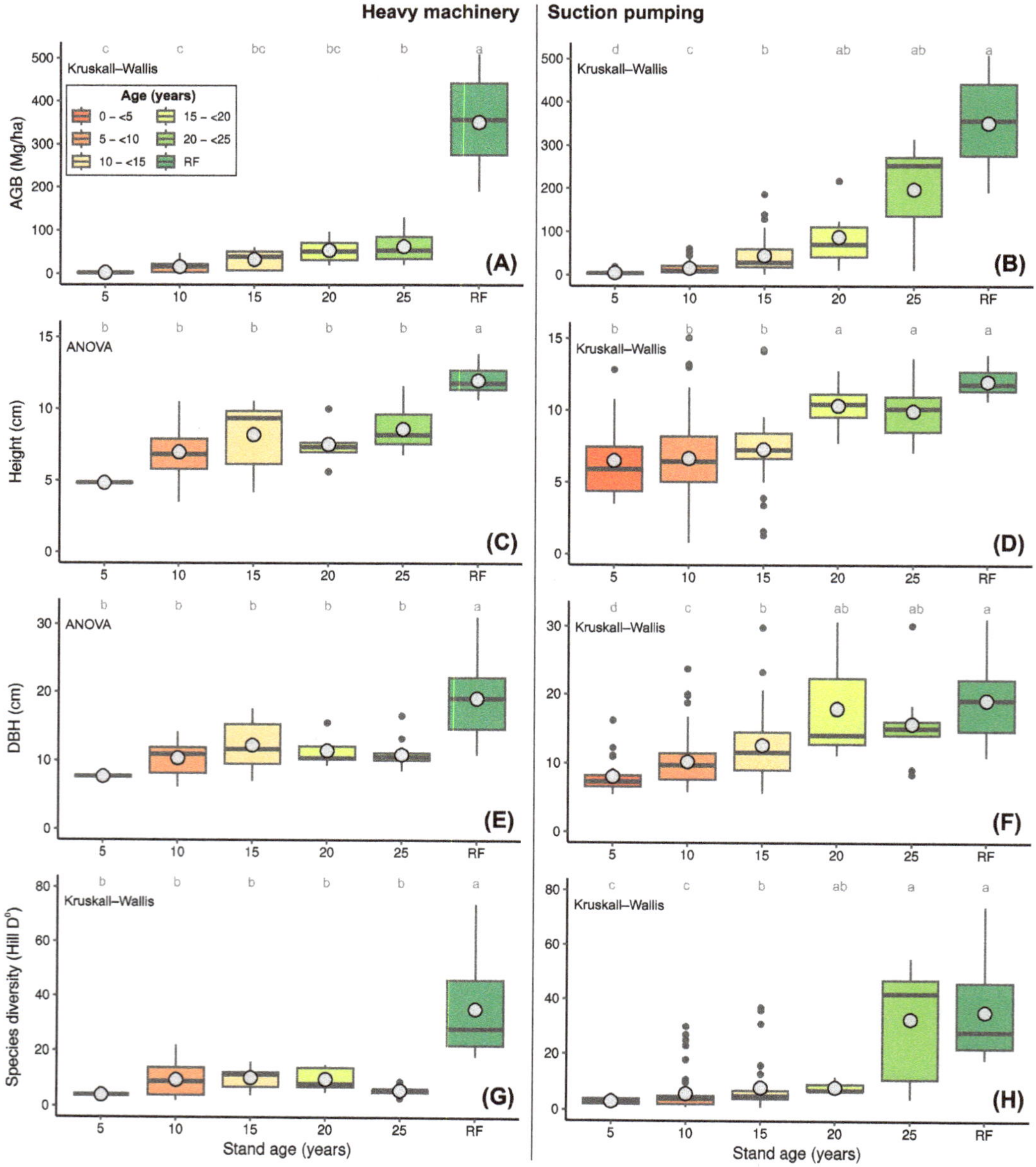

Figure 3. Comparison of AGB, species diversity and forest structure among forest plots in two types of mining activities (heavy machinery and suction pumping) according to five recovery age classes of natural regeneration and in reference forest of tree community across a chronosequence in degraded lands of the Peruvian Amazon. (**A**,**B**) AGB, (**C**,**D**) mean height, (**E**,**F**) mean DBH, and (**G**,**H**) species diversity (Hill ^{0}D). Different letters indicate significant differences among age classes for each variable ($p < 0.05$).

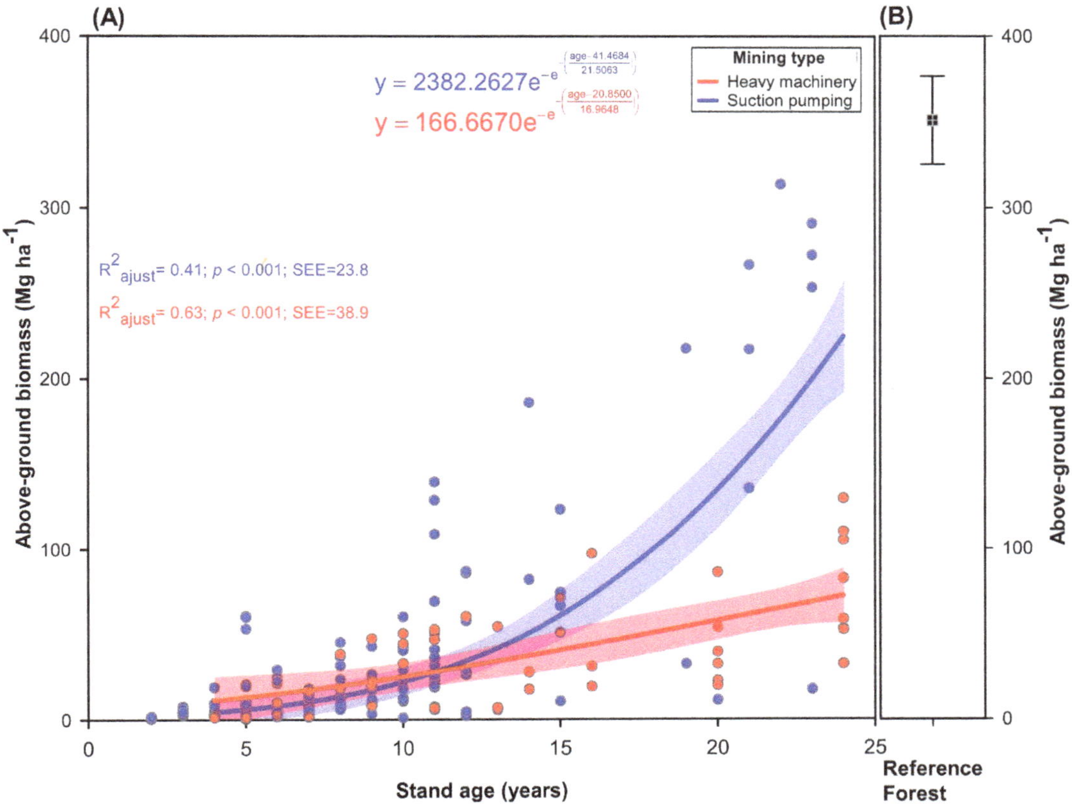

Figure 4. Relationship between aboveground biomass and stand age of tree community in two types of mining activities (heavy machinery and suction pumping) across a chronosequence in degraded lands of the Peruvian Amazon (**A**). Mean AGB (±standard error) of reference forests close to plots in degraded lands (**B**). SEE = standard error of estimate. The blue and red shading around each regression line represent the 95% confidence interval. The solid red circles represent data from a heavy machinery mining. The solid blue circles represent data from a suction pumping mining.

AGB increases recorded in heavy machinery stands (on average ranging from 1.9 to 62.9 Mg ha^{-1}) across the chronosequence were significantly lower (Kruskall–Wallis, $p < 0.001$) than that AGB accumulated in reference forests (351 ± 106 Mg ha^{-1}). However, in pumping mining stands, we found a rapid increase in the mean AGB, on average ranging from 3.8 to 115.7 Mg ha^{-1} depending upon the age of the succession (Figure 3). The AGBs of 20–25-year-old stands (198 ± 115 Mg ha^{-1}) were not significantly different (Kruskall–Wallis, $p > 0.05$) from the reference forest (351 ± 106 Mg ha^{-1}). On average, regenerating communities in pumping mining had reached 56% of reference forest AGB after 20 years, while in heavy machinery it was only 18% (Figure 3A,B). In our study, AGB recovery throughout a chronosequence (2–24-year-old) relative to reference forest AGB, varied from 3.3% to 89.6% (Figures 2 and 4). The resilience of AGB showed a high variation in potential AGB accumulation rates after 10 years of abandonment by goldmining and these patterns differed by mining type, with low rates in heavy machinery mining and high rates in suction pumping (Figure S2).

The vegetation structure (mean height and mean DBH) of secondary forest recovered at a much faster rate than species diversity and AGB, but it varied according to mining type. In >20-year-old regenerating communities in pumping mining stands it reached values of mean DBH and height slightly under and close to those of the reference forest (Kruskall–Wallis, Dun's post hoc test, $p < 0.05$) (Figure 3F). However, regenerating communities in heavy machinery were significantly lower than the reference forest (ANOVA, Bonferroni's post hoc test, $p < 0.05$) (Figure 3E). After 20 years, mean height and DBH in regenerating communities in pumping mining had reached 83% and 82% of reference forest, respectively, while in those of heavy machinery it was only 63% and 60%, respectively (Figure 3A,B).

On the other hand, on average, the number of species (Hill ^{0}D) per plot increased with category of stand age throughout the chronosequence, but the increase was higher in pumping mining stands than in heavy machinery. In pumping mining stands, Hill ^{0}D was highest in plots with a stand age greater than 20 years, and it was statistically indistinguishable from the mean Hill ^{0}D in the reference forest (Kruskall–Wallis, Dun's post hoc test, $p > 0.05$) (Figure 3H). In heavy machinery stands, significantly more Hill ^{0}D were recorded in reference forest than successional forest (Kruskall–Wallis, Dun's post hoc test, $p < 0.01$). There were no differences in Hill ^{0}D values among secondary forests (Kruskall–Wallis, Dun's post hoc test, $p > 0.05$) (Figure 3G).

3.3. Tree Species Dominance and Composition

A total of 303 tree species belonging to 167 genera and 61 families were recorded throughout all successional forests, for a total of 4699 woody trees across all plots. The dominance of tree species, genera, and families changed across the chronosequence and varied significantly according to mining type (Tables 3, S1 and S2). The five most abundant species were more dominant in regenerating areas after heavy machinery mining (IVI value of 27–73.9%) than in areas after suction pumping (IVI value of 59.1–77.5%). In early successional stages (<10 years old), pioneer species were dominant (IVI > 10% [72]). In regenerating areas after suction pumping mining, the dominant species were *Ochroma pyramidale*, *Cecropia engleriana*, and *Cecropia membranacea*, whereas in heavy machinery areas they were *Ochroma pyramidale*, *Trema micrantha*, *Piper aduncum*, and *Vismia macrophylla*. The IVI of pioneer species had a decreasing trend with the increase of stand age. In intermediate secondary succession (10 to <20-year-old), *Ficus insipida*, *Ochroma pyramidale*, and *Inga* spp. were considered dominants in areas of suction pumping mining, whereas in areas recovering from heavy machinery mining they were *Ochroma pyramidale*, *Piper aduncum*, *Miconia minutiflora*, and *Acalypha macrostachya*. On the other hand, in advanced secondary succession, the highest IVI of species differed between mining types. *Sapium marmieri*, *Erythrina poeppigiana*, and *Cecropia membranacea* were the most important species in suction pumping mining, whereas *Tachigali* spp., *Vismia* spp., and *Miconia* spp. were the most important species in heavy machinery. *Virola surinamensis*, *Astrocaryum murumuru*, *Guarea macrophylla*, *Brosimum lactescens*, and *Spondias mombin* were the dominant species in the reference forest plots (Table 3).

In both types of mining, Malvaceae was the dominant plant family in the early successional stages, followed by Urticaceae and Fabaceae in recovering areas of suction pumping, and by Cannabaceae, Fabaceae and Urticaceae in areas of heavy machinery mining. The IVI of Fabaceae and Malvaceae had increasing and decreasing trends, respectively, with the increase in stand age. In areas of heavy machinery mining, Melastomataceae had an increasing trend across the chronosequence. In both types of mining, Fabaceae was the most dominant species in the oldest secondary forest (20–25 years) as well as in the reference forest plots (Table S1).

In the early successional stages, however, Ochroma and Cecropia were the most dominant genera, followed by Trema, Piper, and Vismia in areas affected by heavy machinery mining, and by Andira, Inga, and Tessaria in areas of suction pumping. The IVI of Inga, Ficus, Guazuma, and Malvaceae rose with the increase of stand age in areas affected by suction pumping mining. Meanwhile, the IVI of Miconia, Vismia, and Apeiba increased with the increase of stand age in areas of heavy machinery mining (Table S2).

Table 3. Comparison of five species with the highest importance value index (IVI) of tree community by type of mining activities (heavy machinery and suction pumping) across a chronosequence in degraded lands and reference forest in the Peruvian Amazon.

Forest Stand Age (Years)	Mining Type			
	Suction Pumping		Heavy Machinery	
	Species	IVI (%)	Species	IVI (%)
(a) <5	*Ochroma pyramidale*	38.0	*Ochroma pyramidale*	35.3
	Cecropia engleriana	18.6	*Trema micrantha*	26.6
	Cecropia membranacea	10.2	*Piper aduncum*	10.9
	Trema micrantha	3.6	*Vismia macrophylla*	10.0
	Erythrina poeppigiana	3.4	*Coccoloba mollis*	8.6
(b) 5 to <10	*Ochroma pyramidale*	18.3	*Ochroma pyramidale*	36.1
	Cecropia membranacea	12.1	*Cecropia membranacea*	6.0
	Inga marginata	8.4	*Bauhinia sp1*	2.6
	Cecropia engleriana	6.5	*Erythrina ulei*	2.0
	Tessaria integrifolia	3.7	*Cecropia polystachya*	1.9
(c) 10 to <15	*Ficus insipida*	12.2	*Ochroma pyramidale*	9.9
	Ochroma pyramidale	12.2	*Cecropia engleriana*	4.3
	Cecropia engleriana	7.6	*Apeiba tibourbou*	3.4
	Inga sertulifera	7.3	*Piper aduncum*	2.8
	Cecropia membranacea	5.3	*Bellucia pentamera*	2.8
(d) 15 to <20	*Ochroma pyramidale*	25.8	*Piper aduncum*	10.0
	Ficus insipida	13.1	*Miconia minutiflora*	9.5
	Inga marginata	8.8	*Acalypha macrostachya*	9.2
	Inga thibaudiana	6.5	*Cecropia engleriana*	7.3
	Cecropia membranacea	6.4	*Apeiba membranacea*	5.9
(e) 20 to <25	*Sapium marmieri*	9.4	*Tachigali sp1*	26.2
	Erythrina poeppigiana	5.8	*Miconia tococa*	11.9
	Cecropia membranacea	5.3	*Vismia macrophylla*	10.0
	Pourouma cecropiifolia	3.3	*Vismia baccifera*	8.0
	Guazuma ulmifolia	3.2	*Miconia poeppigii*	7.4
(e) Reference forest	*Virola surinamensis*	3.0		
	Astrocaryum murumuru	2.8		
	Guarea macrophylla	2.4		
	Brosimum lactescens	2.2		
	Spondias mombin	2.1		

Using PERMANOVA in non-metric multidimensional scaling ordinations (nMDS), we found that the tree species composition of natural regeneration in abandoned mining lands differed significantly with mining type ($F = 1.07$, $p < 0.001$). The covariates stand age ($F = 1.12$, $p < 0.001$) and distance from forest edge ($F = 1.13$, $p < 0.001$) had an effect on similarity between assemblages of trees (Table 4), while we found an interactive effect between stand age × distance from forest edge, and between mining type × distance from forest edge (PERMANOVA, $p < 0.05$). But we did not see evidence of an interactive effect between stand age × mining type, and between stand age × distance from forest edge × mining type (PERMANOVA, $p > 0.05$) (Table 4). That would indicate the composition of natural regeneration of tree species at abandoned mining lands changed differently over time and was influenced by distance from the forest edge, and mining type.

PERMANOVA test showed differences in β diversity across mining type groups (Figure 5B). For categories of abandonment time, we only found significant differences in tree species composition between young (<5 years-old) and old stands (20–25 years-old) (Figure 5C).

Table 4. Results of analysis of variance (PERMANOVA) tests on natural regeneration of tree communities across a chronosequence in degraded lands of the Peruvian Amazon. Mining type was treated as fixed effects and stand age (years) and distance from the forest edge (m) were used as covariates. Significant *p*-values ($p < 0.05$) are shown in bold.

Source	df [1]	Sum of Squares	Pseudo-F	*p* (Permutation)
Age	1	5573.4	1.12	**<0.001**
Distance from forest edge (Dist)	1	5637.6	1.13	**<0.001**
Mining type (Min)	1	5300.7	1.07	**<0.001**
Age × Dist	1	5193.6	1.04	**<0.001**
Age × Min	1	5009.9	1.01	0.412
Dist × Min	1	5210.0	1.04	**<0.001**
Age × Dist × Min	1	5066.4	1.02	0.163
Residuals	172	845,150		

[1] df, degree of freedom.

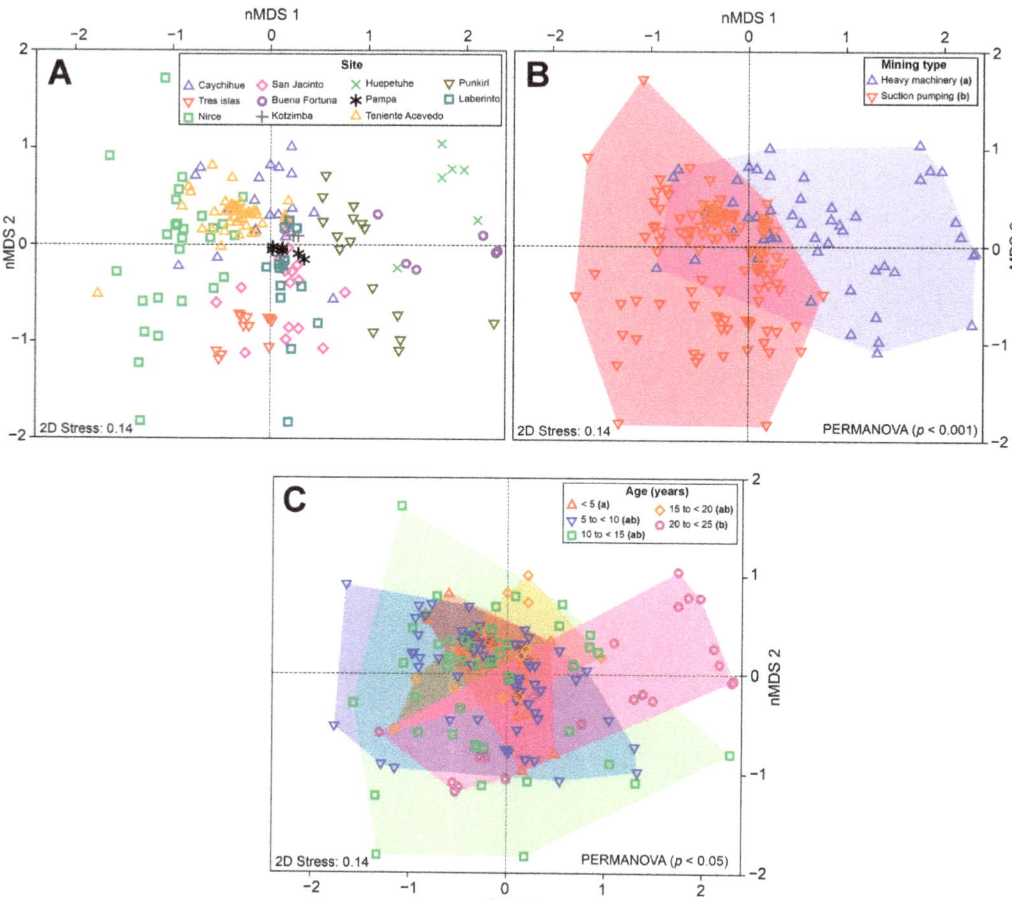

Figure 5. Non-metric multidimensional scaling ordinations (nMDS) of tree community composition (beta diversity) based on Chao–Jaccard similarity between natural regeneration across a chronosequence in degraded lands of the Peruvian Amazon. (**A**) Study sites, (**B**) type of mining, (**C**) stand age. In B and C, hulls overlaid for each variable. In each nMDS, different letters of the legend label indicate statistically significant differences according to pairwise PERMANOVA tests.

4. Discussion

Our study shows for the first time how the type of mining affects the resilience of structure, species diversity, and aboveground biomass of secondary forests following the abandonment of gold mining activity in the Southeastern Peruvian Amazon. To our knowledge, this field-based study provides the first assessment of how the type of mining can limit the recovery of AGB after mining. We sought to understand the relative importance of distance from the forest edge and species diversity in mediating the effects of different types of mining activities (heavy machinery and suction pumping) on AGB. Therefore, for effective restoration/reforestation strategies, these local site variables should be considered [73].

4.1. Effects of Mining Type, Stand Age, and Distance from the Forest on AGB

Our analysis shows that mining type, stand age, and distance from the forest edge had a significant effect on AGB. However, the influence of the distance from the forest edge could vary according to mining type. This is because the distance from the forest edge mediates the effects of species diversity on AGB. The negative effect of the distance from the forest edge was significantly greater in heavy machinery mining due to these lands being subject to high degrees of landscape fragmentation [74]. Meanwhile, in areas recovering from suction pumping mining, tree species diversity had a positive effect on AGB, as found previously in many studies [37,60,75]. This result indicates that these relationships are context-dependent [24], because the effect of the distance from the forest matters significantly in heavy machinery mining but not in suction pumping. This suggests that heavy machinery mining is severely impacting AGB resilience because the AGB accumulation may depend on stand forest age [76], land-use history [37], forest cover [30], soil properties [77,78], proportion of wind-dispersed species [79], and type and severity of disturbance [21,80]. In contrast, the small-scale disturbance caused by suction pumping mining and a relative short distance from a propagule source, in some sites of the study, promotes the establishment of species and increasing species diversity [41,81] and facilitating a faster AGB recovery [80,82] in areas recovering from suction pumping mining. This supports our results in SEM calculations and could explain why the AGB is higher in suction pumping mining forest stands. The positive relationship between AGB and tree species diversity found in suction pumping mining forest stands of the present study was reported by other studies in secondary forests [60,83,84] and old-growth forests [24,58,76,80,85]. Because of forest woody productivity, it is usually positively associated with tree species diversity [85]. Due to the coexistence of functionally different species, this may (1) increase resource partitioning of the environment, (2) enable positive plant–plant interactions and (3) improve growing conditions for establishment and growth of other species [79,83].

4.2. Recovery of AGB, Tree Diversity, and Forest Structure

Understanding the natural regeneration processes on degraded land after human activities is an essential goal of restoration ecology [78]. Our results from the study area support our initial hypothesis that the resilience of ecosystems degraded by mining in the Madre de Dios Amazon would be influenced by the type of mining, time of abandonment, and distance from the forest edge. This effect becomes clearer when we compare our results with reference ecosystems because AGB, effective number of species (Hill ^{0}D), as well as mean height and mean DBH of woody species of natural regeneration differed significantly from reference forests (except for suction pumping forest stands after 20-years of abandonment).

AGB strongly influences ecosystem processes [79] and tropical forests store significant quantities of AGB accumulated in woody plants [86]—64% of global forest AGB [87]. However, the expansion of deforestation and forest degradation in Madre de Dios persists [19,88], resulting in the emission of carbon into the atmosphere [16]. The mean AGB determined in 20–25 years of recovery, 198 ± 115 Mg ha^{-1} in areas of suction pumping mining and 45 ± 78 Mg ha^{-1} in areas recovering from heavy machinery mining in this study is less than the mean AGB reported for tropical Amazon forest from Madre de Dios (230–450 Mg ha^{-1}) [39,86,89]. This suggests a more rapid recovery of AGB in suction pumping mining forest stands in comparison to those of heavy machinery, because >20-year-old suction pumping forest stands store similar amounts of carbon as naturally regenerating forests on abandoned deforested lands (secondary forests from non-mining impacts) of similar age [4,90]. Furthermore, Sharma et al. [29] suggest that secondary forests can accumulate half of the AGB of reference forests within 30 years, which is consistent with our results in suction pumping mining (56% of recovery) but not with heavy machinery mining (18% of recovery). The rapid AGB recovery in suction pumping mining forest stands is consistent with those reported by other studies on tropical secondary forest succession [18,30]. Given that, after 20 years of abandonment, the average AGB accumulation reported by Oberleitner et al. [30] (52% of recovery) and Poorter et al. [18] (60% of recovery) agrees closely with our results (56% of recovery) (Figure S2). Although, there are very few studies of AGB recovery on secondary succession after mining in Amazonian forests, our results are consistent with those reported by Kalamandeen et al. [21] in Guyana, who found woody AGB recovery rates similar to secondary forests recovering from agriculture or pasture in early regenerating sites (2–4 years after abandonment) following gold mining activity. In this way, we found empirical evidence showing that regenerating forests after abandonment by mining activities could represent important carbon sequestration sinks, because secondary forests in the Amazon are a promising way to mitigate climate warming due to their higher productivity and carbon uptake compared to mature forests [4]. Furthermore, when tropical forests are disturbed, AGB is considered the main source of carbon emissions [80].

The rapid recovery of forest structure compared to AGB, species diversity, and composition have been widely reported by previous studies [41,60,79]. In our study, after 15 years, mean height and mean DBH of suction pumping forest stands recovered at a much faster rate than those recovering from heavy machinery mining—88% of DBH and 84% of height versus 58% of DBH and 67% of height, respectively. These differences are likely due to a higher species diversity in suction pumping forest stands, because a higher diversity of species can promote a greater development of the forest structure and increase aboveground light capture, where space and resources are used most effectively [91,92]. Previous studies reported that biomass growth can be negatively influenced by canopy structure [93,94] and forests with more biomass also have a more complex structure [79]. The slower recovery of forest structure generates a slow recovery of AGB in heavy machinery forest stands, because height and tree diameter variation are positively and strongly related to AGB [83]. This can be explained by the presence of low-diameter trees and a lower space optimization by co-occurring tree species in the canopy layer in heavy machinery forest, which impact negatively on biomass accumulation [58,83,91].

Similar to AGB, species diversity also increased with stand forest age [30], but it could vary according to the type of mining. Because only average species diversity in suction pumping forest stands (20–25-year-old) reached a similar level to reference forest at a slower pace than forest structure. These results indicate that heavy machinery mining may have a considerable negative effect on species diversity [21], because suction pumping mining forest stands harbor on average approximately two times more species than heavy machinery stands. Given that species diversity usually starts near zero with no woody plants, due to the elimination of vegetation cover [39,41,42] and severe soil degradation [20,43,44,54] by goldmining, our results highlight the lower impact of suction pumping mining on landscape structure, which promotes an increase over time as new species become established [79], and facilitates the faster recovery across the landscape [80]. The lower species diversity after mining abandonment may be explained by a higher dominance (IVI > 10, Table 3) or monodominance [30,95] of pioneer species in 0–5-year-old stands, due to higher tolerance of certain species to unfavorable soil conditions [95] generated by goldmining activities in the study area. Although monodominance in early secondary succession in Amazonian forests is rare [95], it is usually characterized by low tree species diversity and is linked with low soil fertility [95,96]. This agrees with our results (especially with respect to heavy machinery) because goldmining in the Peruvian Amazon generates extreme soil disturbance [41]. Soil from mine spoils after abandonment is characterized by significantly lower organic matter, cation exchange capacity, and clay content than the reference forests [44,47,54]. On the other hand, the decline in species diversity in the oldest forest stands of heavy machinery mining may be due to an increase in monodominance [96,97] (p.e. *Tachigali* spp., *Miconia* spp. and *Vismia* spp.) along the chronosequence, remoteness from existing forest edges [41,76,97], and a lower soil fertility than suction pumping [54]. In addition, different woody plant species can become monodominant due to smaller natural regeneration community size and may generate stochastic effects on species diversity [98]. As also reported by Oberleitner et al. [30] in tropical secondary forests in Costa Rica, they found an increase in monodominance by a few tree species in the later successional stage. This may explain why suction pumping forest stands have higher tree species diversity and AGB than those of heavy machinery [80].

4.3. *Tree Species Dominance and Composition*

In this study, we found a significant difference in species composition between the two types of mining activities. This may be related to differences in landscape fragmentation and proximity to the sources of propagules [82], as has been observed for other tropical forests [76,79] and secondary forests after mining in the Peruvian Amazon [41]. This difference in species composition and stand structure may have implications for biomass accumulation [30,99], due to the turnover of high wood density species [7,44,60] and the increase in species diversity [24,79,97].

The lower similarity in species composition (<20% in higher successional stages, Table S3) across the chronosequence suggests a slower recovery compared to those of AGB, species diversity and forest structure. This pattern has been observed for other tropical forests [37] and in secondary forests after mining in the Peruvian Amazon [20,39–41]. This highlights the need to promote active restoration strategies to establish both economically and ecologically important valuable native tree species [37]. In our study, the dynamics of the slow recovery of species follow the "initial floristic composition model" hypothesis [100], because the dominance (IVI) is relayed from one floristic group (species, genus or family) to another throughout the chronosequence [97]. This hypothesis states that pioneer species colonize early successional stages which remain for some decades [76], and soon they make suitable conditions for colonization by other species more typical of later successional stages [97]. A similar pattern was found in other secondary successions post mining in the Peruvian Amazon [41,41]. Species typical of the later successional stages are not present in the initial successional stages (<5 years of abandonment), and pioneer species are highly dominant in natural regeneration after goldmining activities.

High IVI values of tree species within a community show their ability to acclimate to their habitat and suggest a high tolerance for degraded environments [99]. Species such as *Ochroma pyramidale*, *Cecropia* spp., *Trema micrantha*, *Tessaria integrifolia*, *Vismia* spp., and *Inga* spp., may represent candidate species for more focused reforestation in degraded lands by gold mining in the study area. However, it could vary according to mining type and the establishment of different species in these severely degraded areas remains largely unstudied. Moreover, medium–long term reforestation/restoration studies in degraded lands by gold mining in the Peruvian Amazon are scarce, with only one study reported on reforestation after 20 years in areas degraded by heavy machinery in Madre de Dios [39]. Therefore, carefully planned field trials are essential for achieving the optimal selection of tree species for restoration or reforestation purposes [21].

Fabaceae is considered the most common family in tropical Amazonian forests [31] and the third largest family of angiosperms [101]. Due to symbiotic nitrogen fixation in Fabaceae species [102], they can drive early ecological succession [21,61] and they are important in secondary forest and successful restoration practices [103]. In our study, the dominance of Fabaceae is similar in advanced successional forest stands and in reference forest, *Inga* species capable of fixing nitrogen were dominant in both mining types. Similar results were found in other studies in degraded lands by gold mining in the Peruvian Amazon [20,39–41]. This is important because nitrogen fixers are known to drive biomass accumulation in recovering secondary forests [21].

4.4. Limitations and Suggestions for Future Research

This study has certain limitations. (1) We used cross-sectional data (chronosequence approach) and future studies need to be longitudinal to confirm the recovery rate of AGB, floristic attributes, and alpha and beta diversity. (2) We used small plots (0.01 to 0.25 ha), which could generate greater heterogeneity in forest structure and greater variation in biomass estimates [104] but not in β-diversity analysis. This is because in heterogeneous vegetation (as in the present study) with a high β-diversity ordination patterns are not sensitive to different plot sizes [105]. Despite these limitations, our results are consistent with other studies in secondary successions [21,41,41,76,79], and provide new insights for a deeper understanding of the natural regeneration recovery of degraded lands by goldmining in the Peruvian Amazon. In future studies, we will analyze differences in functional species traits [106] and in AGB by field measurements of wood density by mining type in chronosequences, in different edaphic conditions. Wood density may change across secondary succession due to changes in species composition; this is usually ignored and could affect AGB estimates [7]. Likewise, tropical forest carbon pools have been affected by soil nutrient availability [30]. This knowledge may be useful for effective strategies to restore forest cover and improve ecosystem services [38] on degraded lands through goldmining in the Peruvian Amazon.

5. Conclusions

Our results clearly showed the dynamics of AGB recovery across a secondary chronosequence after abandonment by goldmining, and the contrasting responses of AGB accumulation between two mining types. This is because the resilience of aboveground biomass, forest structure, and species diversity are affected by type of mining. Successional secondary forest recovery from suction pumping forest stands was more resilient than heavy machinery. After 20 years of abandonment, recovery of AGB, forest structure, and species diversity were higher in suction pumping forest stands than in those of heavy machinery. Moreover, the high AGB accumulation potential of the suction pumping forest stands could play a crucial role in carbon storage and global warming mitigation [107]. Our study disentangles the importance of environmental factors on forest recovery after mining. It helps us understand the resilience of highly degraded ecosystem of the Peruvian Amazon.

Supplementary Materials: The following supporting information can be downloaded at: https://www.mdpi.com/article/10.3390/d16040233/s1, Figure S1: Climate diagram of Puerto Maldonado meteorological station (1950–2023). Values shown in climate diagram correspond to the average month annual temperature (blue line), maximum temperature (red line), minimum temperature (green line), and the total month precipitation (gray bars). Orange line represent the dry season, when precipitation is lower than 100 mm per month; Figure S2: Relationship between aboveground biomass and stand age of tree community in two types of mining activities (heavy machinery and suction pumping) across a chronosequence in degraded lands of the Peruvian Amazon (A). Mean AGB (±standard error) of reference forests close to plots in degraded lands (B). SEE = standard error of estimate; Table S1: Comparison of five families with the highest importance value index (IVI) of tree community by type of mining activities (heavy machinery and suction pumping) across a chronosequence in de-graded lands and reference forest in the Peruvian Amazon; Table S2: Comparison of five genera with the highest importance value index (IVI) of tree community by type of mining activities (heavy machinery and suction pumping) across a chronosequence in degraded lands and reference forest in the Peruvian Amazon; Table S3: Average of floristic similarity (Chao–Jaccard) among forest plots in two types of mining activities (heavy machinery and suction pumping) according to five recovery age classes of natural regeneration and in reference forest of tree community across chronosequence in degraded lands of the Peruvian Amazon.

Author Contributions: Conceptualization, J.G.-Q.; methodology, J.G.-Q., M.H.-M. and S.B.Q.; software, J.G.-Q. and M.H.-M.; validation, J.G.-Q., M.H.-M. and S.B.Q.; formal analysis, J.G.-Q. and M.H.-M.; investigation, J.G.-Q., M.H.-M. and S.B.Q.; resources, V.P.A. and M.H.-M.; data curation, J.G-Q. and M.H-M.; writing—original draft preparation, J.G.-Q.; writing—review and editing, M.H.-M., G.A.A., V.P.A., S.B.Q. and E.E.C.-V.; visualization, J.G.-Q., M.H.-M. and G.A.A.; supervision, G.A.A. and E.E.C.-V.; funding acquisition, V.P.A. and M.H.-M. All authors have read and agreed to the published version of the manuscript.

Funding: The work was supported by Universidad Nacional Amazonica de Madre de Dios.

Institutional Review Board Statement: Not applicable.

Data Availability Statement: The data used are primarily reflected in the article. Other relevant data are available from the corresponding author upon request.

Acknowledgments: Thanks to Erika Sajami Quispe, Laura Cutire Tinta, Laura Ramirez Chávez, Enrrique Pacheco Villanueva, Rembrandt Canahuire Robles, Yoni Fernandez Mamani, Marco Garcia Tueros, Lebana Quispe Sueros, Iván Cruz Herrera, and Jhon Quispe Ampuero for permission to use his field data.

Conflicts of Interest: The authors declare no conflicts of interest. The funders had no role in the design of the study; in the collection, analyses, or interpretation of data; in the writing of the manuscript; or in the decision to publish the results.

References

1. Cooper, D.L.M.; Lewis, S.L.; Sullivan, M.J.P.; Prado, P.I.; ter Steege, H.; Barbier, N.; Slik, F.; Sonké, B.; Ewango, C.E.N.; Adu-Bredu, S.; et al. Consistent patterns of common species across tropical tree communities. *Nature* **2024**, *625*, 728–734. [CrossRef]
2. Houghton, R.A.; Skole, D.L.; Nobre, C.A.; Hackler, J.L.; Lawrence, K.T.; Chomentowski, W.H. Annual fluxes of carbon from deforestation and regrowth in the Brazilian Amazon. *Nature* **2000**, *403*, 301–304. [CrossRef]
3. Mitchard, E.T.A. The tropical forest carbon cycle and climate change. *Nature* **2018**, *559*, 527–534. [CrossRef]
4. Tito, R.; Salinas, N.; Cosio, E.G.; Boza Espinoza, T.E.; Muñiz, J.G.; Aragón, S.; Nina, A.; Roman-Cuesta, R.M. Secondary forests in Peru: Differential provision of ecosystem services compared to other post-deforestation forest transitions. *Ecol. Soc.* **2022**, *27*, art12. [CrossRef]
5. Asner, G.P.; Powell, G.V.N.; Mascaro, J.; Knapp, D.E.; Clark, J.K.; Jacobson, J.; Kennedy-Bowdoin, T.; Balaji, A.; Paez-Acosta, G.; Victoria, E.; et al. High-resolution forest carbon stocks and emissions in the Amazon. *Proc. Natl. Acad. Sci. USA* **2010**, *107*, 16738–16742. [CrossRef]
6. Hall, J.S.; Plisinski, J.S.; Mladinich, S.K.; van Breugel, M.; Lai, H.R.; Asner, G.P.; Walker, K.; Thompson, J.R. Deforestation scenarios show the importance of secondary forest for meeting Panama's carbon goals. *Landsc. Ecol.* **2022**, *37*, 673–694. [CrossRef]
7. Aragón, S.; Salinas, N.; Nina-Quispe, A.; Qquellon, V.H.; Paucar, G.R.; Huaman, W.; Porroa, P.C.; Olarte, J.C.; Cruz, R.; Muñiz, J.G.; et al. Aboveground biomass in secondary montane forests in Peru: Slow carbon recovery in agroforestry legacies. *Glob. Ecol. Conserv.* **2021**, *28*, e01696. [CrossRef]
8. Cherlet, M.; Hutchinson, C.; Reynolds, J.; Hill, J.; Sommer, S.; von Maltitz, G. *World Atlas of Desertification, Mapping Land Degradation and Sustainable Land Management Opportunities*, 3rd ed.; Publication Office of the European Union: Luxembourg, 2018.
9. Walker, L.R.; del Moral, R. *Primary Succession and Ecosystem Rehabilitation*; Cambridge University Press: New York, NY, USA, 2003; ISBN 9780521800761.
10. Prach, K.; Walker, L.R. Differences between primary and secondary plant succession among biomes of the world. *J. Ecol.* **2019**, *107*, 510–516. [CrossRef]
11. Rocha-Nicoleite, E.; Campos, M.L.; Colombo, G.T.; Overbeck, G.E.; Müller, S.C. Forest restoration after severe degradation by coal mining: Lessons from the first years of monitoring. *Brazilian J. Bot.* **2018**, *41*, 653–664. [CrossRef]
12. Kuramoto, J. *Artisanal and Informal Mining in Peru*; International Institute for Enviroment and Development, and World Business Council for Sustainable Development: London, UK, 2001.
13. Stoll, E.; Roopsind, A.; Maharaj, G.; Velazco, S.; Caughlin, T.T. Detecting gold mining impacts on insect biodiversity in a tropical mining frontier with SmallSat imagery. *Remote Sens. Ecol. Conserv.* **2022**, *8*, 379–390. [CrossRef]
14. Álvarez-Berríos, N.; L'Roe, J.; Naughton-Treves, L. Does formalizing artisanal gold mining mitigate environmental impacts? Deforestation evidence from the Peruvian Amazon. *Environ. Res. Lett.* **2021**, *16*, 064052. [CrossRef]
15. Lohbeck, M.; Poorter, L.; Martínez-Ramos, M.; Bongers, F. Biomass is the main driver of changes in ecosystem process rates during tropical forest succession. *Ecology* **2015**, *96*, 1242–1252. [CrossRef]
16. Csillik, O.; Asner, G.P. Aboveground carbon emissions from gold mining in the Peruvian Amazon. *Environ. Res. Lett.* **2020**, *15*, 014006. [CrossRef]
17. Alvarez-Berríos, N.L.; Mitchell Aide, T. Global demand for gold is another threat for tropical forests. *Environ. Res. Lett.* **2015**, *10*, 014006. [CrossRef]

18. Poorter, L.; Bongers, F.; Aide, T.M.; Almeyda Zambrano, A.M.; Balvanera, P.; Becknell, J.M.; Boukili, V.; Brancalion, P.H.S.; Broadbent, E.N.; Chazdon, R.L.; et al. Biomass resilience of Neotropical secondary forests. *Nature* **2016**, *530*, 211–214. [CrossRef]
19. Alarcon-Aguirre, G.; Canahuire Robles, R.R.; Guevara Duarez, F.M.; Rodríguez Achata, L.; Gallegos Chacón, L.E.; Garate-Quispe, J. Dynamics of forest loss in the southeast of the Peruvian Amazon: A case study in Madre de Dios. *Ecosistemas* **2021**, *30*, 2175. [CrossRef]
20. Chambi-Legoas, R.; Ortega Rodriguez, D.R.; Figueiredo, F.d.M.d.; Peña Valdeiglesias, J.; Zevallos Pollito, P.A.; Marcelo-Peña, J.L.; Rother, D.C. Natural Regeneration After Gold Mining in the Peruvian Amazon: Implications for Restoration of Tropical Forests. *Front. For. Glob. Chang.* **2021**, *4*, 594627. [CrossRef]
21. Kalamandeen, M.; Gloor, E.; Johnson, I.; Agard, S.; Katow, M.; Vanbrooke, A.; Ashley, D.; Batterman, S.A.; Ziv, G.; Holder-Collins, K.; et al. Limited biomass recovery from gold mining in Amazonian forests. *J. Appl. Ecol.* **2020**, *57*, 1730–1740. [CrossRef]
22. Ploton, P.; Mortier, F.; Barbier, N.; Cornu, G.; Réjou-Méchain, M.; Rossi, V.; Alonso, A.; Bastin, J.F.; Bayol, N.; Bénédet, F.; et al. A map of African humid tropical forest aboveground biomass derived from management inventories. *Sci. Data* **2020**, *7*, 221. [CrossRef]
23. Liu, C.; Chen, D.; Zou, C.; Liu, S.; Li, H.; Liu, Z.; Feng, W.; Zhang, N.; Ye, L. Modeling Biomass for Natural Subtropical Secondary Forest Using Multi-Source Data and Different Regression Models in Huangfu Mountain, China. *Sustainability* **2022**, *14*, 13006. [CrossRef]
24. Abbasi, U.A.; Gilani, H.; Hussain, K.; Ali, A. Increasing stand stature weakens the positive effects of tree richness and structural imbalance on aboveground biomass in temperate forests: The stand stature hypothesis. *For. Ecol. Manag.* **2023**, *539*, 121040. [CrossRef]
25. Fawcett, D.; Sitch, S.; Ciais, P.; Wigneron, J.P.; Silva-Junior, C.H.L.; Heinrich, V.; Vancutsem, C.; Achard, F.; Bastos, A.; Yang, H.; et al. Declining Amazon biomass due to deforestation and subsequent degradation losses exceeding gains. *Glob. Chang. Biol.* **2023**, *29*, 1106–1118. [CrossRef]
26. Messinger, M.; Asner, G.P.; Silman, M. Rapid assessments of amazon forest structure and biomass using small unmanned aerial systems. *Remote Sens.* **2016**, *8*, 615. [CrossRef]
27. Requena Suarez, D.; Rozendaal, D.M.A.; De Sy, V.; Phillips, O.L.; Alvarez-Dávila, E.; Anderson-Teixeira, K.; Araujo-Murakami, A.; Arroyo, L.; Baker, T.R.; Bongers, F.; et al. Estimating aboveground net biomass change for tropical and subtropical forests: Refinement of IPCC default rates using forest plot data. *Glob. Chang. Biol.* **2019**, *25*, 3609–3624. [CrossRef]
28. Cardozo, E.G.; Celentano, D.; Rousseau, G.X.; Silva, H.R.E.; Muchavisoy, H.M.; Gehring, C. Agroforestry systems recover tree carbon stock faster than natural succession in Eastern Amazon, Brazil. *Agrofor. Syst.* **2022**, *96*, 941–956. [CrossRef]
29. Sharma, S.B.; Kumar, S.; Hegde, N. Biomass and carbon recovery of secondary forest in a Montane Subtropical Forest of North Eastern India. *Trop. Ecol.* **2023**, *64*, 114–121. [CrossRef]
30. Oberleitner, F.; Egger, C.; Oberdorfer, S.; Dullinger, S.; Wanek, W.; Hietz, P. Recovery of aboveground biomass, species richness and composition in tropical secondary forests in SW Costa Rica. *For. Ecol. Manag.* **2021**, *479*, 118580. [CrossRef]
31. Oliveira, L.Z.; de Maçaneiro, J.P.; da Silva, D.A.; Uller, H.F.; de Britto, P.C.; Correia, J.; Piazza, G.E.; Zambiazi, D.C.; Vibrans, A.C.; Fantini, A.C. Structure, Biomass and Diversity of a Late-Successional Subtropical Atlantic Forest in Brazil. *Floresta E Ambient.* **2022**, *29*, e20210095. [CrossRef]
32. Wallis, C.I.B.; Crofts, A.L.; Inamdar, D.; Arroyo-Mora, J.P.; Kalacska, M.; Laliberté, É.; Vellend, M. Remotely sensed carbon content: The role of tree composition and tree diversity. *Remote Sens. Environ.* **2023**, *284*, 113333. [CrossRef]
33. Mora, F.; Jaramillo, V.J.; Bhaskar, R.; Gavito, M.; Siddique, I.; Byrnes, J.E.K.; Balvanera, P. Carbon Accumulation in Neotropical Dry Secondary Forests: The Roles of Forest Age and Tree Dominance and Diversity. *Ecosystems* **2018**, *21*, 536–550. [CrossRef]
34. Rebola, L.C.; Paz, C.P.; Gamarra, L.V.; Burslem, D.F. Land use intensity determines soil properties and biomass recovery after abandonment of agricultural land in an Amazonian biodiversity hotspot. *Sci. Total Environ.* **2021**, *801*, 149487. [CrossRef]
35. Rozendaal, D.; Bongers, F.; Aide, T.; Alvarez-Dávila, E.; Ascarrunz, N.; Balvanera, P.; Becknell, J.M.; Bentos, T.V.; Brancalion, P.H.S.; Cabral, G.A.L.; et al. Biodiversity recovery of Neotropical secondary forests. *Sci. Adv.* **2019**, *5*, eaau3114. [CrossRef]
36. Brienen, R.J.W.; Phillips, O.L.; Feldpausch, T.R.; Gloor, E.; Baker, T.R.; Lloyd, J.; Lopez-Gonzalez, G.; Monteagudo-Mendoza, A.; Malhi, Y.; Lewis, S.L.; et al. Long-term decline of the Amazon carbon sink. *Nature* **2015**, *519*, 344–348. [CrossRef]
37. Ojoatre, S.; Barlow, J.; Jacobs, S.R.; Rufino, M.C. Recovery of aboveground biomass, soil carbon stocks and species diversity in tropical montane secondary forests of East Africa. *For. Ecol. Manag.* **2024**, *552*, 121569. [CrossRef]
38. Chazdon, R.; Letcher, S.G.; van Breugel, M.; Martinez-Ramos, M.; Bongers, F.; Finegan, B. Rates of change in tree communities of secondary Neotropical forests following major disturbances. *Philos. Trans. R. Soc. B Biol. Sci.* **2007**, *362*, 273–289. [CrossRef]
39. Garate-Quispe, J.S.; Canahuire-Robles, R.; Surco-Huacachi, O.; Alarcón-Aguirre, G. Desarrollo estructural y composición florística arbórea en áreas afectadas por minería aurífera en la Amazonía peruana: A 20 años de su reforestación. *Rev. Mex. Biodivers.* **2021**, *92*, 923437. [CrossRef]
40. Alarcón-Aguirre, G.; Sajami Quispe, E.; Vásquez Zavaleta, T.; Ponce Tejada, L.V.; Ramos Enciso, D.; Rodríguez Achata, L.; Garate-Quispe, J. Vegetation dynamics in lands degraded by gold mining in the southeastern Peruvian Amazon. *Trees For. People* **2023**, *11*, 100369. [CrossRef]

41. Garate-Quispe, J.; Velásquez Ramírez, M.; Becerra-Lira, E.; Baez-Quispe, S.; Abril-Surichaqui, M.; Rodriguez-Achata, L.; Muñoz-Ushñahua, A.; Nascimento Herbay, P.; Fernandez-Mamani, Y.; Alarcon-Aguirre, G.; et al. Influence of Distance from Forest Edges on Spontaneous Vegetation Succession Following Small-Scale Gold Mining in the Southeast Peruvian Amazon. *Diversity* **2023**, *15*, 793. [CrossRef]
42. Garate-Quispe, J.S.; de Leon, R.P.; Machaca, M.H.; Laime, E.J.; Ramos, C.N. Growth and survivorship of Vetiveria zizanioides in degraded soil by gold-mining in the Peruvian Amazon. *J. Degrad. Min. Lands Manag.* **2021**, *9*, 3219–3225. [CrossRef]
43. Velásquez Ramírez, M.G.; del Castillo Torres, D.; Guerrero Barrantes, J.A.; Vásquez Bardales, J.; Thomas, E.; Cusi Auca, E.; Chinen Gushiken, M.; Muñoz Diaz, B.; Russo, R.; Corvera Gomringer, R. Soil recovery of alluvial gold mine spoils in the Peruvian Amazon using Stylosanthes guianensis, a promising cover crop. *L. Degrad. Dev.* **2021**, *32*, 5143–5153. [CrossRef]
44. Román-Dañobeytia, F.; Cabanillas, F.; Lefebvre, D.; Farfan, J.; Alferez, J.; Polo-Villanueva, F.; Llacsahuanga, J.; Vega, C.M.; Velasquez, M.; Corvera, R.; et al. Survival and early growth of 51 tropical tree species in areas degraded by artisanal gold mining in the Peruvian Amazon. *Ecol. Eng.* **2021**, *159*, 106097. [CrossRef]
45. Lefebvre, D.; Román-Dañobeytia, F.; Soete, J.; Cabanillas, F.; Corvera, R.; Ascorra, C.; Fernandez, L.E.; Silman, M. Biochar Effects on Two Tropical Tree Species and Its Potential as a Tool for Reforestation. *Forests* **2019**, *10*, 678. [CrossRef]
46. Román-Dañobeytia, F.; Huayllani, M.; Michi, A.; Ibarra, F.; Loayza-Muro, R.; Vázquez, T.; Rodríguez, L.; García, M. Reforestation with four native tree species after abandoned gold mining in the Peruvian Amazon. *Ecol. Eng.* **2015**, *85*, 39–46. [CrossRef]
47. Lefebvre, D.; Williams, A.G.; Kirk, G.J.D.; Paul; Burgess, J.; Meersmans, J.; Silman, M.R.; Román-Dañobeytia, F.; Farfan, J.; Smith, P. Assessing the carbon capture potential of a reforestation project. *Sci. Rep.* **2021**, *11*, 19907. [CrossRef]
48. Garate-Quispe, J.S.; Roca, M.R.G.; Aguirre, G.A. Survival and Growth of Brazil-Nut Seedlings in Tree-Fall Gaps and Forest Understory. *Floresta E Ambient.* **2020**, *27*, e20171168. [CrossRef]
49. MINAM. *Mapa Nacional de Cobertura Vegetal*; Ministerio del Ambiente: Lima, Peru, 2015.
50. Shoo, L.; Freebody, K.; Kanowski, J.; Catterall, C. Slow recovery of tropical old-field rainforest regrowth and the value and limitations of active restoration. *Conserv. Biol.* **2016**, *30*, 121–132. [CrossRef]
51. Araújo, F.; Martins, S.; Meira, J.; Lani, J.; Pires, I. Florística da vegetação arbustivo-arbórea colonizadora de uma área degradada por mineração de caulim, em Brás Pires, MG. *Rev. Árvore* **2005**, *29*, 983–992. [CrossRef]
52. Leiva, J.D. Appropriate technologies and the geosocial evolution of informal, small-scale gold mining in Madre de Dios, Peru. *Extr. Ind. Soc.* **2022**, *12*, 101165. [CrossRef]
53. Caballero, J.; Messinger, M.; Román-Dañobeytia, F.; Ascorra, C.; Fernandez, L.; Silman, M. Deforestation and Forest Degradation Due to Gold Mining in the Peruvian Amazon: A 34-Year Perspective. *Remote Sens.* **2018**, *10*, 1903. [CrossRef]
54. Velásquez-Ramírez, M.; Vega Ruiz, C.; Gomringer, R.; Pillaca, M.; Thomas, E.; Stewart, P.M.; Gamarra Miranda, L.A.; Dañobeytia, F.R.; Guerrero Barrantes, J.A.; Gushiken, M.C.; et al. Mercury in soils impacted by alluvial gold mining in the Peruvian Amazon. *J. Environ. Manag.* **2021**, *288*, 112364. [CrossRef]
55. Wardius, Y.; Hein, S. Terrestrial laser scanning vs. manual methods for assessing complex forest stand structure: A comparative analysis on plenter forests. *Eur. J. For. Res.* **2024**, *143*, 635–649. [CrossRef]
56. Chave, J.; Réjou-Méchain, M.; Búrquez, A.; Chidumayo, E.; Colgan, M.S.; Delitti, W.B.C.; Duque, A.; Eid, T.; Fearnside, P.M.; Goodman, R.C.; et al. Improved allometric models to estimate the aboveground biomass of tropical trees. *Glob. Chang. Biol.* **2014**, *20*, 3177–3190. [CrossRef]
57. APG IV (Angiosperm Phylogeny Group); Chase, M.W.; Christenhusz, M.J.M.; Fay, M.F.; Byng, J.W.; Judd, W.S.; Soltis, D.E.; Mabberley, D.J.; Sennikov, A.N.; Soltis, P.S.; et al. An update of the Angiosperm Phylogeny Group classification for the orders and families of flowering plants: APG IV. *Bot. J. Linn. Soc.* **2016**, *181*, 1–20. [CrossRef]
58. Su, L.; Heydari, M.; Omidipour, R.; Soheili, F.; Cheraghi, J.; Manuel Villa, P.; Prévosto, B. Stand structural diversity and elevation rather than functional diversity drive aboveground biomass in historically disturbed semiarid oak forests. *For. Ecol. Manag.* **2023**, *543*, 121139. [CrossRef]
59. Zanne, A.E.; Lopez-Gonzalez, G.; Coomes, D.A.; Ilic, J.; Jansen, S.; Lewis, S.L.; Miller, R.B.; Swenson, N.G.; Wiemann, M.C.; Chave, J. Data from: Towards a worldwide wood economics spectrum [Dataset]. *Dryad*. 2009. [CrossRef]
60. Aryal, D.R.; De Jong, B.H.J.; Sánchez-Silva, S.; Haas-Ek, A.; Esparza-Olguin, L.; Ochoa-Gaona, S.; Ghimire, R.; Morales-Ruiz, D.E. Biomass recovery along a tropical forest succession: Trends on tree diversity, wood traits and stand structure. *For. Ecol. Manag.* **2024**, *555*, 121709. [CrossRef]
61. Galván-Cisneros, C.M.; Sánchez Montaño, L.R.; Ojeda-Rodríguez, A.E.; Meira-Neto, J.A.A. Structures of tropical dry forests in the Andes: Forest conservation, composition and the role of fabaceae and myrtaceae. *Cerne* **2023**, *29*, e-1033189. [CrossRef]
62. Curtis, J.; McIntosh, R. An upland forest continuum in the prairie-forest border region of Wisconsin. *Ecology* **1951**, *32*, 476–496. [CrossRef]
63. Kindt, R.; Coe, R. *Tree Diversity Analysis. A Manual and Software for Common Statistical Methods for Ecological and Biodiversity Studies*; World Agroforestry Centre: Nairobi, Kenia, 2005.
64. Chao, A.; Kubota, Y.; Zelený, D.; Chiu, C.H.; Li, C.F.; Kusumoto, B.; Yasuhara, M.; Thorn, S.; Wei, C.L.; Costello, M.J.; et al. Quantifying sample completeness and comparing diversities among assemblages. *Ecol. Res.* **2020**, *35*, 292–314. [CrossRef]
65. Li, D. hillR: Taxonomic, functional, and phylogenetic diversity and similarity through Hill Numbers. *J. Open Source Softw.* **2018**, *3*, 1041. [CrossRef]

66. Melo, A. *CommEcol: Community Ecology Analyses*, R package version 1.7.1; R Foundation for Statistical Computing: Vienna, Austria, 2021.
67. Niku, J.; Warton, D.I.; Hui, F.K.C.; Taskinen, S. Generalized Linear Latent Variable Models for Multivariate Count and Biomass Data in Ecology. *J. Agric. Biol. Environ. Stat.* **2017**, *22*, 498–522. [CrossRef]
68. Rosseel, Y. lavaan: An R Package for Structural Equation Modeling. *J. Stat. Softw.* **2012**, *48*, 1–36. [CrossRef]
69. R Core Team. *R: A Language and Environment for Statistical Computing*; R Foundation for Statistical Computing: Vienna, Austria, 2024.
70. Muthén, L.; Muthén, B. *Mplus User's Guide*, 8th ed.; Muthén & Muthén: Los Angeles, CA, USA, 2017.
71. Jain, A.; Li, X.H.; Chen, W.N. Similarities and differences in gut microbiome composition correlate with dietary patterns of Indian and Chinese adults. *AMB Express* **2018**, *8*, 104. [CrossRef]
72. Liu, F.; Tan, C.; Yang, Z.; Li, J.; Xiao, H.; Tong, Y. Regeneration and growth of tree seedlings and saplings in created gaps of different sizes in a subtropical secondary forest in southern China. *For. Ecol. Manag.* **2022**, *511*, 120143. [CrossRef]
73. Van Der Sande, M.T.; Powers, J.S.; Kuyper, T.W.; Norden, N.; Salgado-Negret, B.; Silva De Almeida, J.; Bongers, F.; Delgado, D.; Dent, D.H.; Derroire, G.; et al. Soil resistance and recovery during neotropical forest succession. *Philos. Trans. R. Soc. B Biol. Sci.* **2023**, *378*, 20210074. [CrossRef] [PubMed]
74. Sánchez-Cuervo, A.M.; Santos, L.; Dallmeier, F.; Garate, P.; Bravo, A.; Vanthomme, H. Twenty years of land cover change in the southeastern Peruvian Amazon: Implications for biodiversity conservation. *Reg. Environ. Chang.* **2020**, *20*, 8. [CrossRef]
75. de Almeida Valente, F.D.; de Castro, M.F.; Lustosa Filho, J.F.; de Carvalho Gomes, L.; Neves, J.C.L.; da Silva, I.R.; de Oliveira, T.S. Native multispecies and fast-growing forest root biomass increase C and N stocks in a reclaimed bauxite mining area. *Environ. Monit. Assess.* **2023**, *195*, 129. [CrossRef] [PubMed]
76. de Oliveira, C.C.; Martins, F.R.; Souza, B.C.; de Sa Barretto Sampaio, E.V.; Loiola, M.I.B.; Soares, A.A. Resilience and successional trends of woody vegetation in seasonally dry tropical forests. *For. Int. J. For. Res.* **2023**, *96*, 740–753. [CrossRef]
77. Kim, S.; Zang, H.; Mortimer, P.; Shi, L.; Li, Y.; Xu, J.; Ostermann, A. Tree species and recovery time drives soil restoration after mining: A chronosequence study. *L. Degrad. Dev.* **2018**, *29*, 1738–1747. [CrossRef]
78. Damptey, F.; Birkhofer, K.; de la Riva, E.G. Soil Properties and Biomass Attributes in a Former Forest Restoration. *Land* **2020**, *9*, 209. [CrossRef]
79. Poorter, L.; Craven, D.; Jakovac, C.C.; Van Der Sande, M.T.; Amissah, L.; Bongers, F.; Chazdon, R.L.; Farrior, C.E.; Kambach, S.; Meave, J.A. Multidimensional tropical forest recovery. *Science* **2021**, *374*, 1370–1376. [CrossRef]
80. Uscanga, A.; Bartlein, P.J.; Silva, L.C.R. Local and Regional Effects of Land-Use Intensity on Aboveground Biomass and Tree Diversity in Tropical Montane Cloud Forests. *Ecosystems* **2023**, *26*, 1734–1752. [CrossRef]
81. Li, T.; Yang, H.; Yang, X.; Guo, Z.; Fu, D.; Liu, C.; Li, S.; Pan, Y.; Zhao, Y.; Xu, F.; et al. Community assembly during vegetation succession after metal mining is driven by multiple processes with temporal variation. *Ecol. Evol.* **2022**, *12*, e8882. [CrossRef]
82. Velasco-Murguía, A.; del Castillo, R.F.; Rös, M.; Rivera-García, R. Successional pathways of post-milpa fallows in Oaxaca, Mexico. *For. Ecol. Manag.* **2021**, *500*, 119644. [CrossRef]
83. Mensah, S.; Noulèkoun, F.; Salako, V.K.; Lokossou, C.S.M.J.; Akouété, P.; Seifert, T.; Glèlè Kakaï, R. Structural and taxonomic diversity predict above-ground biomass better than functional measures of maximum height in mixed-species forests. *Appl. Veg. Sci.* **2023**, *26*, e12732. [CrossRef]
84. Calderón-Balcázar, A.; Cárdenas, C.D.; Díaz-Vasco, O.; Fandiño, E.; Márquez, T.; Pizano, C. Biomass and carbon stocks of four vegetation types in the Llanos Orientales of Colombia (Mapiripán, Meta). *Trees For. People* **2023**, *12*, 100380. [CrossRef]
85. Kohyama, T.I.; Sheil, D.; Sun, I.-F.; Niiyama, K.; Suzuki, E.; Hiura, T.; Nishimura, N.; Hoshizaki, K.; Wu, S.-H.; Chao, W.-C.; et al. Contribution of tree community structure to forest productivity across a thermal gradient in eastern Asia. *Nat. Commun.* **2023**, *14*, 1113. [CrossRef]
86. Sousa, T.R.; Schietti, J.; Ribeiro, I.O.; Emílio, T.; Fernández, R.H.; ter Steege, H.; Castilho, C.V.; Esquivel-Muelbert, A.; Baker, T.; Pontes-Lopes, A.; et al. Water table depth modulates productivity and biomass across Amazonian forests. *Glob. Ecol. Biogeogr.* **2022**, *31*, 1571–1588. [CrossRef]
87. Santoro, M.; Cartus, O.; Carvalhais, N.; Rozendaal, D.M.A.; Avitabile, V.; Araza, A.; De Bruin, S.; Herold, M.; Quegan, S.; Rodríguez-Veiga, P.; et al. The global forest above-ground biomass pool for 2010 estimated from high-resolution satellite observations. *Earth Syst. Sci. Data* **2021**, *13*, 3927–3950. [CrossRef]
88. Alarcon-Aguirre, G.; Mamani Mamani, M.; Canahuire-Robles, R.R.; Vasquez Zavaleta, T.; Peña Valdeiglesias, J.; Diaz Revoredo, J.; Rodríguez Achata, L.; Ramos Enciso, D.; Garate-Quispe, J. Forest Loss Related to Brazil Nut Production in Non-Timber Forest Product Concessions in a Micro-Watershed in the Peruvian Amazon. *Remote Sens.* **2023**, *15*, 5438. [CrossRef]
89. Asner, G.; Knapp, D.; Martin, R.; Tupayachi, R.; Anderson, C.; Mascaro, J.; Sinca, F.; Chadwick, D.; Sousan, S.; Higgins, M.; et al. *La Geografía del Carbono en Alta Resolución del Perú*; Stanford: Lima, Peru, 2014.
90. Gutierrez, N.; Alvarez, C.; Riveros, J. *Aguilar-Amuchastegui Estimación del Carbono en la Biomasa aérea de los bosques de la Región de Madre de Dios*; WWF: Lima, Peru, 2014.
91. Noulèkoun, F.; Birhane, E.; Mensah, S.; Kassa, H.; Berhe, A.; Gebremichael, Z.M.; Adem, N.M.; Seyoum, Y.; Mengistu, T.; Lemma, B.; et al. Structural diversity consistently mediates species richness effects on aboveground carbon along altitudinal gradients in northern Ethiopian grazing exclosures. *Sci. Total Environ.* **2021**, *776*, 145838. [CrossRef]

92. Yuan, Z.; Ali, A.; Ruiz-Benito, P.; Jucker, T.; Mori, A.S.; Wang, S.; Zhang, X.; Li, H.; Hao, Z.; Wang, X.; et al. Above- and below-ground biodiversity jointly regulate temperate forest multifunctionality along a local-scale environmental gradient. *J. Ecol.* **2020**, *108*, 2012–2024. [CrossRef]
93. Silva, J.L.d.S.; Araujo, R.A.d.; Esteves, V.P.P.; Loebmann, D.G.d.S.W.; Vicente, L.E.; Paschoal, J.P.; Morgado, C.d.R.V. Analysis of vegetation recovery in areas impacted by bauxite mining in the Amazon Forest. *Clean Technol. Environ. Policy* **2021**, *23*, 1617–1640. [CrossRef]
94. Martínez-garcía, E.; Nilsson, M.B.; Laudon, H.; Lundmark, T.; Fransson, J.E.S.; Wallerman, J.; Peichl, M. Drought response of the boreal forest carbon sink is driven by understorey—Tree composition. *Nat. Geosci.* **2024**, *17*, 197–204. [CrossRef]
95. Gris, D.; Casagrande, J.C.; Marques, M.R.; Oldeland, J.; Damasceno-Júnior, G.A. Periodic flooding and edaphic factors shape Erythrina fusca dominance in riparian forests in the Pantanal wetland. *Trop. Ecol.* **2024**, *2024*, 1–15. [CrossRef]
96. Makelele, I.A.; Verheyen, K.; Boeckx, P.; Cizungu Ntaboba, L.; Mujinya Bazirake, B.; Ewango, C.; Bauters, M. Afrotropical secondary forests exhibit fast diversity and functional recovery, but slow compositional and carbon recovery after shifting cultivation. *J. Veg. Sci.* **2021**, *32*, e13071. [CrossRef]
97. Poorter, L.; Amissah, L.; Bongers, F.; Hordijk, I.; Kok, J.; Laurance, S.G.W.; Lohbeck, M.; Martínez-Ramos, M.; Matsuo, T.; Meave, J.A.; et al. Successional theories. *Biol. Rev.* **2023**, *98*, 2049–2077. [CrossRef]
98. Staude, I.R.; Weigelt, A.; Wirth, C. Biodiversity change in light of succession theory. *Oikos* **2023**, *2023*, e09883. [CrossRef]
99. Adinugroho, W.C.; Prasetyo, L.B.; Kusmana, C.; Krisnawati, H.; Weston, C.J.; Volkova, L. Recovery of Carbon and Vegetation Diversity 23 Years after Fire in a Tropical Dryland Forest of Indonesia. *Sustainability* **2022**, *14*, 6964. [CrossRef]
100. Bombino, G.; D'Agostino, D.; Marziliano, P.A.; Pérez Cutillas, P.; Praticò, S.; Proto, A.R.; Manti, L.M.; Lofaro, G.; Zimbone, S.M. A Nature-Based Approach Using Felled Burnt Logs to Enhance Forest Recovery Post-Fire and Reduce Erosion Phenomena in the Mediterranean Area. *Land* **2024**, *13*, 236. [CrossRef]
101. Liu, H.; Ye, Q.; Lundgren, M.R.; Young, S.N.R.; Liu, X.; Luo, Q.; Lin, Y.; Ye, N.; Hao, G. Phylogeny and climate explain contrasting hydraulic traits in different life forms of 150 woody Fabaceae species. *J. Ecol.* **2024**, *112*, 741–754. [CrossRef]
102. Zhang, Y.; Fu, Y.; Xian, W.; Li, X.; Feng, Y.; Bu, F.; Shi, Y.; Chen, S.; van Velzen, R.; Battenberg, K.; et al. Comparative phylogenomics and phylotranscriptomics provide insights into the genetic complexity of nitrogen-fixing root-nodule symbiosis. *Plant Commun.* **2024**, *5*, 100671. [CrossRef]
103. Balieiro, F.d.C.; Costa, C.A.; de Oliveira, R.B.; de Oliveira, R.; Donagemma, G.K.; de Andrade, A.G.; Capeche, C.L. Carbon stocks in mined area reclaimed by leguminous trees and sludge. *Rev. Arvore* **2017**, *41*, e410610. [CrossRef]
104. Saatchi, S.; Marlier, M.; Chazdon, R.L.; Clark, D.B.; Russell, A.E. Impact of spatial variability of tropical forest structure on radar estimation of aboveground biomass. *Remote Sens. Environ.* **2011**, *115*, 2836–2849. [CrossRef]
105. Otypková, Z.; Chytry, M. Effects of plot size on the ordination of vegetation samples. *J. Veg. Sci.* **2006**, *17*, 465–472. [CrossRef]
106. Chai, Y.; Yue, M.; Wang, M.; Xu, J.; Liu, X.; Zhang, R. Plant functional traits suggest a change in novel ecological strategies for dominant species in the stages of forest succession. *Oecologia* **2016**, *180*, 771–783. [CrossRef]
107. Chen, X.; Luo, M.; Larjavaara, M. Effects of climate and plant functional types on forest above-ground biomass accumulation. *Carbon Balance Manag.* **2023**, *18*, 5. [CrossRef]

diversity

MDPI

Article

Arrested Succession on Fire-Affected Slopes in the Krummholz Zone and Subalpine Forest of the Northern Limestone Alps

Marta De Giuli [1], Markus Winkler [2], Thomas Deola [1], Julia Henschel [1], Oliver Sass [2], Peter Wolff [1] and Anke Jentsch [1,2,*]

1 Department of Disturbance Ecology and Vegetation Dynamics, University of Bayreuth, 95448 Bayreuth, Germany; marta.degiuli@uni-bayreuth.de (M.D.G.); thomas.deola@uni-bayreuth.de (T.D.); peter.wolff@uni-bayreuth.de (P.W.)

2 Department of Geomorphology, University of Bayreuth, 95448 Bayreuth, Germany; markus.c.winkler@uni-bayreuth.de (M.W.); oliver.sass@uni-bayreuth.de (O.S.)

* Correspondence: anke.jentsch@uni-bayreuth.de

Abstract: Fire in the Northern Alps is comparatively rare. Yet, previous human-ignited fire events in subalpine forests up to the treeline have triggered severe fire damage to vegetation and soil. Here, we investigate post-fire vegetation dynamics in the Northern Limestone Alps about 80 years after disturbance. We observed higher species richness in burned compared to unburned vegetation and clearly distinct floristic communities emerging after fire-driven forest removal, with several alpine specialist species uniquely found in the burned subalpine sites. The functional composition of vegetation was also distinct, with higher relative forb cover in burned plots. This difference was likely driven by disturbance-related environmental changes, such as increased light availability, offering safe sites for subalpine and alpine species. Due to a general lack of tree encroachment, we consider this a case of arrested succession after fire. We conclude that the recovery of fire-affected subalpine forests is modulated by complex interactions of climatic and biotic filters producing extreme site conditions, controlling the recolonization of the disturbed areas by forest species while providing safe sites for the establishment of a rich subalpine and alpine low-statured flora. The coupling of disturbance and abiotic filters makes high-elevation treeline ecotones very vulnerable to climate change.

Keywords: vegetation dynamics; wildfire; recovery; alpine communities; calcareous alps; arrested succession; microclimate; floristic survey; *Pinus mugo*; extreme site; diversity; high-elevation ecosystem

1. Introduction

The high montane regions of the European Alps are unfamiliar with fire as a disturbance [1], with reconstructed fire frequencies in northern Tyrol in the range of hundreds of years [2,3]. Soil and microclimate variability, vegetation types, land use changes, and the high heterogeneity of alpine relief resulting in landscape fragmentation are key factors in shaping fire regimes at high elevations [4–6], These factors themselves are closely related to vegetation dynamics [2,7,8]. In these alpine systems, disturbance regimes are often associated with climatic events such as extreme temperatures, wind, and ice, as well as geomorphological events such as avalanches, rockslides, and permafrost, rather than fire [9]. However, fire is known to occur occasionally in the European Alps, triggering successional dynamics [10–14] and amplifying many of these geomorphological disturbances, with observed increases in soil erosion rates [15,16], debris flows [16], and avalanche occurrences [17] after fire. In alpine regions, in particular, alterations to compounding disturbances are expected with climate change [18], especially with changes such as the projected temperature increase, the lengthening of the growing season, and change in snow cover [19–21]. These projected changes are affecting the aridity and flammability of areas already prone to fire: southern aspects [10,22]. Despite the growing body of research on the effects of fire on mountain forest ecosystems [23–26], the effects of the introduction of fire on krummholz vegetation have been mostly neglected.

Krummholz vegetation, composed of dwarfed or stunted vegetation formed by the continued pressure of wind and low temperatures, is found in treeline landscapes in the subalpine and subarctic zones [27]. Due to the short stature of the woody vegetation in this system, any unlikely fire consumes whole individuals from trunk to treetop, drastically altering resource ratios and the nature of the limiting factor, i.e., moving the system from nutrient limitation to water limitation [28]. Studies globally have found long-term effects on post-fire vegetation dynamics in these krummholz systems, with burned areas reported to more closely resemble the species composition of subalpine grasslands than their original communities [29–32]. However, predicting post-fire forest recovery versus regime shift to open montane grasslands remains an ecological challenge.

In the Northern Limestone Alps, krummholz vegetation is mostly composed of *Pinus mugo* Turra individuals disrupted by vertical, gravity-dependent mechanical disturbance, yielding a heterogeneous pattern, often with minimal understory community (Figure 1). The removal of this layer of vegetation, either by fire or other disturbances, causes a successional cascade with the recruitment of fast-growing pioneer forbs and graminoids peaking within the first couple of years after disturbance, followed by slower recruitment of woody vegetation [14]. These successional dynamics vary in speed and magnitude with elevation and moisture availability—often affected by slope aspect [12,33]. When conditions are particularly extreme, successional dynamics may pause or arrest at a stage prior to that which existed pre-disturbance, raising questions about the system's ability to return to the former steady state, here, to subalpine forest dynamics.

Figure 1. Treeline landscapes with krummholz patches forming woody patches on rocky outcrops, i.e., Ammergebirge, or dense thickets on shallow soils, i.e., Wetterstein (@A.Jentsch).

The conservation potential of these open subalpine areas is significant both as a refugia for species that are unable to outcompete others while tracking their climatic and biotic niches and as clusters of high diversity [29,34]. Soon after disturbance, these open sites provide habitat to important fauna as well, with the Bavarian Alps homing three endangered species of grouse, namely *Tetrao urogallus* L., *Tetrao tetrix* L., and *Tetrastes bonasia* L., which need a mix of open woodland and grassland to survive [35,36]. In part due to this, the Arnspitze massif in the Northern Limestone Alps has been a designated nature protection area (Naturschutzgebiet) since 1942, ironically five years before the first fire. Drastic changes in vegetation and character could impact the region's ability to effectively serve this purpose of nature protection, making the understanding of these dynamics more pressing.

Given the importance of nature conservation of these open krummholz stands and the lack of studies on their post-fire recovery, we aim to investigate the vegetation dynamics and their potential drivers 75 years after the fire in the Northern Limestone Alps. Here, we included halted regeneration and emerging plant species diversity. First, we tested whether the plant communities on burned slopes were compositionally distinct from those on unburned slopes and whether they had higher species diversity. We expected the removal of woody, shading vegetation and the opening of new niches by fire to alter the biotic and abiotic filters, enabling the colonization of novel species. Second, we used species elevational range data to compare the 'elevational niche space' of burned and unburned slopes, anticipating a preferential encroachment of alpine specialists or lowland colonizer species in burned areas. We further disentangled these community differences by comparing the environmental indicator values of the species. Third, we looked more closely at the potential environmental drivers of these observed differences, including soil depth and *P. mugo* cover as a proxy for light availability.

2. Materials and Methods

2.1. Study Site

Our study area is in the Wetterstein mountain range in the Northern Limestone Alps on the southern and eastern slopes of the Große Arnspitze (2196 m a.s.l.; 47.40° N, 11.22° W), hereafter referred to as Arnspitze. The peak is part of the Arnspitze massif located on the border between Austria and Germany, approximately 20 km northwest of Innsbruck, Austria (Figure 2). It lies between the suboceanic, cool, and humid northern edge of the European Alps and the subcontinental, warm, and dry Tyrolean Inn Valley. The average annual temperature in Innsbruck (574 m) and Mittenwald (915 m) is 8.6 °C and 6.5 °C, respectively, and near zero at the elevation of the summit. The precipitation in the area has a distinct summer maximum, characterized by frequent, intense thunderstorms, and a mean annual precipitation of 850 mm and 1430 mm in Innsbruck and Mittenwald, respectively, mirroring the expected elevational trend [10]. The mean annual precipitation for the massif is approximated between 1500 and 1750 mm [37,38]. The area is characterized by Wettersteinkalk limestone, a very pure limestone with a $CaCO_3$ content of 95.6% [39], which slows the rate of soil formation [40] and leads to karstification. The rapid drainage of the karstified limestone coupled with the high annual evaporation results in very little available surface water and a very dry environment [37,41].

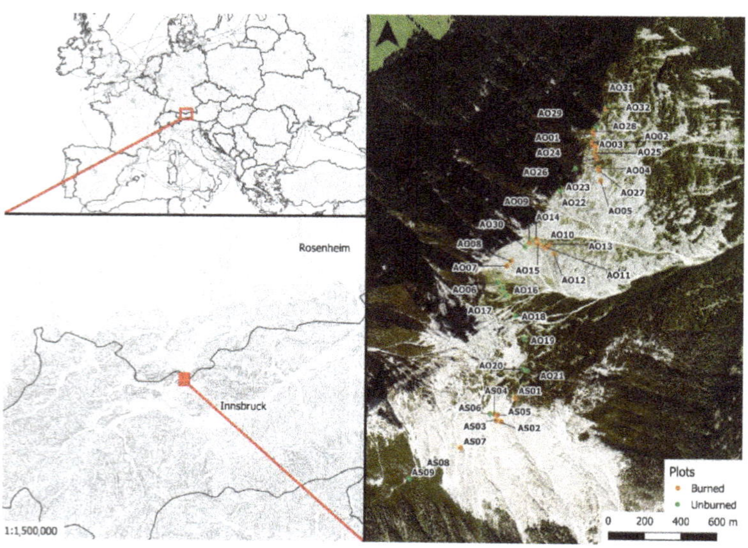

Figure 2. Location of the Arnspitze massif, Germany (left) and of the 41 sampling plots. Source: ESRI World Imagery.

The treeline ecotone in the Northern Limestone Alps is at an elevation of 1800–2400 m, lower than in the central Swiss Alps, in part due to the cooler conditions at this northern fringe. It is characterized by a mix of open stands of *Picea abies* (L.) H.Karst. sometimes mixed with *Larix decidua* Mill. (*Piceion excelsae* Pawłowski et al. 1928) and *P. mugo* krummholz vegetation, partly forming dense thickets (*Erico-Pinion mugo* Leibundgut 1948). On these slopes, the krummholz vegetation is observed ranging from 1200 m to the ridges at 2050 m of elevation [10], and the rocky slopes without scree cover are occupied by small statured vegetation including dwarf shrubs such as *Dryas octopetala* L. and *Daphne striata* Tratt., graminoids, forbs, and ferns. These taxa mostly belong to *Seslerion caeruleae* Br.-Bl. in Br.-Bl. et Jenny, with some components of *Potentillion caulescentis* Br.-Bl. in Br.-Bl. et Jenny 1926, the latter presumably spread from nearby cliffs untouched by the fires.

In May 1946, the southern slope of the Arnspitze burned for about 11 days [42], consuming 5–6 ha of forest (dominated by *P. abies*) and approximately 100 ha of krummholz stands (dominated by *P. mugo*) at elevations of 1200–2000 m (Figure 3). Large swaths of the slope were left stripped of vegetation and humus soil [43]. Prior to the fire, the slope had an even cover of either *P. mugo* at higher elevations or forest at lower elevations (compare Figure 3a,b; see Sass, 2019). The effect of severe fires such as this on geomorphological processes has been subject to several investigations [43,44], demonstrating that the measured erosion rates on the burned, southern slope are about 10 times higher than on unburned slopes [35]. Two more, probably less severe forest fires were discovered in the area using radiocarbon dating [42]: one about 1266–1012 BCE, the other 1206–1292 CE [43].

Figure 3. (**a**) Pre-fire eastern slope of the Arnspitze (Wetterstein, Northern Limestone Alps) harboring subalpine forest and krummholz vegetation (@R.Rischreiter, before 1946) versus (**b**) post-fire Eastern slope of the Arnspitze 80 years later devoid of woody vegetation exposing bare rock (@A.Jentsch in 2023).

2.2. Sampling Design

A total of 41 plots were established over the span of two years 2020–2021 during the months of July and August: 32 on the eastern slope and 9 on the southern slope of Arnspitze. Each 3 × 3 m square plot was aligned so that the upper edge was parallel to the contour line, thus pointing toward the summit or ridge. The plots were located between 1800 m and 1950 m of elevation, in areas that were safely accessible by the samplers. The positions of the plots can be seen in the map (Figure 2).

The 20 plots established in 2020 were all located in the 1946–1947 fire scar, while the 21 plots established in 2021 were partly in the fire scar, partly in the adjacent intact krummholz pine or forested stands (fire history status seen in Figure 2). A total of 26 plots were established in the fire scar, hereafter 'burned plots', and 15 in intact krummholz or forested sites, hereafter 'unburned plots'. The placement of the burned plots was random within the former krummholz vegetation (taken from historical photographs), while the unburned plots were preferentially selected to be characteristic of the unaffected slope, both floristically and geophysically.

2.3. Soil Depth

To determine the soil thickness in each plot, a 1 m long metal rod with 5 cm markings was pushed into the soil until it encountered resistance. Measurements were taken using 5 cm intervals, rounding up (i.e., 12.8 cm of soil recorded as 15 cm), with shallow soils less than 5 cm in depth categorized as 2.5 cm deep. A total of five of these measurements were taken in each plot, one in the geometric center of the square and once along each diagonal, halfway between the center and the corner. To account for potential within-plot variation in soil depth, the median soil depth per plot was used for all subsequent analyses.

2.4. Species Composition

Vascular plant species composition and cover were assessed in each plot. Nomenclature was checked and updated according to Plants of the World Online [45]. Plant individuals were identified in the field with 186 identified to the species level, 37 only to the genus level, and 3 remained unidentified. For completeness, the entire species list was only used in the species diversity and composition analyses. Species cover in percent was assessed using the Londo scale [46], a decimal scale with an appropriate fine grain to approximate cover separate from abundance.

Ecological indicator values (EIVs) for each taxon identified to the species level were extracted from both Ellenberg and Landolt repositories [47,48]. Ellenberg EIVs are commonly used bioindicator values with a long-standing tradition in Central Europe. To balance this with our study site location (in the Alps), we also included Landolt EIVs which were formulated for Switzerland in particular and may have a more accurate representation of the indicator values for the species we observed at higher elevations. We focused on the following EIVs: temperature, light, moisture, moisture variability, nutrients, humus content, reaction/soil pH, root depth, dominance in situ, and soil aeration.

We tested whether burned and unburned sites differed in their species richness, plant community composition, species' elevational range, and species' ecological indicator values. 'e determined the alpha, beta, and gamma diversity across all plots and between fire tories (burned vs. unburned). We calculated the Shannon index [49] as the alpha rsity metric rather than the Simpson index to obtain a balanced weight of rare species, the diversity() function in the *vegan* R package. We used Whittaker's multiplicative iversity [50], which is determined as follows:

$$\beta = \frac{S}{m}$$

he number of species in a composite community, and m is the mean number of plot. We used the total number of unique species present in each plot as the ess, testing the significance of this difference with a t-test using the function

whether the two types of plots were occupied by distinctly different plant compared the species composition of each plot. We tested this distinctness OVA using distance matrices with the function adonis2() in the *vegan* sualized these results with a non-multidimensional scaling (NMDS) ay–Curtis dissimilarity in species space using the metaMDS() function kage.

abundance of each functional group—specifically forbs and 1 the difference between the relative cover of each group in the ots with a t-test using the function t_test() in R. Relative cover was tional group cover in a plot divided by the total vegetative cover

ity composition as a function of elevation, we tested for a nge of the species present in each site type, as well as the species occurring uniquely in each. The elevational range data were obtained from the Swiss National Databank of Vascular Plants [51]. We tested for differences in maximum elevation, minimum elevation, range breadth (maximum elevation − minimum elevation), and relative elevational distance of the plot elevation from the species' minimum elevation (1900 − minimum elevation)/range breadth) for species in each site type with a *t*-test using the function t_test() in R, and plotted using the helper package *ggbreak* [52].

2.6. Krummholz Species Recruitment

To investigate the edge changes in the present krummholz vegetation, we used freely available orthophotos provided by the Land Tyrol (© TIRIS geoinformation system), comparing pictures taken in 1974, 2001, and 2016 with UAV images taken in situ (2021).

3. Results

3.1. Effect of Fire on Species Composition and Plant Functional Types

Across all 41 plots, the Shannon diversity index was 2.5 ± 0.8, while the average species diversity in each plot (±SD) was 31.7 ± 10.8, beta diversity (total number of species/mean number of species per site) was 89, and gamma diversity (total number of species across all plots) was 226. Species richness varied significantly between burned and unburned plots (Figure 4b, $p = 0.05$), with burned plots having the greatest species richness per plot (average species/plot ± SD; 35.1 ± 8.8 vs. 25.8 ± 11.5). This pattern also carries across the alpha, beta, and gamma diversity metrics, where we see burned plots having a higher alpha Shannon index and gamma diversity but lower beta diversities, 4.2 vs. 2.8, 174 vs. 136, and 3.9 vs. 6.7, respectively.

The community composition was significantly different between the burned and unburned sites (*p*-value < 0.001) (Figure 4a,b). Species unique to burned plots included *Carex curvula* All., *Ranunculus alpestris* L., *Rumex scutatus* L., and *Pedicularis kerneri* Dalla Torre. Species unique to unburned plots included *Alchemilla flabellata* Buser, *Calamagrostis varia* (Schrad.) Host, and *Trollius europaeus* L. We observed a clustering of plots by fire history with burned plots occupying a more circular cluster than unburned plots (orange area on the NMDS, Figure 4a). This result is consistent with our earlier observation of beta and gamma diversity, indicating a more uniform community composition among burned plots than among unburned plots.

We were interested in looking at these compositional differences from a perspective of functional type—specifically, differences in forb and graminoid cover (Figure 4c). Our results indicate a significant difference in forb, graminoid, shrub, and tree cover in fire-affected slopes (Figure 4c; $p < 0.001$, $p = 0.035$, $p = 0.003$, $p < 0.001$, respectively), with relative forb and graminoid cover higher in burned plots (0.64 ± 0.20 vs. 0.20 ± 0.15 and 0.24 ± 0.12 vs. 0.12 ± 0.12, respectively). We observed an inversion of this pattern for trees and shrubs, where burned plots had hardly any tree or shrub cover compared to the 0.52 ± 0.24 and 0.16 ± 0.12 of the unburned plots (Figure 4c). The top three dominant taxa in each of these categories are highlighted in Table 1. Excluding trees, of which there were only three species present at our site, we observed two species that were dominant in both burn categories: *Dryas octopetala* (Figure 5) and *Rhododendron hirsutum*. This hints at either a slow homogenization of the communities 80 years after disturbance or the ability of these taxa to colonize the disturbed terrain in an effective manner. The large differences in mean plot covers of these taxa, coupled with the clear distinction of the communities seen in Figure 4a and their unique functional composition, provide more support to the latter suggestion.

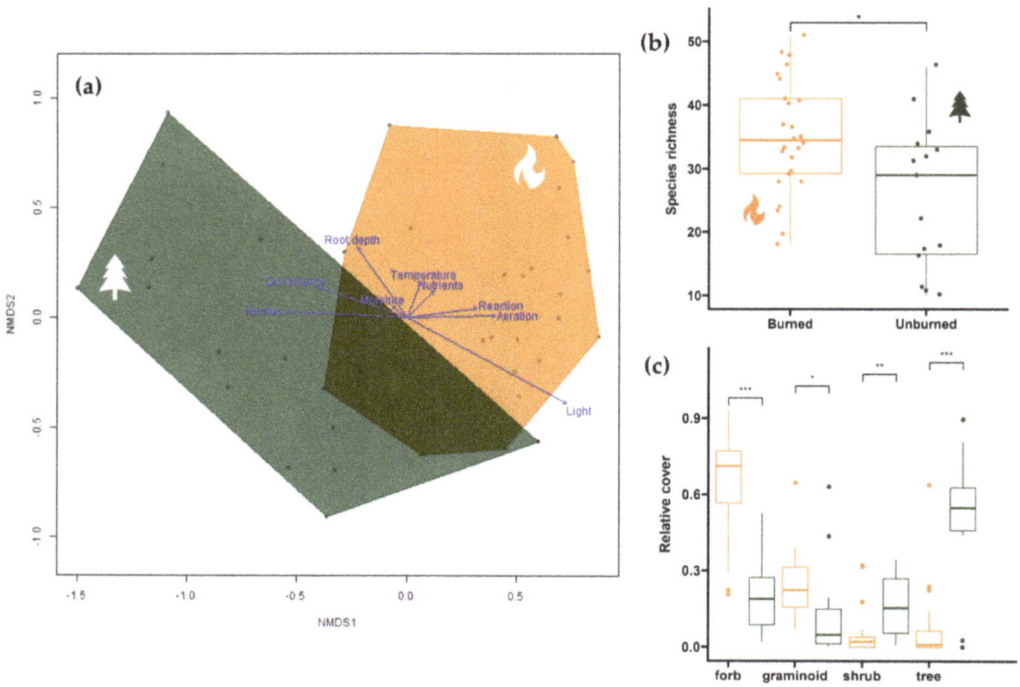

Figure 4. Analysis of community composition of the burned and unburned plots: (**a**) NMDS ordination using Bray–Curtis dissimilarity of plots in species space. Convex hulls show burned (orange) and unburned (green) plots. Blue vectors are the Landolt ecological indicator values of the species in the analysis. (**b**) Comparison of species richness and (**c**) comparison of the relative cover of functional types (forb, graminoid, and tree) of plots by fire history (orange: burned, green: unburned). Significance levels indicated by "***" = 0.001, "**" = 0.01, and "*" = 0.05.

Table 1. Top three dominant taxa of each functional type (forb, graminoid, shrub, and tree) by mean plot cover (percentage) in burned and unburned plots.

	Forb	Graminoid	Shrub	Tree
Burned	*D. octopetala* (14.4%)	*Nardus stricta* L. (11.7%)	*Rhododendron hirsutum* L. (2%)	*P. abies* (5.2%)
	Crepis foetida L. (11%)	*Carex humilis* Leyss. (5.75%)	*Vaccinium vitis-idea* L. (1.3%)	*L. decidua* (3%)
	Gymnocarpium robertianum (Hoffm.) Newman (10%)	*C. curvula* (4.85%)	*Vaccinium myrtillus* L. (0.7%)	*P. mugo* (1%)
Unburned	*Viola biflora* L. (9.1%)	*Alpagrostis alpina* (Scop.) P.M. Peterson, Romasch., Soreng & Sylvester (20.0%)	*Rhododendron ferrugineum* L. (15.2%)	*P. mugo* (54.2%)
	Erica carnea L. (7.3%)	*Carex sempervirens* Vill. (19.6%)	*R. hirsutum* (12.3%)	*P. abies* (23.1%)
	D. octopetala (5.4%)	*Luzula campestris* (L.) DC. (15.0%)	*Vaccinium uliginosum* L. (6.8%)	*L. decidua* (14.3%)

Figure 5. Characteristic open patch krummholz ecotone species, *Dryas octopetala* and *Globularia cordifolia*, with *Pinus mugo* in the background, covering shallow soil in the Northern Limestone Alps (@A.Jentsch in 2023).

The difference in relative cover of the functional types is likely driven by a change in species diversity after fire, where forb species richness is significantly (p = 0.005) higher in burned plots (27.9 ± 7.6) than in unburned plots (18.2 ± 10.6), as is graminoid species richness (p = 0.029) between burned plots (5.2 ± 1.7) and unburned plots (3.37 ± 2.1). However, this increase in richness does not seem to translate to a greater number of rare or endangered species, as we see slightly fewer endangered taxa present in burned plots. As seen in Figure 6, this does not insinuate a larger number of ruderal species in burned plots, and the main ruderal taxa present (*C. varia*) is only present in unburned plots.

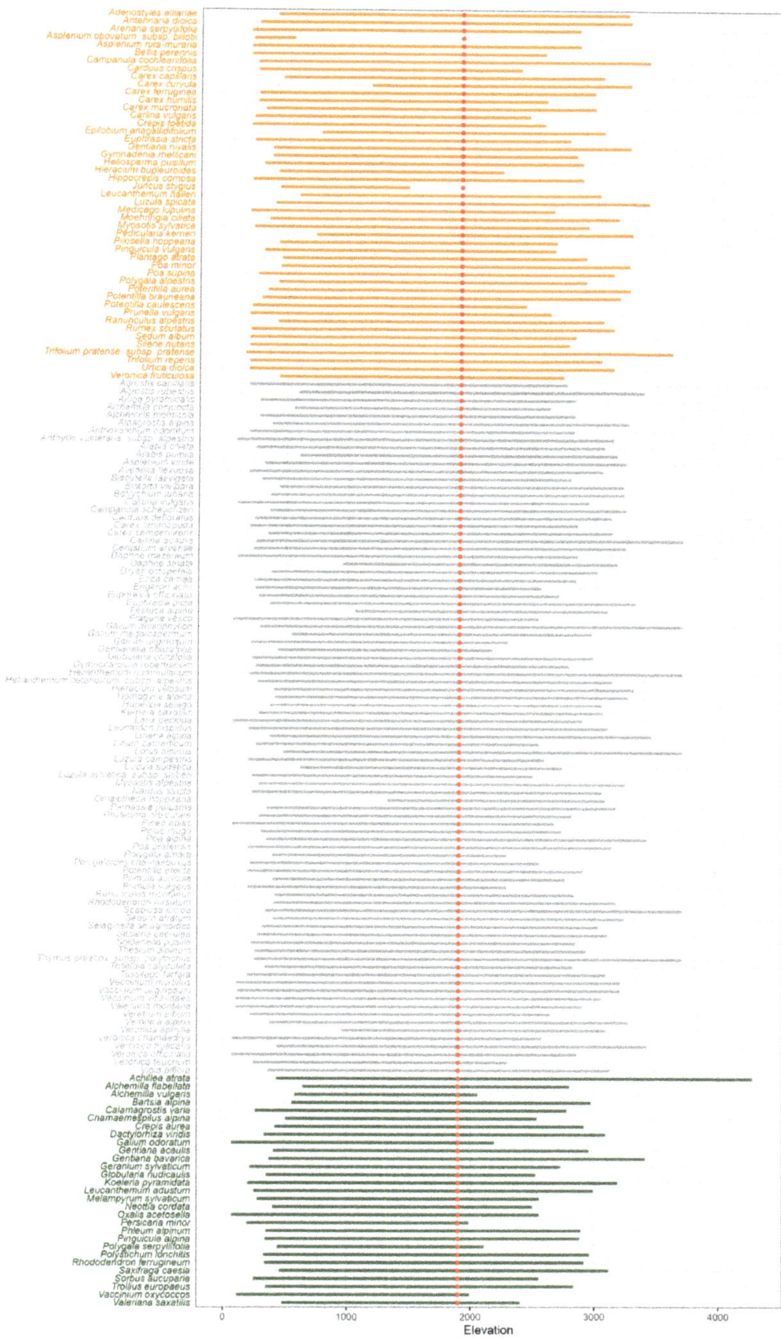

Figure 6. Elevation ranges of taxa identified to the species level found uniquely in burned plots (orange), unburned plots (green), or common to both (gray). The red dotted line indicates the elevation of the sampling plots.

3.2. Elevational Ranges of Colonizing Forbs and Graminoids

The elevational ranges of all the species present in both the burned and unburned plots were very similar (Figure 6). While the species in unburned plots had generally lower values (Figure 7a–d), we did not find a significant difference (Figure 7a–c) in minimum elevation (323 m ± 109; 324 m ± 188), maximum elevation (2948 m ± 428; 2936 m ± 387), and elevation range breadth (2625 m ± 438; 2612 m ± 404) between the species uniquely present in burned or unburned plots. These results seem to indicate that the species growing in burned areas do not have distinctly different elevational niches from those growing in unburned areas. This is further supported by the lack of significant difference in the relative distance of plot elevation from the range minima between the species in burned and unburned plots (64.0% ± 43.8; 61.5% ± 10.9), as seen in Figure 7d.

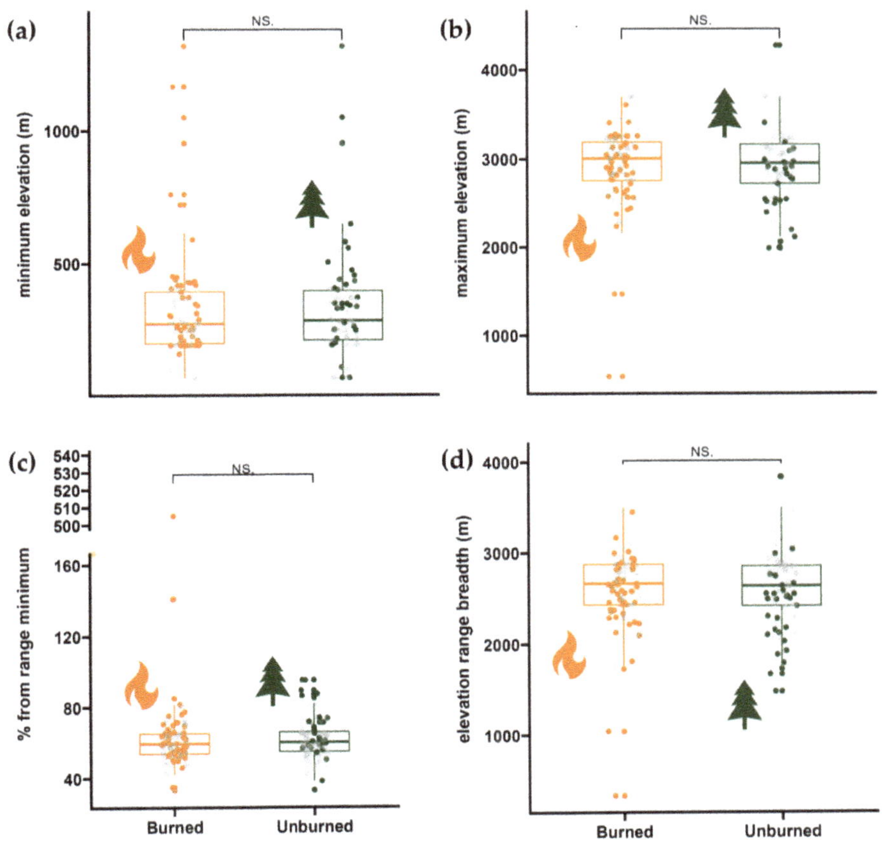

Figure 7. (**a**–**d**) Comparison of range extents of all species found in the burned and unburned plots; species unique to each fire history are in orange and green, respectively, while species common to both fire histories are in gray: (**a**) minimum limit of each species present; (**b**) maximum limit of each species; (**d**) the elevation extent (calculated as maximum elevation − minimum elevation) for each species; (**c**) location and relative elevational distance of the plot (1900 m) from the minimum elevation limit of each species (calculated as (1900 − minimum elevation)/elevation extent); NS indicates non-significant.

The same trends were seen in the elevational ranges of the subset of species *unique* to each fire history (burned vs. unburned): They did not vary significantly in minimum elevation (334 m ± 206; 334 m ± 194), maximum elevation (2853 m ± 573; 2748 m ± 417), elevation range breadth (2519 m ± 556; 2404 m ± 425), or relative elevational distance of plot location from range minima (71.9% ± 43.8; 66.9% ± 10.9). These results indicate that species colonizing the burned sites come from similar elevational ranges rather than representing a net upward shift of valley-dwelling species or a downward shift of alpine species. This is likely due to abiotic filters preventing their immigration and establishment. The 29 species unique to unburned plots include *A. flabellata* and *T. europaeus*, but also *C. varia* (a grass that commonly thrives after disturbance). The 47 species unique to burned plots include subalpine–alpine specialists, such as *Antennaria dioica* (L.) Gaertn., *C. curvula*, *R. alpestris*, *Gentiana nivalis* L., and *P. kerneri*, as well as the two outliers *Asplenium obovatum ssp. billotii* (F.W.Schultz) O.Bolòs, Vigo, Masalles & Ninot, and *Jucus stygius* L., which have documented elevational ranges of 200–450 m and 870–1470 m, respectively [53].

3.3. Effect of Fire on Community Ecologic Indicator Values

The temperature indicator value of the species supports the results found above, with inconclusive results due to the mismatch between the temperature Ellenberg and Landolt indicator values between species present in burned and unburned plots (Table 2). Similarly, the species indicative of nutrient availability (both Ellenberg and Landolt) did not vary significantly between burned and unburned plots. The species present in the burned area had significantly higher soil aeration (3.07, 2.61, $p < 0.001$), light (3.80, 3.46, $p < 0.001$), and pH (3.45, 3.28, $p = 0.024$) Landolt indicator values (Table 2). This contrasted with their significantly lower dominance in situ (2.36, 2.59, $p < 0.001$), humus content (3.00, 3.53, $p < 0.001$), moisture (2.75, 2.85, $p = 0.043$) Landolt indicator values compared to unburned plots. These results are robust to the exclusion of trees in the analyses, which, due to their very different rooting architecture and physiological requirements, could cause skewing of the data. Species present in burned plots are thus indicative of dryer, rockier, more basic soils that receive a high amount of sunlight. Taken with the inconclusive results of the temperature indicator value, this implies that the species present in these communities vary in their physical requirements rather than elevational provenance.

Table 2. Table of significance in variation in Ellenberg and Landolt indicator values for communities in burned and unburned plots. Significance levels indicated by "****" = 0.0001, "**" = 0.01, and "*" = 0.05; ns and grayed-out rows indicate non-significant results.

Indicator	Source	Mean Burned	Mean Unburned	Statistic	*p*-Value	Significance
Aeration	Landolt	3.07	2.61	5.09	<<0.001	****
Dominance in situ	Landolt	2.36	2.59	−3.99	<<0.001	****
Humus content	Landolt	3.00	3.53	−6.53	<<0.001	****
Light	Ellenberg	7.44	6.91	5.50	<<0.001	****
	Landolt	3.80	3.46	5.74	<<0.001	****
Moisture	Ellenberg	4.71	4.93	−2.44	0.015	*
	Landolt	2.75	2.85	−2.02	0.043	*
Moisture variability	Landolt	1.50	1.57	−1.45	0.148	ns
Nutrients	Ellenberg	2.96	3.02	−0.57	0.571	ns
	Landolt	2.29	2.28	0.12	0.902	ns
Reaction/Soil pH	Ellenberg	6.61	6.12	2.81	0.005	**
	Landolt	3.45	3.28	2.27	0.024	*
Root depth	Landolt	1.82	1.90	−1.08	0.279	ns
Temperature	Ellenberg	2.91	3.05	−1.64	0.102	ns
	Landolt	2.35	2.26	2.03	0.043	*

These results can be seen visually affecting the community composition NMDS in Figure 4a, with the burned community shifting along the light indicator value vector and the unburned community shifting along the humus content and dominance in situ indicator value vectors.

3.4. *Krummholz Species Recruitment*

We observed no clear evidence of tree regeneration (both individual species and combined) in the burned sites. Changes in woody canopy cover in time occurred only in unburned areas (Figure 8), visible in the filling of canopy gaps, with no discernable change in the canopy boundary. This is supported by the lack of tree cover in burned plots, where the mean tree cover per plot was less than 0.08% (Figure 4c).

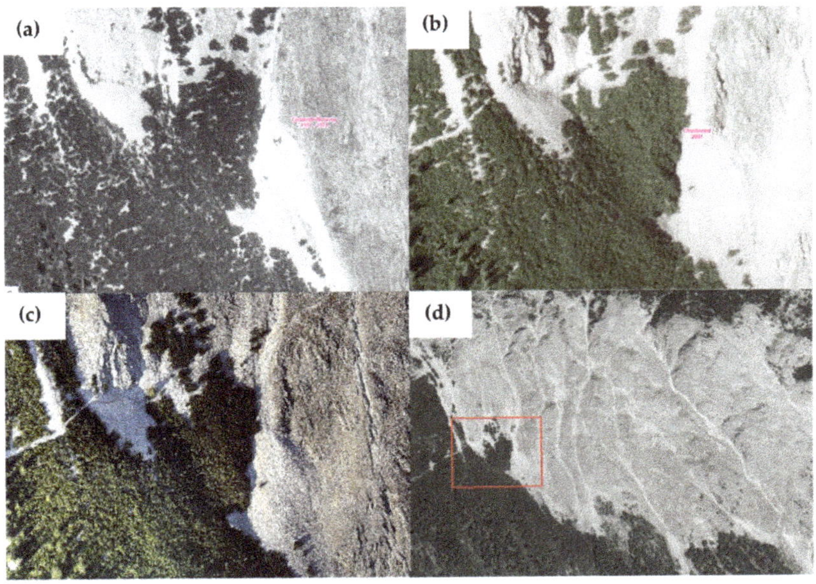

Figure 8. Patterns of *P. mugo* growth on a section of the southern slope of the Arnspitze at approximately 1600 m a.s.l. in (**a**) 1974, (**b**) 2001, (**c**) 2020, and (**d**) overview (section of interest outlined in red) data: (**a**,**b**,**d**) orthophotos (LBA.TIROL.GV.AT, DATA.TIROL.GV.AT), (**c**) orthomosaic created from project UAV data.

3.5. *Potential Drivers of Post-Fire Regime Shift*

Evidence from plot proxy data and measured environmental data suggests that the trend of increased forb diversity after fire is driven mostly by increasing light availability and partly by decreasing soil depth. As a proxy of light extinction, we recorded *P. mugo* cover, which was strongly related to fire history with burned plots having no cover and unburned plots an average of 32% cover ($p < 0.001$). Forb richness varied significantly with *P. mugo* cover (Figure 9a). Plots with low *P. mugo* cover (<25%) had significantly higher forb diversity than plots with higher cover (28.3 ± 9.0; 18.6 ± 8.8; $p = 0.018$), but we observed no significant variation in graminoid diversity with *P. mugo* cover (4.8 ± 1.9; 4.1 ± 2.2; $p > 0.05$). While forb richness responded to *P. mugo* cover, we observed no significant response in the relative cover of forbs and graminoids (Figure 9b).

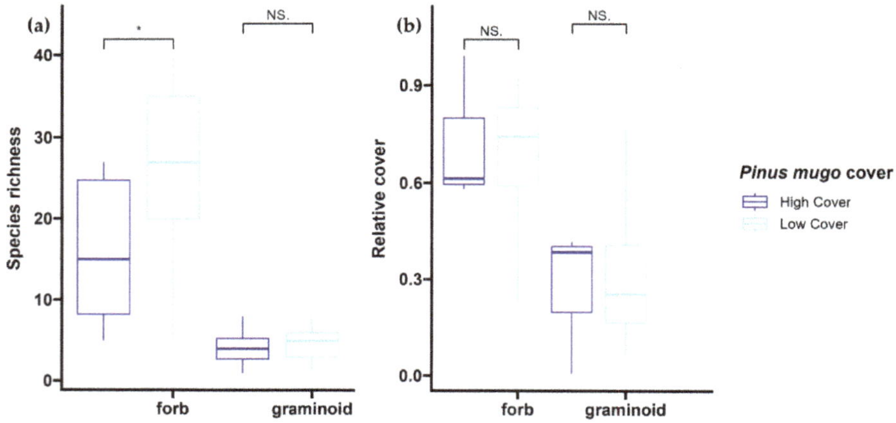

Figure 9. (**a**) Comparison of plots with high vs. low *Pinus mugo* cover in species richness and (**b**) relative cover by the functional group of plots grouped by *P. mugo* cover class. Significance levels indicated by "*" = 0.05; NS indicates non-significant.

Though burned and unburned plots did not vary significantly in elevation, aspect, or slope, they showed distinctly different community compositions (Figure 4a) and forb species relative cover (Figure 4c). Soil depth and variability also changed significantly as a function of fire history, with 27.4 cm ± 18.6 mean soil depth in burned plots compared to 41.2 cm ± 21.0 in unburned plots. Forb species diversity varied significantly with soil depth, with diversity decreasing with an increase in the substrate (Figure 10a). Plots with the shallowest soils had the greatest richness per plot (30.8 ± 8.2), significantly greater than the mid-depth (19.8 ± 9.0; $p = 0.007$) and deepest soils (21.4 ± 5.2, $p = 0.009$). However, we did not observe a significant difference in species diversity between the deeper two soil depths (Figure 10b).

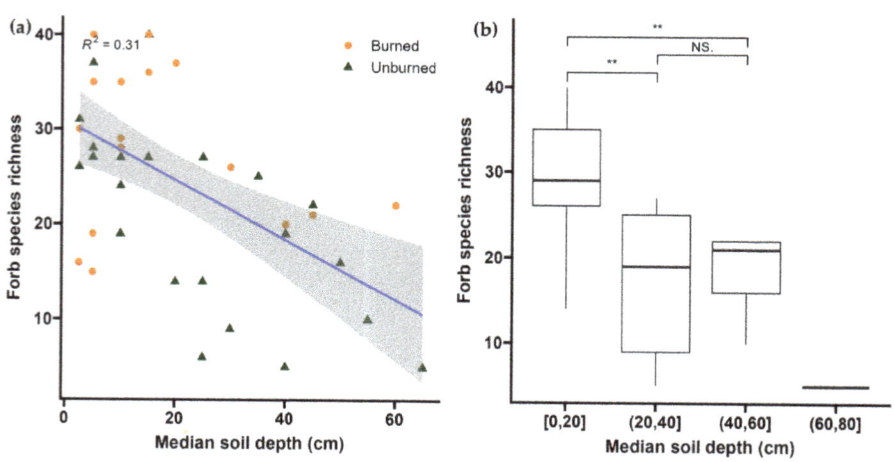

Figure 10. Forb species richness by (**a**) median soil depth and (**b**) median soil depth class. Significance levels indicated by "**" = 0.01; NS indicates non-significant.

4. Discussion

Understanding the dynamics of fire recovery in the subalpine and alpine belt is becoming increasingly important as these zones are becoming hotspots for change [18,20] and priorities for biodiversity conservation [34]. The long-term fire recovery of subalpine krummholz—including evidence of arrested succession—is not well studied and is an issue that needs to be monitored to help address pressing ecological and safety concerns in the mountains. These studies need to be held in the context of changing climate and disturbance regimes [54], shifting snow [20,21] and growing periods [20], as well as water availability affected by changing temperature and summer precipitation regimes [19], increased fire risk and natural fire occurrence [55], and the subsequent risk of soil erosion [15,16] and avalanche occurrence [17]. Critically, these environmental changes may have already altered baseline climatic conditions necessary for the complete post-fire recovery of subalpine systems and the recruitment of currently absent woody vegetation. These halted successional subalpine krummholz dynamics can trigger a regime shift to open, high-elevation tundra [31,56] generating novel habitats for biodiversity of high nature value.

In the Northern Limestone Alps, fire seems to have a distinct impact on the plant community in the krummholz zone: entirely removing woody, krummholz vegetation—here *P. mugo*—and shifting the community toward a herbaceously dominated, species-rich, subalpine grassland. This is a more homogeneous community, as seen by the lower beta diversity and closer clustering of the burned plots (compare Figure 4a). This homogenization toward species-rich subalpine grassland or tundra has also been observed in other studies, where previously krummholz or shrubby subalpine vegetation gained richness while losing woody dominance and spatial heterogeneity after fire [29,31]. This plant species diversity is gained mostly through forb and graminoid diversity, which is significantly higher in burned plots than in unburned plots (Figure 4c). The ecological indicator values of the species present in the communities also indicate a shift toward plant species that are associated with drier, brighter sites with shallower soils in burned areas (Table 2), likely driven by the removal of the *P. mugo* biotic filter by fire, providing more light and freeing substrate for potential herbaceous colonizers.

Though this shift in species composition could also open niches for lowland disturbance specialists or ruderal invasives [57,58], we did not see a variation in the elevation ranges of the species between burned and unburned sites (Figure 6) or the presence of "weeds" in burned plots (Figure 4). By contrast, certain alpine specialist species were found only in burned plots, such as *C. curvula*, *P. kerneri*, *Plantago atrata* Hoppe, *Veronica fruticans* Jacq., and *Poa supina* Schrad., while others such as *Festuca alpina* Suter, *D. striata*, and *Veronica aphylla* L. were observed growing in both burned and unburned plots. This, coupled with other species found only in unburned sites such as *Gentiana acaulis* L., *Gentiana bavarica* L., and *Globularia nudicaulis* L., which appeared to be unable to colonize the burned sites, seems to indicate that the new community is not composed of species novel to these elevations, rather a new grouping of species already present at these elevations.

The homogeneity and richness of the post-fire communities give us a potential insight into the duration of recovery. Our survey was conducted just under 80 years after the latest occurrence of fire on these slopes. Though these systems are known to take decades or even centuries to recover [3], the complete lack of *P. mugo* seedlings indicates that the herbaceous vegetative recovery is temporally different from that of woody species. This observation is consistent with other studies that observed little to no woody species regeneration within the first half-century after fire and reconstructed a recovery time of over 150 years for the restocking of woody krummholz vegetation [29–31]. However, sites recovering from fire in subalpine forests in the Valais in the Central Alps see a significant rewooding after only 10 years, with aspect playing a significant role in the successful reestablishment of seedlings: northern aspects had a 17-fold increase in stem density compared to southern aspects 19 years after fire [14,33].

Potential drivers of this arrested succession could be three-fold. Here, we have shown a lack of tree seedling recruitment in burned areas (Figure 8), likely due to prolonged periods with frozen soil and the potential of extreme temperatures, in line with Walker et al. 2010 [59]. We have also uncovered that limited substrate availability (in the form of soil depth) is a key driver of forb diversity and can thus be key to supporting a viable herbaceous recovery while preventing the establishment of *P. mugo* seedlings [60]. The third factor, not discussed in this paper, that prevents the establishment of seedlings could be linked to browsing or grazing pressure from ungulates [61].

Climate change is bringing other abiotic stressors into the picture, though with predictions for the Eastern Alps indicating increases in summer and autumn temperatures as well as in the frequency and severity of heatwaves [62,63]. Beniston et al. (2004) [62] forecast a shift in precipitation expected in the northern Eastern Alps from large summer events to spring and autumn rainfall. Combined with forecasted earlier snowmelt dates [20,21,64], these stressors will expose vegetation to longer, likely more intense, dry conditions, leading to increased fire risk in much of this zone [10,22]. While in the past, fire management strategies may not have been considered necessary in these areas, the projected increase in fire risk coupled with the long-term impacts of fire on vegetation analyzed in this study, as well as on other abiotic factors [10–18], necessitates bringing this management strategy to the forefront.

5. Conclusions

Albeit potentially temporary, these post-fire subalpine herbaceous communities are vital for the protection of diversity in our mountains and can be useful to help mitigate the negative effects of our changing alpine climates by acting as refugia for an array of threatened and endangered species. To further understand the dynamics of recovery in these systems and decipher whether the community shift we have observed is a plateau in the successional return to the previously woody krummholz vegetation or rather an alternate subalpine grassland steady state, it is important to look at the system as a whole—combining biotic and abiotic factors and placing this one site within the context of many others. Studies in other such sites are scarce (but see [14] for more forested sites in the European Alps and [30,31] for sites in Japan and the USA, respectively), and a thorough analysis of the abiotic elements present in these extreme post-fire habitats is lacking (see [42]). Given this gap, we recommend further investigation into understanding the causes of *P. mugo* regeneration failure in these scenarios with a focus on micrometeorological factors and seed, seedling, and sapling physiology, as well as the impact of ungulate browsing. Management of these areas should include these results in their assessment of new conservation goals, making sure to balance the ability of fire to create more open systems with the likelihood that these changes may be permanent.

Author Contributions: Conceptualization, A.J. and O.S.; field data collection, T.D., J.H., A.J., M.W. and P.W.; formal analysis, M.D.G.; writing—original draft preparation, M.D.G. and A.J.; writing—review and editing, M.D.G., T.D., A.J., O.S., M.W. and P.W. All authors have read and agreed to the published version of the manuscript.

Funding: We acknowledge a small grant by the Bayreuth Center of Ecology and Environmental Research (BayCEER) at the University of Bayreuth, Germany and the Open Access Publishing Fund of the University of Bayreuth.

Institutional Review Board Statement: Not applicable.

Data Availability Statement: Data will be made available upon request.

Conflicts of Interest: The authors declare no conflicts of interest. The funders had no role in the design of the study; in the collection, analyses, or interpretation of data; in the writing of the manuscript; or in the decision to publish the results.

References

1. Breckle, S.-W.; Lawlor, G.; Lawlor, D.W.; Walter, H.; Breckle, S.-W.; Walter, H. *Walter's Vegetation of the Earth: The Ecological Systems of the Geo-Biosphere*, 4th ed.; Springer: Berlin/Heidelberg, Germany, 2002; ISBN 978-3-540-43315-6.
2. Genries, A.; Muller, S.D.; Mercier, L.; Bircker, L.; Carcaillet, C. Fires Control Spatial Variability of Subalpine Vegetation Dynamics during the Holocene in the Maurienne Valley (French Alps). *Écoscience* **2009**, *16*, 13–22. [CrossRef]
3. Sass, O.; Heel, M.; Leistner, I.; Stöger, F.; Wetzel, K.; Friedmann, A. Disturbance, Geomorphic Processes and Recovery of Wildfire Slopes in North Tyrol. *Earth Surf. Process. Landf.* **2012**, *37*, 883–894. [CrossRef]

4. Carcaillet, C.; Ali, A.A.; Blarquez, O.; Genries, A.; Mourier, B.; Bremond, L. Spatial variability of fire history in subalpine forests: From natural to cultural regimes. *Ecoscience* **2009**, *16*, 1–12. [CrossRef]
5. Zumbrunnen, T.; Menéndez, P.; Bugmann, H.; Conedera, M.; Gimmi, U.; Bürgi, M. Human impacts on fire occurrence: A case study of hundred years of forest fires in a dry alpine valley in Switzerland. *Reg. Environ. Chang.* **2012**, *12*, 935–949. [CrossRef]
6. Leys, B.; Carcaillet, C. Subalpine fires: The roles of vegetation, climate and, ultimately, land uses. *Clim. Chang.* **2016**, *135*, 683–697. [CrossRef]
7. Higuera, P.E.; Brubaker, L.B.; Anderson, P.M.; Hu, F.S.; Brown, T.A. Vegetation mediated the impacts of postglacial climate change on fire regimes in the south-central Brooks Range, Alaska. *Ecol. Monogr.* **2009**, *79*, 201–219. [CrossRef]
8. Leys, B.; Finsinger, W.; Carcaillet, C. Historical range of fire frequency is not the Achilles' heel of the Corsican black pine ecosystem. *J. Ecol.* **2014**, *102*, 381–395. [CrossRef]
9. Tinner, W.; Conedera, M.; Ammann, B.; Lotter, A.F. Fire Ecology North and South of the Alps since the Last Ice Age. *Holocene* **2005**, *15*, 1214–1226. [CrossRef]
10. Sass, O.; Wetzel, K.-F.; Friedmann, A. Landscape Dynamics of Sub-Alpine Forest Fire Slopes in the Northern Alps-Preliminary Results. *Z. Für Geomorphol. Suppl.* **2006**, *142*, 207–227.
11. Moser, B.; Gimmi, U.; Wohlgemuth, T. Ausbreitung Des Erdbeerspinats Blitum Virgatum Nach Dem Waldbrand von Leuk, Wallis (2003). *Bot. Helv.* **2006**, *116*, 179–207. [CrossRef]
12. Moser, B.; Temperli, C.; Schneiter, G.; Wohlgemuth, T. Potential Shift in Tree Species Composition after Interaction of Fire and Drought in the Central Alps. *Eur. J. For. Res.* **2010**, *129*, 625–633. [CrossRef]
13. Wohlgemuth, T.; Hester, C.; Jost, A.-R.; Wasem, U.; Moser, B. Dynamik Der Wiederbewaldung Im Waldbrandgebiet von Leuk (Wallis) | Recruitment Dynamics Following the Forest Fire near Leuk (Valais). *Schweiz. Z. Forstwes.* **2010**, *161*, 450–459. [CrossRef]
14. Wohlgemuth, T.; Moser, B. Zehn Jahre Vegetationsdynamik Auf Der Waldbrandfläche von Leuk (Wallis). *Schweiz. Z. Forstwes.* **2018**, *169*, 279–289. [CrossRef]
15. Shakesby, R.A. Post-Wildfire Soil Erosion in the Mediterranean: Review and Future Research Directions. *Earth-Sci. Rev.* **2011**, *105*, 71–100. [CrossRef]
16. Mandrone, G.; Vacha, D.; Chicco, J. Post-Wildfire Erosion Rates and Triggering of Debris Flows: A Case Study in Susa Valley (Bussoleno). *E3S Web Conf.* **2023**, *415*, 1–4. [CrossRef]
17. Sass, O.; Heel, M.; Hoinkis, R.; Wetzel, K.-F. A Six-Year Record of Debris Transport by Avalanches on a Wildfire Slope (Arnspitze, Tyrol). *Z. Geomorphol.* **2010**, *54*, 181–193. [CrossRef]
18. Seidl, R.; Thom, D.; Kautz, M.; Martin-Benito, D.; Peltoniemi, M.; Vacchiano, G.; Wild, J.; Ascoli, D.; Petr, M.; Honkaniemi, J.; et al. Forest Disturbances under Climate Change. *Nat. Clim. Chang.* **2017**, *7*, 395–402. [CrossRef] [PubMed]
19. Auer, I.; Böhm, R.; Jurkovic, A.; Lipa, W.; Orlik, A.; Potzmann, R.; Schöner, W.; Ungersböck, M.; Matulla, C.; Briffa, K.; et al. HISTALP—Historical Instrumental Climatological Surface Time Series of the Greater Alpine Region. *Int. J. Climatol.* **2007**, *27*, 17–46. [CrossRef]
20. Gobiet, A.; Kotlarski, S.; Beniston, M.; Heinrich, G.; Rajczak, J.; Stoffel, M. 21st Century Climate Change in the European Alps—A Review. *Sci. Total Environ.* **2014**, *493*, 1138–1151. [CrossRef]
21. Klein, G.; Vitasse, Y.; Rixen, C.; Marty, C.; Rebetez, M. Shorter Snow Cover Duration since 1970 in the Swiss Alps Due to Earlier Snowmelt More than to Later Snow Onset. *Clim. Chang.* **2016**, *139*, 637–649. [CrossRef]
22. Cansler, C.A.; McKenzie, D.; Halpern, C.B. Area Burned in Alpine Treeline Ecotones Reflects Region-Wide Trends. *Int. J. Wildland Fire* **2016**, *25*, 1209. [CrossRef]
23. Dupire, S.; Curt, T.; Bigot, S.; Fréjaville, T. Vulnerability of Forest Ecosystems to Fire in the French Alps. *Eur. J. For. Res.* **2019**, *138*, 813–830. [CrossRef]
24. Moris, J.V.; Vacchiano, G.; Ravetto Enri, S.; Lonati, M.; Motta, R.; Ascoli, D. Resilience of European Larch (*Larix decidua* Mill.) Forests to Wildfires in the Western Alps. *New For.* **2017**, *48*, 663–683. [CrossRef]
25. Maringer, J.; Ascoli, D.; Küffer, N.; Schmidtlein, S.; Conedera, M. What Drives European Beech (*Fagus sylvatica* L.) Mortality after Forest Fires of Varying Severity? *For. Ecol. Manag.* **2016**, *368*, 81–93. [CrossRef]
26. Frejaville, T.; Curt, T.; Carcaillet, C. Bark Flammability as a Fire-Response Trait for Subalpine Trees. *Front. Plant Sci.* **2013**, *4*, 466. [CrossRef] [PubMed]
27. Körner, C. The Cold Range Limit of Trees. *Trends Ecol. Evol.* **2021**, *36*, 979–989. [CrossRef] [PubMed]
28. Jentsch, A.; White, P. A Theory of Pulse Dynamics and Disturbance in Ecology. *Ecology* **2019**, *100*, e02734. [CrossRef] [PubMed]
29. Douglas, G.W.; Ballard, T.M. Effects of Fire on Alpine Plant Communities in the North Cascades, Washington. *Ecology* **1971**, *52*, 1058–1064. [CrossRef]
30. Tomback, D.F. Post-Fire Regeneration of Krummholz Whitebark Pine: A Consequence Of Nutcracker Seed Caching. *Madroño* **1986**, *33*, 100–110.
31. Takahashi, K. Damage of Alpine Vegetation by the 2009 Fire on Mt. Shirouma, Central Japan: Comparison between Herbaceous Vegetation and Pinus Pumila Scrub. *Landsc. Ecol. Eng.* **2012**, *8*, 123–128. [CrossRef]
32. Kalas, M.M.; Berg, C. Regeneration of High Montane Plant Communities in the "Nationalpark Kalkalpen" (Northern Alps) after Fire Events. In Proceedings of the 5th Symposium for Research in Protected Areas, Mittersill, Austria, 10–12 June 2013.
33. Gauye, C.; Bernard, C.; Moser, B.; Rigling, A.; Wohlgemuth, T. L'importance de l'exposition Sur Le Reboisement Suite à Deux Incendies de Forêt En Valais. *Schweiz. Z. Forstwes.* **2023**, *174*, 238–242. [CrossRef]

34. Geppert, C.; Bertolli, A.; Prosser, F.; Marini, L. Red-Listed Plants Are Contracting Their Elevational Range Faster than Common Plants in the European Alps. *Proc. Natl. Acad. Sci. USA* **2023**, *120*, e2211531120. [CrossRef] [PubMed]
35. Schweiger, A.-K.; Nopp-Mayr, U.; Zohmann, M. Small-Scale Habitat Use of Black Grouse (*Tetrao tetrix* L.) and Rock Ptarmigan (Lagopus Muta Helvetica Thienemann) in the Austrian Alps. *Eur. J. Wildl. Res.* **2012**, *58*, 35–45. [CrossRef]
36. Scridel, D.; Tenan, S.; Brambilla, M.; Celva, R.; Forti, A.; Fracasso, I.; Volcan, G.; Dorigatti, E.; Anderle, M.; Marchesini, A. Early-Succession Secondary Forests Following Agropastoral Abandonment Are Key Winter Habitats for the Conservation of a Priority Bird in the European Alps. *Eur. J. For. Res.* **2022**, *141*, 1029–1043. [CrossRef]
37. Baumgartner, A.; Reichel, E.; Weber, G. *Der Wasserhaushalt Der Alpen: Niederschlag, Verdunstung, Abfluß u. Gletscherspende Im Gesamtgebiet d. Alpen Im Jahresdurchschnitt Für d. Normalperiode 1931–1960; Mit 68 Tab*; R. Oldenbourg: München, Germany, 1983; ISBN 978-3-486-27251-2.
38. Schiechtl, H.; Stern, R.; Meisel, K. Karte Der Aktuellen Vegetation von Tirol 1:100,000. XI. Teil: Blatt 2 Lechtaler Alpen-Wetterstein. *Grenoble* **1987**, *XXX*, 25–48.
39. Hüttl, C. *Steuerungsfaktoren und Quantifizierung der Chemischen Verwitterung auf dem Zugspitzplatt (Wettersteingebirge, Deutschland): Mit 77 Tabellen*; Münchener Geographische Abhandlungen Reihe B; Geobuch-Verl: München, Germany, 1999; ISBN 978-3-925308-51-2.
40. Sass, O.; Haas, F.; Schimmer, C.; Heel, M.; Bremer, M.; Stöger, F.; Wetzel, K. Impact of Forest Fires on Geomorphic Processes in the Tyrolean Limestone Alps. *Geogr. Ann. Ser. Phys. Geogr.* **2012**, *94*, 117–133. [CrossRef]
41. Korch, O.; Friedmann, A. Vegetation und Vegetationsdynamik auf dem Zugspitzplatt (Bayerische Alpen) Natur- und Kulturlandschaft im hochalpinen Raum als Produkt natürlicher, anthropogener und zoogener Einflüsse. *Polarforschung* **2016**, *86*, 35–45. [CrossRef]
42. Sass, O. *Waldbrände in den Nordtiroler Kalkalpen: Verbreitung-Geschichte-Regeneration*; Innsbrucker Geographische Studien; Geographie Innsbruck: Innsbruck, Austria, 2019; ISBN 978-3-901182-44-0.
43. Heel, M. Waldbrände in den Nördlichen Kalkalpen: Raumzeitliche Verteilung und Beispiele Lokaler Auswirkungen. Ph.D. Thesis, Universität Augsburg, Augsburg, Germany, 2016.
44. Stine, M.B.; Butler, D.R. Effects of Fire on Geomorphic Factors and Seedling Site Conditions within the Alpine Treeline Ecotone, Glacier National Park, MT. *Catena* **2015**, *132*, 37–44. [CrossRef]
45. Plants of the World Online. Facilitated by the Royal Botanic Gardens, Kew. Available online: http://www.plantsoftheworldonline.org/ (accessed on 17 May 2024).
46. Londo, G. The decimal scale for releves of permanent quadrats. *Vegetatio* **1976**, *33*, 61–64. [CrossRef]
47. Ellenberg, H.; Weber, H.E.; Düll, R.; Wirth, V.; Werner, W. *Zeigerwerte von Pflanzen in Mitteleuropa*; Scripta Geobotanica; 3. Durchgesehene Auflage.; Verlag Erich Goltze GmbH & Co KG: Göttingen, Germany, 2001; ISBN 978-3-88452-518-0.
48. Landolt, E. *Flora Indicativa: Ökologische Zeigerwerte und Biologische Kennzeichen zur Flora der Schweiz und der Alpen*; 2. Völlig Neu Bearbeitete und Erweiterte Aufl.; Haupt: Bern, Switzerland, 2010; ISBN 978-3-258-07461-0.
49. Hill, M.O. Diversity and evenness: A unifying notation and its consequences. *Ecology* **1973**, *54*, 427–473. [CrossRef]
50. Whittaker, R.H. Evolution and measurement of species diversity. *Taxon* **1972**, *21*, 213–251. [CrossRef]
51. Jutzi, M.; Vilpert, M.; Juillerat, P.; Eggenberg, S. Swiss National Databank of Vascular Plants. Version 1.15. Swiss National Biodiversity Data and Information Centres—Infospecies.ch. 2023. Occurrence Dataset. Accessed via GBIF.org. Available online: https://doi.org/10.15468/7jffp3 (accessed on 1 December 2023).
52. Xu, S.; Chen, M.; Feng, T.; Zhan, L.; Zhou, L.; Yu, G. Use Ggbreak to Effectively Utilize Plotting Space to Deal with Large Datasets and Outliers. *Front. Genet.* **2021**, *12*, 774846. [CrossRef]
53. Hess, H.E.; Landolt, E.; Hirzel, R. *Flora der Schweiz und Angrenzender Gebiete*; 1: Pteridophyta bis Caryophyllaceae; 2. Durchges. Aufl.; Birkhäuser: Basel Stuttgart, Germany, 1976; ISBN 978-3-7643-0843-8.
54. Wohlgemuth, T.; Jentsch, A.; Seidl, R. *Disturbance Ecology*; Springer Nature: Berlin/Heidelberg, Germany, 2022; Volume 32, ISBN 3-030-98756-6.
55. Wastl, C.; Schunk, C.; Leuchner, M.; Pezzatti, G.B.; Menzel, A. Recent Climate Change: Long-Term Trends in Meteorological Forest Fire Danger in the Alps. *Agric. For. Meteorol.* **2012**, *162*, 1–13. [CrossRef]
56. Daskalova, G.N.; Myers-Smith, I.H.; Bjorkman, A.D.; Blowes, S.A.; Supp, S.R.; Magurran, A.E.; Dornelas, M. Landscape-Scale Forest Loss as a Catalyst of Population and Biodiversity Change. *Science* **2020**, *368*, 1341–1347. [CrossRef] [PubMed]
57. Haider, S.; Palm, S.; Bruelheide, H.; de Villemereuil, P.; Menzel, A.; Lachmuth, S. Disturbance and Indirect Effects of Climate Warming Support a Plant Invader in Mountains. *Oikos* **2022**, *2022*, e08783. [CrossRef]
58. Lembrechts, J.J.; Pauchard, A.; Lenoir, J.; Nuñez, M.A.; Geron, C.; Ven, A.; Bravo-Monasterio, P.; Teneb, E.; Nijs, I.; Milbau, A. Disturbance is the key to plant invasions in cold environments. *Proc. Natl. Acad. Sci. USA* **2016**, *113*, 14061–14066. [CrossRef] [PubMed]
59. Walker, L.R.; Wardle, D.A.; Bardgett, R.D.; Clarkson, B.D. The Use of Chronosequences in Studies of Ecological Succession and Soil Development. *J. Ecol.* **2010**, *98*, 725–736. [CrossRef]
60. Šenfeldr, M.; Treml, V. Which Generative Reproduction Characteristics Determine Successful Establishment of the Subalpine Shrub *Pinus mugo*? *J. Veg. Sci.* **2020**, *31*, 403–415. [CrossRef]
61. Simon, A.; Katzensteiner, K.; Gratzer, G. Drivers of Forest Regeneration Patterns in Drought Prone Mixed-Species Forests in the Northern Calcareous Alps. *For. Ecol. Manag.* **2019**, *453*, 117589. [CrossRef]

62. Beniston, M. The 2003 heat wave in Europe: A shape of things to come? An analysis based on Swiss climatological data and model simulations. *Geophys. Res. Lett.* **2004**, *31*, L02202. [CrossRef]
63. Beniston, M. August 2005 intense rainfall event in Switzerland: Not necessarily an analog for strong convective events in a greenhouse climate. *Geophys. Res. Lett.* **2006**, *33*, L05701. [CrossRef]
64. Beniston, M.; Keller, F.; Goyette, S. Snow pack in the Swiss Alps under changing climatic conditions: An empirical approach for climate impacts studies. *Theoret. Appl. Climatol.* **2003**, *74*, 19–31. [CrossRef]

diversity

MDPI

Article

Succession as a Natural Tool for Restoration of Oak—Lime Forests on Aspen-Covered Clearcuts

Alina Nasibullina *, **Katharina Tiebel and Sven Wagner**

Institute of Silviculture and Forest Protection, Dresden University of Technology, Pienner Str. 8, 01737 Tharandt, Germany; katharina.tiebel@tu-dresden.de (K.T.); sven.wagner@tu-dresden.de (S.W.)
* Correspondence: alina.nasibullina@tu-dresden.de

Abstract: The genus *Quercus*, including species like pedunculate oak (*Quercus robur* L.), can play a key role in maintaining climate-resistant mixed forests due to its broad ecological spectrum and drought tolerance. Unfortunately, in some parts of Europe, clearcutting has drastically reduced the oak population. An example of this event is our survey of heritage Oak—Lime forests in European Russia, which were transformed into pure aspen stands. The aim of our study was to provide forecasts and silvicultural recommendations for the passive restoration of these forests. We took a chronosequence approach to assess changes associated with natural succession over 60 years. In our survey of the development of oaks, limes and accompanying tree species (aspen, birches, maples, elms), we used 190 plots ranging across a wide spectrum of forest disturbance due to clearcutting. We demonstrate that aspen reproduce rapidly by root suckers after cutting and occupy more than 60% of the space. But the dominance of aspen decreases continuously from the age of 30, and then the lime trees begin to dominate. Oak does not show successful natural regeneration. Therefore, we recommend planting oak seedlings or sowing acorns, i.e., active restoration, in combination with the natural restoration of lime.

Keywords: chronosequence; generalized additive model; rejuvenation; *Quercus robur* L.; *Tilia cordata* Mill.; *Populus tremula* L.

1. Introduction

The pedunculate oak (*Quercus robur* L.) is predicted to adapt to the expected climatic changes [1,2]. Due to its wide ecological scope and great drought tolerance, oak can become an essential component of climate-resilient mixed forests [3,4]; thus, it is regarded as an available option to meet the challenges of future climate change [5–7]. Also, the genus oak, with its great species abundance [8,9], might be suitable for multifunctional forest management, including the production of timber [10–12].

The natural range of pedunculate oak covers all of western Europe—with the exception of most of the Iberian Peninsula—Scandinavia, central Europe, Turkey, the Republics of Belarus, Lithuania, Latvia, Estonia, Ukraine and much of the territory of the European part of Russia [13]. Oak forests are also naturally present in the central Volga region of Russia [14–16], but the degradation of oak forests in these areas has been progressing over a long period of time. The oak forests in this region have been actively exploited for the past 300 years, and the development of shipbuilding by Peter the Great (1672–1725) depleted large volumes of oak timber [17]. In later years, other anthropogenic factors such as wars, the construction of a large hydroelectric power station and, of course, the lack of forest management reduced the oak population by 1/3 [18]. Another noteworthy reason is the extremely hard winters of the region, such as 1941–1942 and especially 1978–1979, when the temperature dropped down to −50 to −54 °C. As an example, about half of all oaks in the Republic of Tatarstan (a part of the central Volga region [19]) died after the winter of 1978–1979 [1].

As reported by Petrov [20] and Puryaev et al. [18], the oak stands in Tatarstan have been shown to regenerate 'unsatisfactorily' with certain harvesting methods (logging of the thickest oaks, i.e., exploitation, or clearcutting). 'Unsatisfactory' here refers to altered species composition, species ratios, and the performance of individual trees. Compared to the harvested stands, successional stands appear to be of less ecological and economic value; i.e., they are degraded. Despite the problem of oak forest loss, there are very few studies on forest restoration after clearcutting [21–23].

In order to significantly alter conditions for valuable oak-lime stands to be harvested in the future, (i) foresters need to know exactly how different harvesting methods affect species regeneration and stand composition over a long period of time [24]. We could not cover this aspect of sustainability in the present study; however, we looked into the question of (ii) how to bring back existing degraded stands to a higher level of economic and ecological value and what restoration strategy should be developed and made available for restoration of the oak-lime forests.

An intuitively compelling and cost-effective strategy for restoration could be seen in the deliberate use of residual structures of the clearcuts, which are sometimes called 'legacies' [25]. As in many cases of degraded forests, in the oak-lime forests of Tatarstan, some remnant structures of the former diverse forests may exist after the last clearcut, i.e., in the recent stands. In particular, there may be some oak trees left at forest margins, lime trees may resprout after the last cut, and here and there, elm trees or maple trees survived in an understory feature. These legacies could be taken as an initial step to upgrading the forest during succeeding decades by passive restoration.

In contrast to studies on thinning, most research works on clearcuts succession focus mainly on relatively short monitoring periods and the initial consequences arising directly from clearcuts and not on the long-term effects [26–28]. The lack of understanding of a long-term series can be partially overcome by taking chronosequences of current forest stands systems [29]. Chronosequence is a term used by Jenny [30]; however, the method has been used for a large number of systems since the very beginning of ecological science [31]. In relation to vegetation dynamics, the chronosequence *sensu stricto* represents a 'space–time substitution': a spatial representation of a chronological sequence of vegetation changes divided into states of development in which environmental factors besides time are not assumed to play a role [31]. Thus, for a successful chronosequence analysis, the sampling method should measure most of the natural variability of species diversity at each stage of development and reduce as much as possible the variation of environmental factors that could affect the vegetation development derived from the chronosequence [32].

We applied the chronosequence approach to record the changes associated with natural succession and tree species development in formally oak-lime forests [14] over the 60 years post-clearcutting in order to provide forecasts and silvicultural recommendations for these and other similar areas with degraded oak forests.

2. Materials and Methods

2.1. Study Area

The Republic of Tatarstan is located in the east of European Russia between 47° and 54° north latitude and 54° and 57° east longitude [33]. Typical for this region is a temperate continental climate with hot summers and moderate cold winters, but heavy frosts of up to −52 °C can also occur. In summer, it is usual for temperatures to climb up to 40 °C. The average annual precipitation in the region ranges from 360 to 510 mm. The snow level in forests is usually between 40 and 45 cm but can reach 100 cm during high-snow winters [34].

Tatarstan is mostly characterized by a flat terrain. Mountainous uplands can be seen only in the south of the region. The entire territory of the region is conventionally divided into three parts (Figure 1).

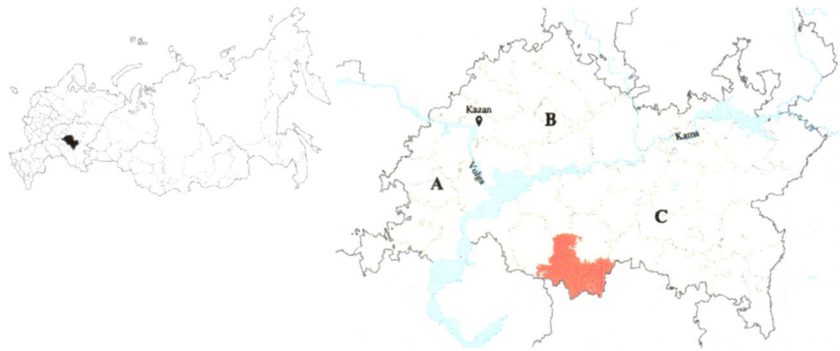

Figure 1. (**Left**) The inserted map of Russia shows the location of the Republic of Tatarstan in black. (**Right**) Map layout of Tatarstan with three parts of the territory (A, B, C with river borders in-between) and the research district Nurlat (red colored).

We conducted our measurements within Zakamye (C), in the Nurlat forestry district (Figure 1), as the problem of transformation of oak-lime forests to the aspen (*Populus tremula* L.) stand after clearcut is heavier here. The area of Zakamye forests of the Republic of Tatarstan is 634,233 ha. The aspen forests in Zakamye grow in the area of 189,854 ha, which is about 30% of the total area of the forest stands [35]. The Nurlat forestry district is the department with the largest concentration of aspen in the Tatarstan Republic (with aspen showing 37% of forest stock). The total forest area of the Nurlat forestry district is 83,140 ha. In addition to the dominant aspen (*P. tremula*) and very rare oaks (*Q. robur*), there are also individual trees of maple (*Acer platanoides* L.), birch (*Betula pendula* Roth), lime (*Tilia cordata* Mill.) and elm (*Ulmus glabra* Huds.) admixed.

Typically, the forests in this region are managed by clearcutting with a rotation period of up to 40 years.

Soils of these region are diverse—from different gray forest soils mostly typical for the forest sites to various types of nutrient-rich chernozems on the border to the agricultural land [36].

2.2. Site Selection

2.2.1. Site Measurements

Two measurement sessions were conducted in August 2020 and in August 2021. In an application of the chronosequence approach to successional sites, several prerequisites have to be met to yield sound and unbiased data.

According to the forestry department of Nurlat (personal communication with foresters, August 2020), no operations of tending, thinning and planting had been carried out in the study area in the past 62 years since the last clearcut (including the year of our last measurement (2021)). However, the residue from clearcutting was pushed together and left on the ground, forming clumps. By this 'no-intervention behavior' of forest management, an unspoilt forest dynamic after clearcut should be secured.

The sampling procedure must be the same for all plots; this is described below.

The choice of locations for the plots was made in a three-step procedure, which is outlined below. *First*, a series of 'years after clearcutting' (further on termed 'age') was defined by evaluating the administration's taxation data (this means a pre-stratification). In order to depict the succession dynamics in the time series, short intervals were to be selected for the years directly after clearcutting, and these intervals were to be extended with increasing plot age. The aim was therefore to establish 5-year intervals between 10 and 30 years and 10-year intervals from 30 to 60 years so that the study would cover 10 age groups. We aimed for at least 15 plots for each age to adequately represent the variation in forest structure.

Since a sufficient number of stands was not available for each desired age after clearcutting, we had to deviate somewhat from the target, and the following ten different ages were chosen: 1, 3, 6, 10, 15, 20, 30, 42, 52, and 62 years after the last clearcut. These years guaranteed a satisfactory number of stands to choose from.

Secondly, stands that had a signature 'aspen-dominant' taxation data (this means again a pre-stratification) were selected to exclude all other forest types, which enabled areas that depicted the problem of tree species to change from heritage oak-lime forests to aspen-covered clearcuts. Approx. 40% of the Nurlat district is 'aspen-dominated'. In this way, we chose an experimental design that allowed us to investigate whether succession was a natural tool for restoring oak-lime forests on aspen-dominated clearcuts. In addition, the stands of each age were distributed over the entire forest enterprise area to represent the variation in soil/site classes from different gray forest soils to various types of chernozems as well as possible. According to the World Reference Base for Soil Resources [37], we differentiate two zones: (1) phaeozems and (2) chernozems.

Third, no more than five plots within a single chosen stand were allowed to be established. This sampling within a stand was performed by placing plots randomly on the stands area, securing a minimum distance of 100 m apart to each other. All together, the design aims to follow a stratified random sampling approach in which only the stratum 'aspen-dominant' defined by the forest taxation is considered.

In the temperate deciduous forest zone in general, and thus also in Tatarstan, catastrophic events such as those known from coniferous forests—like snow breakage, windthrow, wildfire or even insect calamities—are unlikely. This is particularly true for the younger forest stages, because the vitality of the trees there is particularly high. An exception could be snow pressure in younger and very dense stands, which can occur during wet snow events in autumn or spring when foliage is present. In Nurlat, smaller or larger events such as those just mentioned have not occurred in recent decades, which is why we focused on human-made clearcuts.

All over, 190 plots were established and measured (Figure 2). For all single plots, we recorded and analyzed the geoposition (GARMIN GPSMap 64s, UTM coordinates), forest residues, dead wood, leaf litter and vegetation survey. Vegetation surveys included the determination of dominant species in the herb, shrub and tree layers. Changes in soil types and microclimatic conditions between the different age groups were not part of the measurements.

Each single plot (12 m radius) was divided into 3 concentric sample circles (sc): sc0 with a 2 m radius, sc1 with a 6 m radius and sc3 with a 12 m radius.

Starting with sc0 (tree height (h) from 20 cm; diameter at breast height (dbh) up to 6.99 cm), we named the species, determined the type of their reproduction, organized the young trees in three height classes (B0: 20–50 cm; B1: 50–180 cm; B2: from 180 cm), counted the trees of each height class, marked trees with chalk, and noted browsing and bark damage when counting.

In sc1 (dbh 7.00–29.99 cm) and sc2 (dbh $\geq$ 30 cm), we named the tree species, measured the dbh, and noted crown and bark damages. In addition to the measurements listed above, tree height was taken using a Vertex clinometer (Haglöf Sweden AB, Långsele, Sweden) on two randomly selected trees of each tree species starting at a dbh of 7 cm in each sc; exceptionally thick trees were excluded from the height measurements.

While all diameters were measured once a tree exceeded 7 cm, height was taken on two randomly selected trees of each tree species starting at a dbh of 7 cm in each plot. In this way, it was aimed to determine an average height of the trees by species and age.

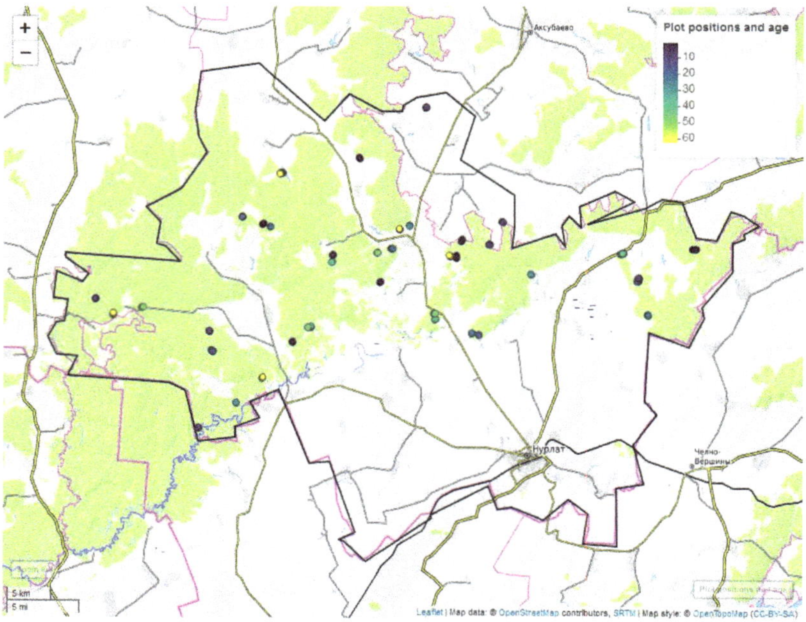

Figure 2. Plots distribution within the Nurlat forestry district. Neighbors when positioned in the same stand obscure some of the plots.

2.2.2. Stand's History

A map from 1927 [16] was used to collect information about the predominant species on a particular plot in that year. Only oak, lime and aspen-dominated stands are shown on that map. No other tree species like elm, maple or birch were dominant in those years. However, on the plots of our investigation, ages 15 or older, no lime was dominant in 1927, reducing the levels of the dominant species to two. Since no information on the immediate pre-harvest condition, i.e., before the last clearcut, was available for the individual stands, we used the 1927 stand map. We were able to assign a dominant tree species from 1927 to each individual plot and incorporate this information into a model (see Section 2.3) to explain the influence of this part of the stand history.

2.3. *Statistics*

2.3.1. Dbh Density Distributions

Dbh distributions for different age classes were established by kernel (Gaussian) density estimates. The bandwidth for the estimates is the standard deviation of the smoothing kernel. The integral under each individual distribution is always '1', and thus the graphs of the distributions of different species are comparable. However, the density plots will not represent the ratio of stem numbers or basal area of the different tree species to each other on the plots. An attempt was therefore made to calculate the development of the proportion of species in the basal area during succession separately (see Section 2.3.3).

2.3.2. Height Development Model

Based on the height measurements in the single plots (see Section 2.2), we established a height development model over time for five key tree species (oaks were excluded; for an explanation, see Section 3.2): aspen, birch, maple, elm, and lime. To do so, we followed a generalized additive model approach with integrated smoothness estimation (gam). Gams have the advantage of being able to integrate non-linear effects on the growth level, such as the age trend, into a linear model using non-parametric smoothers [38].

Within this procedure, the smooth terms were represented using penalized regression splines. Our model consisted of two parts, the first part being the generalized linear model and the second part being the additive smoother. The smoother was established to cover the effect of the time since clearcut, which essentially was a covariate. The generalized linear part of the model included species as a fixed factor. The dependent variable in the model was the height for each species on the plot level. The distributions of the dependent variable were observed carefully. In all species, Gaussian distribution best represented the height on plot level, and thus the generalized approach was implemented with an identity link. Residuals of the models were checked for heteroscedasticity [39]. However, in no case was it necessary to adjust the model to deviate from assumptions.

The model formula for height as dependent variable, species as fixed factor and time as covariate thus was

$$height_i = \alpha + \beta_1 species_i + f(X_i) + \varepsilon_i$$

with *height* being the height of trees of a species on the single plot, α being the intercept, β_1 being the parameter of the fixed effect, i.e., species, $f(X)$ being the smoother on time, and ε being the error term.

2.3.3. Species Proportion during Succession

The proportion of species in mixtures was calculated using the proportion of the basal area of species in the total basal area of a single plot, taking into account the six most important species and excluding, e.g., goat willow (*Salix caprea* L.), alder (*Alnus glutinosa* L.) and hazel (*Coryllus avellana* L.), as these species did not exhibit relevant basal area proportions. The mean value of the species shares of all plots of a certain age gives the mean share of a species at that age.

2.3.4. Categorical Logistic Regression for Species Proportion

'Species' represents a response variable that falls in a non-ordered finite set of categories. Accordingly, multinomial data are given [40,41]. Multinomial logistic regression was used to model nominal outcome variables in which the log odds of the outcomes were modeled as a linear combination of the predictor variables. Thus, a multinomial log-linear model for categorical data was used to analyze species proportion as a function of age of clearcut sites and of the dominant species in 1927. A model with the dominant species in 1927 as a 2-level factor {oak, aspen} and years after clearcut, i.e., 'age' as a continuous covariate, was built. The difference between an ordinary logit model and the multinomial logit is that the latter models the choice of *each* category as a function of the characteristics of the observation.

In a multinomial logit model, the sum of all individual probabilities was restricted to 1:

$$p_{i,j} = \frac{e^{\eta_{i,j}}}{1 + \sum\limits_{i=2}^{J} e^{\eta_{i,j}}}$$

with $p_{i,j}$ as the probability of the ith observation falling into category j. J is the number of categories with the baseline category 1, which is aspen. $\eta_{i,j}$ is the linear predictor.

The final model for the linear predictor in our investigation is shown below:

$$\eta_{i,j,p} = \beta_1 + \beta_{i,j,p} domspec1927_p + \beta_{i,j} age + \varepsilon_{i,j,p}$$

with β_1 as the reference parameter for the category 'aspen' in combination with the dominating species in 1927, which is 'aspen' at the age 0, $\beta_{i,j,p}$ as the parameter for the jth category and the pth dominating species of that site in 1927 and $\beta_{i,j}$ as the parameter for the age effect. $\varepsilon_{i,j}$ is the error term.

Probabilities for scenarios based on the observed data were built. These data were used to plot predicted probabilities. Sampling distributions were simulated (i.e., for confidence intervals) and could be used to predict probabilities for two scenarios and their first differences.

Statistics were performed using the open-source software R version 4.2.0 (R Core Team, 2022), packages 'MNLpred' [42], 'nnet' [43] and 'MASS' [44]. The graphs were created with the package 'ggplot2' [45].

3. Results

3.1. Dbh Development

Our analysis showed the development of dbh in six key tree species—aspen, birch, elm, maple, lime and oak—at six different age groups (the dbh values from other age classes are less than 7 cm and are not presented here) and thus represented the chronological trend among tree species (Figure 3).

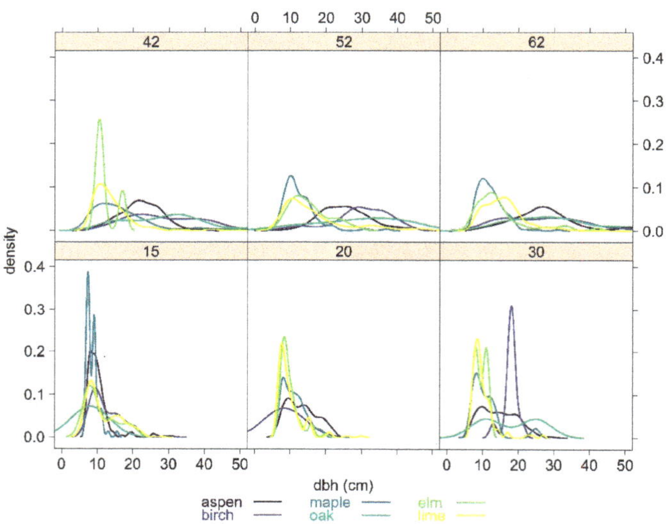

Figure 3. Density distribution of the diameter at breast height (dbh) of the tree species.

All tree species showed dbh larger than 7 cm from the age of 15 onwards.

The diameter distributions of the tree species showed an increasing dispersion of the diameters with age in general. From the age of 15 onwards, the modal of the distribution for aspen shifted from 8 cm (at age 15) to 20 cm (age 42) to 26 cm (age 62). Some aspen trees exceeded the dbh of 45 cm at the age of 62.

This trend was similar but not as pronounced in birch. In elm and maple, this kind of shift to larger dbh by time was almost absent. In lime, the trend to larger dbh was detectable, and it was combined with a small proportion of trees having diameters larger than 30 cm at the age of 62. In oak, the diameter distribution was highly dispersed from the age of 30, and oak had some trees of more than 60 cm at the age of 62.

3.2. Height Development of Key Tree Species

We established a height age relation via a generalized additive model (gam; see Section 2.3.2).

Oaks were excluded from this analysis because no oaks were found in plots younger than 30 years (with the exception of some oaks in 3-year-old areas (see Section 3.5)), and large oaks in older plots were stocky in shape.

All our key species—maple, elm and lime, except birch—showed a significant difference in growth relative to aspen (Table 1). There was also a tight relation between height and age with significant parameters of the model for aspen, maple, birch and lime. However, elm showed constant height over the entire chronosequence, and thus the smoothers of the gam for this species were not significant.

Table 1. Parameters of generalized additive model (gam) for aspen, maple, elm, birch and lime.

Parametric Coefficients:		Fixed Effect	Parameter	Standard Error	t Value	Pr (>|t|)
	β_0	intercept	20.167	0.333	60.635	$<2 \times 10^{-16}$ ***
	β_1	maple	−7.086	0.597	−11.872	$<2 \times 10^{-16}$ ***
	β_2	elm	−9.411	0.986	−9.546	$<2 \times 10^{-16}$ ***
	β_3	birch	0.844	0.794	1.063	0.288
	β_4	lime	−4.257	0.497	−8.574	$<2 \times 10^{-16}$ ***
Approximate significance of smooth terms:			edf	Ref. df.	F	*p*-value
	β_5	age:aspen	2.998	3.498	93.390	$<2 \times 10^{-16}$ ***
	β_6	age:maple	2.662	3.161	8.088	2.28×10^{-5} ***
	β_7	age:elm	1.000	1.000	0.160	0.690
	β_8	age:birch	1.803	2.158	24.130	$<2 \times 10^{-16}$ ***
	β_9	age:lime	1.816	2.246	39.701	$<2 \times 10^{-16}$ ***

Note: the intercept refers to gradient difference released for variant with aspen; Pr (>|t|) is the *p*-value for each test, *** *p*-value < 0.001.

Figure 4 shows that the height–age curves for aspen, birch, maple, elm and lime begin at a height of about 11 m at age 15. Beyond this age, aspen, birch, and lime clearly showed increasing height with age. At age 62, the trees reached a height of 25 m for aspen and birch and 20 m for lime, while maple did not appear to exceed 15 m, and elm stagnates at a height of 12 m. Hazel stayed at a constant height of about 3.3 m throughout the entire observation period.

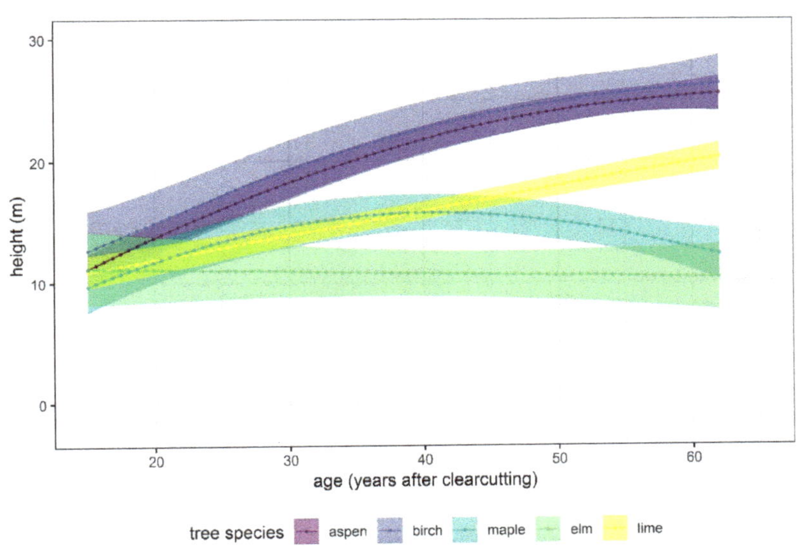

Figure 4. Height development of key tree species over time (age). The representation of aspen is overshadowed by that of birch, and it runs almost identically in the upper part of the graph.

3.3. *Basal Area Development*

Figure 5 shows the total basal area development of the basal area (m^2/ha) for all tree species over the course of 62 years.

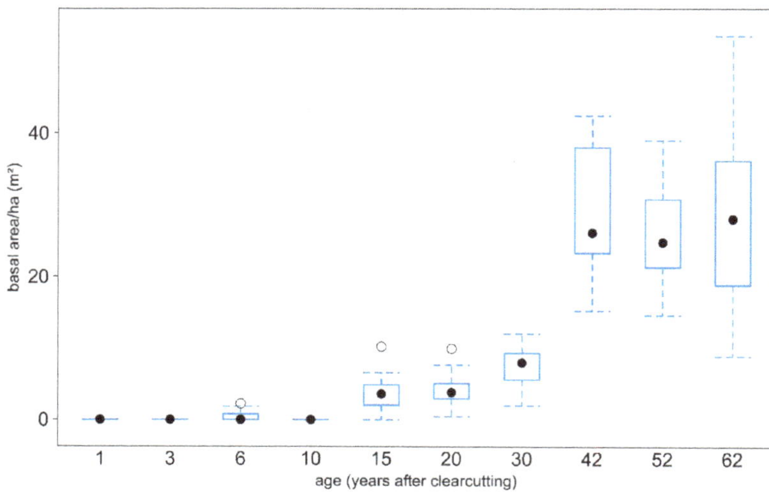

Figure 5. Basal area of all tree species in the tree layer by age.

The variance of the basal area values as indicated by the boxplots is due to multiple plots of the same age, which vary in basal area. It is obvious that—apart from a certain basal area at the age of six—the basal area built up from the age of 15. At age 42, the median basal area settles at slightly less than 30 m^2 and remains there at ages 52 and 62.

3.4. *Development of Tree Species Proportion*

We established the graphs of the development of tree species proportion from basal area over time, starting at the age of 15 years (Figure 6).

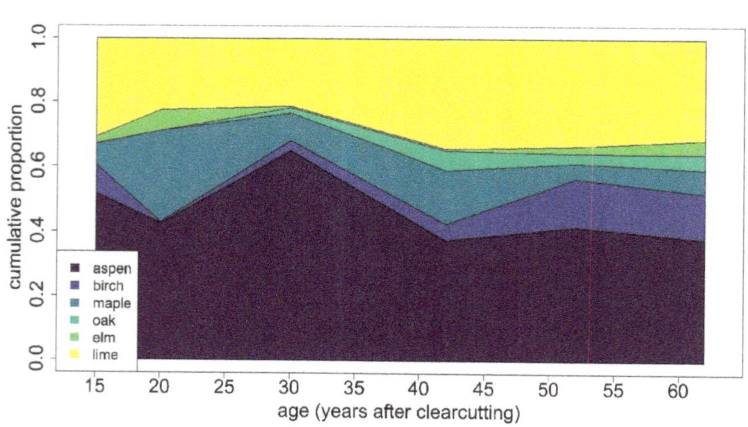

Figure 6. Chronology of proportions of tree species by basal area weighting.

It can be seen that the proportions of the different species varied with time and that species with low proportions in basal area like elm and oak, but also birch, had periods with no proportion at all. From the age of 30 years onwards, there was a decrease in the aspen proportion and an increase in the lime proportion in general. In addition, maple seemed to vanish, whereas birch and oak showed an increase very slowly over time. In elm, any kind of trend is difficult to detect. This graph does not show the influence of the dominant species in 1927.

To justify the observations described above, we established a multinomial log-linear model for categorical data with the predictor variables 'years after clearcutting' and 'dominant species in 1927' (Figure 7).

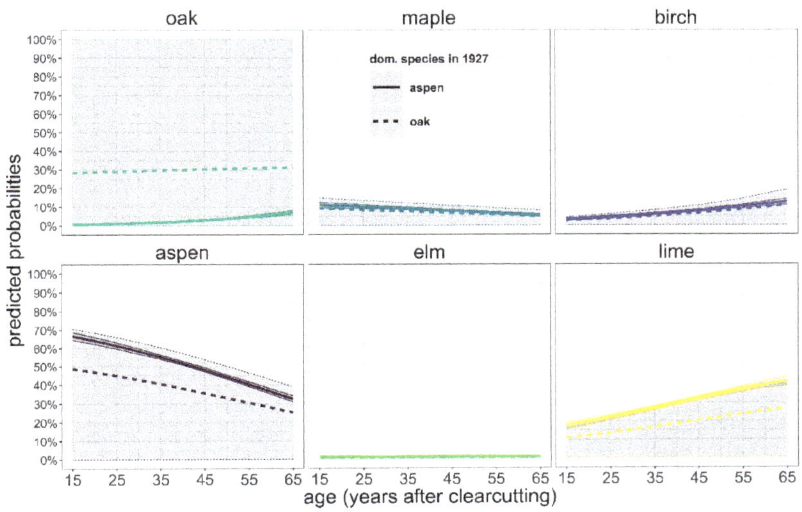

Figure 7. Model predictions for proportion of species in basal area in dependence of age and the dominating species in 1927. Also shown are 90% confidence intervals (gray shading). The broad confidence bands in the graphs are those of the variant with oak dominance in 1927.

The model predictions clearly confirmed the above-mentioned trends and showed the importance of the dominant species in 1927. If aspen was the dominant species 100 years ago, then aspen and lime today have higher proportions in the stands than in the case where oak dominated the stands in 1927. Birch, maple and elm showed no response to the dominating species in 1927. Oak seemed to benefit from a stand history in which 100 years ago oak dominated. However, due to the low numbers of plots with an oak-dominated history and with oaks in the recent stands, these findings regarding the effect of the dominating species in 1927 showed broad confidence bands and are thus hardly significant.

3.5. Rejuvenation Layer

Figure 8 shows how the density of young plants (number/ha) varies over time. Immediately after clearcutting, aspen had an average regeneration density of about 52,000 plants ha^{-1}, lime had about 14,000 plants ha^{-1}, and maple had about 4000 plants ha^{-1}. Birch and elm had about 1000 regeneration plants ha^{-1}. Aspen was dominated by root suckers, lime by stem shoots, birch and maple by seedlings, and elms used both. Aspen was represented at the age of 62 with about 220 regeneration plants ha^{-1}, while maple and lime had 410 regeneration plants ha^{-1} each.

While aspen, maple and lime were continuously present in the regeneration stratum, birch fails after about 20 years after clearcutting. Elm also dropped out after about 15 years but began to show regeneration again at age 52 (about 200 plants ha^{-1}). Oak was never present in the regeneration stratum except for 3-year-old areas (150 plants ha^{-1}).

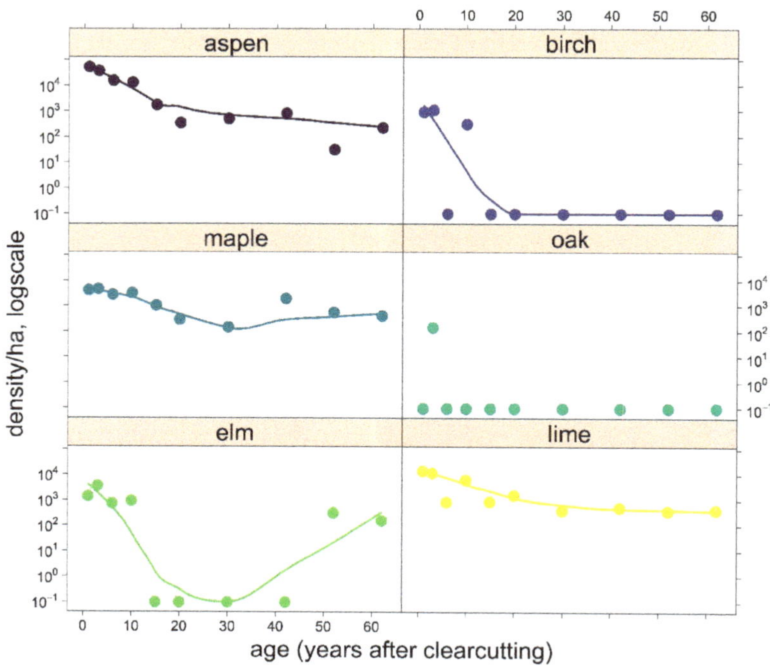

Figure 8. Relation between regeneration density (log-transformed) of key tree species and age as time after clearcutting (smoothed).

4. Discussion

4.1. Rejuvenation Layer

We observed a rejuvenation layer in all our plots and could take data for six key tree species: *P. tremula*, *B. pendula*, *A. platanoides*, *U. glabra*, *Q. robur* and *T. cordata*.

In the first 10 years after clearcutting, we could see the 'explosion' of pioneer species such as birch and especially aspen, which produce more than 50,000 root suckers ha^{-1}, which is common for aspen, especially on rich soils [46–51]. Initially, limes had about 14,000 sprouts ha^{-1} and, normally, the capacity to vegetative propagation did not seem to reduce with the stem age [52,53]. In the rejuvenation layer, we noticed a good presence of maples (4000 plants ha^{-1}) and elms (1000 plants ha^{-1}), which were mostly reproduced by seeds. According to the work of Collin et al. [54], despite the generally good capacity of *U. glabra* for setting seed or producing root suckers, they do not sucker and do not often produce viable resprouts from the stump—which may be a reason for the later low numbers of elms in these forests (clearcutting). *A. platanoides* has the trait of high seed production and the ability to produce root shoots as well as resistance to a variety of environmental conditions, and it has a high growth rate [55,56]. The germination of maple seeds and the shade tolerance of seedlings permits them to survive for long periods of time—estimated to be more than 20 years [56]—and its abundance remains stable throughout all the stages of our investigations.

Of course, the regeneration layer of all trees becomes less over time, and from the age of 20 years, there are very few young trees of all species (with dbh less than 7 cm). And this tendency is common for forest ecosystems, as rejuvenation is dependent on light availability [57–59].

Our main objective was to detect the rejuvenation steps for oaks. But unfortunately, we could only find some young oaks in 3-year-old areas.

4.2. Dbh and Height Development of Key Tree Species

The diameter distributions of our six key tree species *P. tremula, B. pendula, A. platanoides, U. glabra, Q. robur* and *T. cordata* (Figure 3) showed a displacement of the modal value with age in aspen, birch and lime, while the dbh of maple and elm did not show significant changes over time. The graph in Figure 3 demonstrates that the aspens grew rather fast and left other trees to grow up under their canopy. Aspen, due to its fast reproduction by root suckers [51,60], overtakes other pioneer trees like shade-intolerant birch, reproducing by seeds [61–63]. Also, the fact that birch trees were not found in all plots may indicate that they did not compete with aspen on these nutrient-rich soils [50]. In addition, we noted that for oak, no changes in diameter were observed from the age of 42 onwards. The spontaneous occurrence of quite mature oaks in the stands from 30 years onwards could be due to various reasons, which will be discussed in detail below in Section 4.5.

Figure 4 indicates that within 45 years, these stands developed from mono-layered into two-layered ones, having the fast-growing aspen and birch in the upper layer and maple with elm in the bottom layer. Lime, at the age of 62, seemed to reach for the upper layer, being 5 m lower than the pioneers.

4.3. Basal Area by Age

Basal area growth estimation is a tool that describes forest dynamics over time (i.e., growth, mortality, reproduction and associated stand-level changes) and therefore is widely used in forest management to provide inventory updates as well as to predict potential future yields and explore forest alternatives [64].

Our results showed (Figure 5) that the basal area of our plots increased with the age of the trees, peaking at the age of 42, and it had a tendency to level out after this age. The basal area was on average 30 m^2 for the 42-year-old plots and 30 m^2 for the older plots. In order to explain the constant total basal area, it can be assumed that lime and birch (Figure 3), as well as the undergrowing elms and maples, became thicker, thus creating a relatively constant living basal area, while the aspen was dying off (verified observations of the death of aspen from the age 30 due to stem rot [65–67]). Details on the proportions of tree species over the course of time can be found in Section 4.4.

Therefore, the asymptotic increase in stand basal area was as expected [68,69], while the absolute asymptote value ($\approx$30 m^2) is species and site-specific.

4.4. Development of Tree Species Proportion

It is well known that clearcutting affects the species distribution and species changes in the forest ecosystems [22,70–72]. In general, the structure and function of ecosystems depend on solar energy/radiation, water, nutrients, disturbance intensity and plant compositions [73], and these influence the biotic distribution within an ecosystem. Disturbances influence the establishment and the development of forests by affecting the distribution of space availability [74]. A forest stand's succession course is determined by the interaction and relative influence of allogenic (e.g., soil characteristics and climate change) and autogenic factors (e.g., stand characteristics and species life cycle characteristics), and their relative importance can change over the course of stand development [75]. In our measurements of the development of main tree species after such a disturbance as clearcutting, we analyzed the interactions of the autogenic factors (natural succession and proportion changes) over the course of 60 years (Figure 6).

The measurements of the basal area indicated that soon after harvesting, the aspens reproduced rapidly through root suckers and occupied more than 60% of the area at the age of 30. The fast reproduction of aspen on the clearcuts was also observed in several other studies, as mentioned above. Light-demanding species (aspen) generally grow faster than shade-tolerant species (lime) during the first few years [58]. This dominance decreases continuously from the age of 30 years onwards. We mentioned previously that studies in Tatarstan had already shown that *P. tremula* started to die from the age of 30, which was due to the high incidence of stem rot [65–67]. At the same age, maple was also declining over time, while lime was gaining proportion and was increasing from 40 years to about half of the basal area (Figure 6). *T. cordata* survived the hard conditions after clearcutting due to its ability to form sprouts [52,76] and its high shade tolerance. When aspen started to die, the lime trees began to gain dominance. The slow loss of maple could be also due to natural causes. Maple trees are shade tolerant when young, but as the trees age, more light is required for optimal growth [55], which is why they could no longer compete with the fast-growing aspen and lime trees.

Once again, we could see that the birches were almost absent from the plots between the ages of 20 and 40, but their proportion in basal area in older plots was higher (Figure 6). It is most likely that some old birches were leaved during the clearcutting process, which is a common practice in Tatarstan (personal communication with foresters, August 2020), and these birches recovered once the aspen died off. Also, the oak was almost completely missing up to the age of 30. This event is discussed in the next section.

4.5. The Oak–Paradox

In our study, we demonstrated a tendency of lime to return to these historical oak-lime forests [14–16] if the stands were not cut after 40 years and had the opportunity for natural development.

We can see that oak was represented in these areas, but only by a few individuals, at the age of 30 years and older (Figure 6). So, the question arises: where did these oaks come from, and did they regenerate naturally after clearcutting?

First, from personal communication with foresters (2020), it is known that some of these oaks were left in the felling process for two reasons: some individuals were too big for the harvesters (setting mainly for aspen) or these oaks were not registered for harvesting. Thus, these rare oaks were unintentionally but systematically left in the clearcut; i.e., these oaks are what we may call 'retention trees'. Assuming that oaks which rejuvenated after clearcutting cannot grow to 20 cm dbh in 30 years—especially in extreme competition with pioneer tree species—we may cautiously conclude that the oak retention trees were those in the sample which had an age of more than 30 years. *Second*, in the older plots, there were actually some oaks with a diameter of less than 20 cm; here, we are talking about 14 trees in the 190 plots. Thus, in all likelihood, there were a few oaks that rejuvenated into the aspen stands after all, i.e., after clearcutting (most likely by vectors e.g., animals [77]). We classified this form of regeneration from seed from the fact that they are very thin—e.g., with 9 cm at age 52 or with 15 cm at age 62. However, since the diameter distributions in oak were definitely not bimodal (exception: age 30), we unfortunately cannot make a clear assignment to either origin, i.e., retention vs. zoochorous. We can, however, see from the multinomial log-linear model and Figure 7 that the oak proportion in a particular plot related to the historical tree species assemblage of that plot; i.e., oaks were almost exclusively present in plots today when there were oaks dominating on the same site 100 years ago. Thus, oak seemed unable to arrive on plots within 60 years after the last clearcut when it was not present there in close vicinity 100 years ago.

4.6. Feasibility

Therefore, we could observe that leaving some seed-producing oaks on clearcuts seemed to be an insufficient practice for natural regeneration (we observed some young oaks in 3-year-old plots only; see Figure 8). The practical forestry approach in Tatarstan of planting oaks in combination with clearcutting in order to maintain or increase the area under oak cultivation harbors considerable risks for the oaks: without tending management, competition from the natural regeneration of maple, hazel and lime can threaten the existence of the oaks [78].

On the other hand, the rotation period of 40 years practiced in the aspen-dominated areas perpetuates the dominance of the root-sucker-capable aspen and, as we have shown, eliminates the oaks.

Studies mainly focusing on forest regeneration showed that the seedling establishment phase was the most sensitive regeneration stage of oaks [79] with pressure of competition [80,81] and herbivory [82,83]. According to different Tatarian sources [84–87], there was a high occurrence of moose (*Alces alces* L.) and roe deer (*Capreolus pygargus* Pallas) in combination with a very variable population of jay (*Garrus glandarius* L., most important vector of acorns in European forests [77,88]). The abundance of the jay in these forests in general depended on the presence of oaks (and acorns) [89]. Thus, the browsing pressure on the few oaks regenerating from seed in combination with the strong competition of aspen and lime trees seemed to be a problem for the natural regeneration of oak. Based on our measurements and observations, these forests formerly containing a characteristic proportion of oak then regressed to lime stands instead of Oak—Lime stands.

However, the fact that we recorded the presence of healthy mature oaks and at least some vital younger ones proved that oaks still had a chance for a living in the forests of Tatarstan [14–16]. This allows and may even demand the reintroduction of oak seedlings by planting or direct seeding into these areas, as has been demonstrated elsewhere in various restoration studies [90–94], i.e., enrichment planting.

The challenge here is to determine the most appropriate times and methods for oak regeneration. In other forests of the Volga region, it has been observed that the understory of maple and elm plays an important role in the development of oaks up to the age of 20–40 years [95]. However, as maple and elm are often affected by frost, more frost-resistant limes play an important role in this area [96], while aspen is rather a competitor of oak [97,98]. As the requirements of oaks, i.e., the company by lime trees and the die-off of aspen from the age of 30, are met (as our study shows), we only need the artificial introduction of oaks. The enrichment planting or sowing should take place in the 30–40-year-old stands when the aspen begins to die back. To convert the stands and achieve 1000–3500 oaks/ha, 3000–10,000 acorns should be sown, depending on the germination percentage [91]. Suitable germination sites, such as a thick layer of moss and litter, and sufficient water in the topsoil, are required for sowing [13,99]. These are not always available, so planting might be the better alternative. In addition, oak is a light-demanding tree species [13,99]. More than 20% of the light should be available; otherwise, oak will not rejuvenate or planted trees will die [13]. If planting is planned in the stands, on the one hand, the method established in the central Volga region of planting 2-year-old oak seedlings with a row spacing width of 6–10 m (a distance of 0.75 m within a row) on cleared areas [100] can be adopted for the stands. However, when planting in the stands, it is not always possible to plant in rows, as the overstory trees interrupt the rows, and there is root competition. Furthermore, rejuvenation can die due to competition for water and nutrients with the overstory trees, in particular lime sprouts [78]. Attention must also be paid to the light conditions under the canopy. In order to create seed tree initials for the future, small groups of oaks could be planted in gaps where, for example, aspens have died or need to be removed by thinning. There are fewer difficulties when it comes to the planting of groups in the gaps as in rows. The planting of small groups, for example 50–100 groups with 8–30 plants [101–103], should be dispersed over the stands.

Moreover, it is also known that regular thinning has a positive effect on better establishment of the light-demanding oaks [100,104,105]. Depending on the strong competition from pioneer tree species in the canopy and the understory, at least two tending interventions per decade are required in favor of oak [78,106–108].

The combined type of restoration (planting oaks under the natural succession of accompanying tree species and supportive tendings) may be not only ecologically sustainable but also cost effective [100].

5. Conclusions

To research the prospect of Oak—Lime forests restoration through an exclusively natural regeneration (i.e., passive restoration), we effectively implemented the chronosequence method and identified the regeneration processes 62 years after clearcutting. *P. tremula*, *B.*

pendula, *A. platanoides*, *U. glabra*, and *T. cordata* successfully survived clearcutting and were well presented in the rejuvenation layer. We also demonstrated the shift in proportions among these tree species during those years. Limes (and individual birch trees) overcame the high spread of aspen, as aspen started dying naturally after 30 years, whereas maples and elms found their place in the understory. Given that we found no natural regeneration (passive restoration) in oak, we suggest that the best approach is to reintroduce oak seedlings or acorns (active restoration) in combination with the natural restoration of limes and other accompanying tree species to recover the historical Oak—Lime forest (combined restoration).

Author Contributions: Conceptualization, A.N. and S.W.; methodology, S.W. and K.T.; software, S.W.; validation, A.N., K.T. and S.W.; formal analysis, A.N. and S.W.; investigation, A.N.; resources, A.N.; data curation, S.W.; writing—original draft preparation, A.N.; writing—review and editing, S.W. and K.T.; visualization, S.W.; supervision, S.W.; project administration, A.N.; funding acquisition, S.W. All authors have read and agreed to the published version of the manuscript.

Funding: This work was supported by the Federal Ministry of Food and Agriculture (Bundesministerium für Ernährung und Landwirtschaft, BMEL) [grant number 28I03001].

Institutional Review Board Statement: Not applicable.

Data Availability Statement: The original contributions presented in the study are included in the article, further inquiries can be directed to the corresponding author.

Acknowledgments: Our thanks go to the Federal Ministry of Food and Agriculture (Bundesministerium für Ernährung und Landwirtschaft, BMEL) for its financial support and to the research team of the ASTAT project (Development of sustainable forest management and utilization concepts for forest stands characterized by aspen and initiation of research networks in Ukraine, in Baltic States, the Republic of Kyrgyzstan and selected CIS countries) and especially to Anna Moosmann for her organizational expertise. Also, we are very grateful to the Ministry of Forestry of the Republic of Tatarstan for their support of forest measurements and to all foresters of the Nurlat forest district for great assistance. Many thanks to Natalia L. Blatt for providing valuable cartography, and a special gratitude to Robert Schlicht for statistical input.

Conflicts of Interest: The authors declare that they have no conflicts of interest.

References

1. Askeyev, O.; Tishin, D.; Sparks, T.; Askeyev, I. The effect of climate on the phenology, acorn crop and radial increment of pedunculate oak (*Quercus robur*) in the middle Volga region, Tatarstan, Russia. *Int. J. Biometeorol.* **2005**, *49*, 262–266. [CrossRef] [PubMed]
2. Schroeder, H.; Nosenko, T.; Ghirardo, A.; Fladung, M.; Schnitzler, J.-P.; Kersten, B. Oaks as Beacons of Hope for Threatened Mixed Forests in Central Europe. *Front. For. Glob. Change* **2021**, *4*, 670797. [CrossRef]
3. Albert, M.; Nagel, R.-V.; Nuske, R.; Sutmöller, J.; Spellmann, H. Tree Species Selection in the Face of Drought Risk—Uncertainty in Forest Planning. *Forests* **2017**, *8*, 363. [CrossRef]
4. Pretzsch, H.; Bielak, K.; Block, J.; Bruchwald, A.; Dieler, J.; Ehrhart, H.-P.; Kohnle, U.; Nagel, J.; Spellmann, H.; Zasada, M.; et al. Productivity of mixed versus pure stands of oak (*Quercus petraea* (MATT.) LIEBL. and *Quercus robur* L.) and European beech (*Fagus sylvatica* L.) along an ecological gradient. *Eur. J. For. Res.* **2013**, *132*, 263–280. [CrossRef]
5. Bolte, A.; Ammer, C.; Löf, M.; Madsen, P.; Nabuurs, G.-J.; Schall, P.; Spathelf, P.; Rock, J. Adaptive forest management in central Europe: Climate change impacts, strategies and integrative concept. *Scand. J. For. Res.* **2009**, *24*, 473–482. [CrossRef]
6. Lindner, M.; Maroschek, M.; Netherer, S.; Kremer, A.; Barbati, A.; Garcia-Gonzalo, J.; Seidl, R.; Delzon, S.; Corona, P.; Kolström, M.; et al. Climate change impacts, adaptive capacity, and vulnerability of European forest ecosystems. *For. Ecol. Manag.* **2010**, *259*, 698–709. [CrossRef]
7. Puettmann, K.; Messier, C. Simple Guidelines to Prepare Forests for Global Change: The Dog and the Frisbee. *Northwest Sci.* **2020**, *93*, 209. [CrossRef]
8. Brändle, M.; Brandl, R. Species richness of insects and mites on trees: Expanding Southwood. *J. Anim. Ecol.* **2001**, *70*, 491–504. [CrossRef]
9. Manos, P.S.; Stanford, A.M. The Historical Biogeography of Fagaceae: Tracking the Tertiary History of Temperate and Subtropical Forests of the Northern Hemisphere. *Int. J. Plant Sci.* **2001**, *162*, S77–S93. [CrossRef]

10. Löf, M.; Brunet, J.; Filyushkina, A.; Lindbladh, M.; Skovsgaard, J.P.; Felton, A. Management of oak forests: Striking a balance between timber production, biodiversity and cultural services. *Int. J. Biodivers. Sci. Ecosyst. Serv. Manag.* **2016**, *12*, 59–73. [CrossRef]
11. Attocchi, G. *Silviculture of Oak for High-Quality Wood Production*; Swedish University of Agricultural Sciences: Uppsala, Sweden, 2015; Available online: https://core.ac.uk/reader/77127980 (accessed on 21 June 2023).
12. von Lüpke, B. Silvicultural methods of oak regeneration with special respect to shade tolerant mixed species. *For. Ecol. Manag.* **1998**, *106*, 19–26. [CrossRef]
13. Rumiantsev, M.; Lukyanets, V.; Musienko, S.; Mostepanyuk, A.; Obolonyk, I. Main problems in natural seed regeneration of pedunculate oak (*Quercus robur* L.) stands in Ukraine. *For. Stud.* **2018**, *69*, 7–23. [CrossRef]
14. Bohn, U.; Gollub, G.; Hettwer, C.; Neuhäuslová, Z.; Raus, T.; Schlüter, H. Karte Der Natürlichen Vegetation Europas. Bundesamt für Naturschutz. 2000/2003. Available online: https://de.scribd.com/doc/14698027/karte-der-naturlichen-vegetation-europas (accessed on 17 April 2023).
15. Gavrilov, N. Растительный миръ ср. и н. Поволжья и Заволжья. Sankt Petersburg. 1901. Available online: http://tat-map.ru/Tematich/Selsk_lesn_hoz/rast_mir-sr-pov.jpg (accessed on 24 August 2023).
16. Kapustin, A. Forest map of Tatarstan 1:420 K [Карта лесов Татарской Республики 1:420K]. Kazan. 1927. Available online: http://retromap.ru/1419279 (accessed on 23 August 2023).
17. Istomina, E.G. The ship forests of european Russia as a resource for regional development in the eighteenth and nineteenth centuries [Корабельные леса европейской России как ресурс регионального развития в XVIII-XIX вв]. *Вестник Рггу Серия Литературоведение Языкознание Культурология* **2016**, *10*, 106–119.
18. Puryaev, A.S.; Saripov, I.N.; Petrov, W.A. The oak forests of the Middle Volga region: State, reproduction and conservation [Дубравы Среднего Поволжья: состояние, воспроизводство и сохранение]. *Лесохоз Информ* **2019**, *3*, 190–198.
19. Yakowlev, I.A.; Yekowlev, A.S. Oak forests of the middle Volga region (history, causes of degradation and current state) [Дубравы среднего поволжья (история, причины деградации и современное состояние)]. Марийский государственный технический университет, Yoshkar-Ola, Отчет о НИР. 1999. Available online: http://oaks.forest.ru/region/sredvolga/1.html (accessed on 22 January 2024).
20. Petrov, V.A. Ecological and silvicultural features of natural regeneration in disturbed oak forests in the Chuvash Republic [Эколого-лесоводственные особенности естественного возобновления в расстроенных дубравах Чувашской Республики]. Kazan State University, Kazan. 2004. Available online: https://earthpapers.net/ekologo-lesovodstvennye-osobennosti-estestvennogo-vozobnovleniya-v-rasstroennyh-dubravah-chuvashskoy-respubliki (accessed on 29 June 2023).
21. Seifert, J.R.; Selig, M.F.; Douglass, J.F.; Morrissey, R.C. Natural Oak Regeneration Following Clearcutting on the Hoosier National Forest. FNR-260; Purdue University: West Lafayette, IN, USA, 2005; 12. Available online: https://www.in.gov/isda/files/FNR-260-Oak-Regen-Following-Clearcut.pdf (accessed on 29 June 2023).
22. Kovács, B.; Tinya, F.; Németh, C. Ódor Unfolding the effects of different forestry treatments on microclimate in oak forests: Results of a 4-yr experiment. *Ecol. Appl.* **2020**, *30*, e02043. [CrossRef] [PubMed]
23. Mölder, A.; Sennhenn-Reulen, H.; Fischer, C.; Rumpf, H.; Schönfelder, E.; Stockmann, J.; Nagel, R.-V. Success factors for high-quality oak forest (*Quercus robur*, *Q. petraea*) regeneration. *For. Ecosyst.* **2019**, *6*, 49. [CrossRef]
24. Yeagle, J.A.; Groninger, J.W. Long-term effects of clearcutting on tree species composition in an oak-hickory forest. In *General Technical Report SRS-92*; Department of Agriculture, Forest Service, Southern Research Station: Asheville, NC, USA, 2006; pp. 538–540. Available online: https://www.srs.fs.usda.gov/pubs/gtr/gtr_srs092/gtr_srs092-127-yeagle.pdf (accessed on 29 June 2023).
25. Mitchell, R.; Franklin, J.; Palik, B.; Kirkman, L.; Smith, L.; Engstrom, T. Natural Disturbance-Based Silviculture for Restoration and Maintenance of Biological Diversity. In *Final report to the National Commission of Science for Sustainable Forestry*; 2002; 120. Available online: https://www.nrs.fs.usda.gov/pubs/jrnl/2003/nc_2003_mitchell_001.pdf (accessed on 29 June 2023).
26. Hédl, R.; Šipoš, J.; Chudomelová, M.; Utinek, D. Dynamics of herbaceous vegetation during four years of experimental coppice introduction. *Folia Geobot.* **2017**, *52*, 83–99. [CrossRef]
27. Tinya, F.; Kovács, B.; Aszalós, R.; Tóth, B.; Csépányi, P.; Németh, C.; Ódor, P. Initial regeneration success of tree species after different forestry treatments in a sessile oak-hornbeam forest. *For. Ecol. Manag.* **2020**, *459*, 117810. [CrossRef]
28. Vild, O.; Roleček, J.; Hédl, R.; Kopecký, M.; Utinek, D. Experimental restoration of coppice-with-standards: Response of understorey vegetation from the conservation perspective. *For. Ecol. Manag.* **2013**, *310*, 234–241. [CrossRef] [PubMed]
29. Ewald, J.; Hédl, R.; Chudomelová, M.; Petřík, P.; Šipoš, J.; Vild, O. High resilience of plant species composition to coppice restoration—A chronosequence from the oak woodland of Gerolfing (Bavaria). *Tuexenia* **2018**, *38*, 61–78. [CrossRef]
30. Jenny, H. Factors of soil formation: A system of quantitative pedology, First edition. In *McGraw-Hill Publications in the Agricultural Sciences*; McGraw-Hill Book Company, Inc.: New York, NY, USA, 1941.
31. Pickett, S.T.A. Space-for-Time Substitution as an Alternative to Long-Term Studies. In *Long-Term Studies in Ecology*; Springer: New York, NY, USA, 1989; pp. 110–135. [CrossRef]
32. Mitchell, R. *A Comparison of Three Natural Succession Chronosequence Case Studies from the South Island, New Zealand to Select Predictable Indices for Evaluating Restoration Success*; University of Canterbury: Christchurch, New Zealand, 2005.

33. Chernov, V.I. формирование хозяйственно-ценных насаждений осины (Populus tremula L.) в лесах республики Татарстан [Formation of economically useful aspen (Populus tremula L.) Stands in Forests of the Republic of Tatarstan]. Kazan State University, Kazan, Russia. 2015; Available online: https://dissercat.com/content/formirovanie-khozyaistvenno-tsennykh-nasazhdenii-osiny-populus-tremula-l-v-lesakh-respubliki (accessed on 22 June 2022).
34. Bemmann, A.; Gasisullin, A.H.; Wagner, S.; Puryaev, A. Wald und Forstwirtschaft in der Republik Tatarstan. *Forst Holzwirtsch.* **2015**, *141*, 1022–1023.
35. Garipov, N.R.; Puryaev, A.S. Structure of aspen forests of Zakamye of the Republic of Tatarstan. *For. Inf.* **2017**, *4*, 19–27.
36. Minnihanov, R.N. Atlas of Territories of the Republic of Tatarstan [Атлас земель Республики Татарстан]. Kazan, Russia: ФГУП 'Госземкадастрсъемка'-ВИСХАГИ. 2005. Available online: https://rusneb.ru/catalog/000200_000018_RU_NLR_bibl_979409/ (accessed on 16 May 2023).
37. IUSS Working Group WRB. World Reference Base for Soil Resources. In *International Soil Classification System for Naming Soils and Creating Legends for Soil Maps*, 4th ed.; International Union of Soil Sciences (IUSS): Vienna, Austria, 2022.
38. Uhl, E.; Ammer, C.; Spellmann, H.; Schoelch, M.; Pretzsch, H. Growth and growth resilience to stress of Silver fir and Norway spruce. *Allg. Forst Und Jagdztg.* **2013**, *184*, 278–292.
39. Zuur, A.; Ieno, E.; Walker, N.; Saveliev, A.; Smith, G. *Mixed Effects Models and Extensions in Ecology With R*; Springer: New York, NY, USA, 2009; Volume 1. [CrossRef]
40. Faraway, J.J. *Extending the Linear Model with R*; Taylor & Francis: Abingdon, UK, 2006; p. 345.
41. Venables, W.N.; Ripley, B.D. *Exploratory Multivariate Analysis. Modern Applied Statistics with S*; Springer: New York, NY, USA, 2002; pp. 301–330. [CrossRef]
42. Neumann, M. MNLpred: Simulated Predicted Probabilities for Multinomial Logit Models. 2021. Available online: https://cran.r-project.org/web/packages/MNLpred/index.html (accessed on 20 November 2023).
43. Ripley, B.; Venables, W. nnet: Feed-Forward Neural Networks and Multinomial Log-Linear Models. 2023. Available online: https://cran.r-project.org/web/packages/nnet/index.html (accessed on 20 November 2023).
44. Ripley, B.; Venables, B.; Bates, D.M.; Hornik, K.; Gebhardt, A.; Firth, D. MASS: Support Functions and Datasets for Venables and Ripley's MASS. 2023. Available online: https://cran.r-project.org/web/packages/MASS/index.html (accessed on 20 November 2023).
45. Wickham, H.; Chang, W.; Henry, L.; Takahashi, K.; Wilke, C.; Woo, K.; Yutani, H.; Dunnington, D.; van den Brand, T. ggplot2: Create Elegant Data Visualisations Using the Grammar of Graphics.12 October 2023. Available online: https://cran.r-project.org/web/packages/ggplot2/index.html (accessed on 20 November 2023).
46. Demakov, Y.P.; Nureeva, T.V.; Puryaev, A.S.; Krasnov, V.G. Economic basis and an experience of plantation forest growing in the central Volga region. *Сибирский Лесной Журнал* **2018**, *2*, 3–14. [CrossRef]
47. Gigunova, S.N.; Fedorov, N.I.; Mikhaylenko, O.I. Restoration succession in pine and birch deforested areas of the Southern Urals central part [Восстановительные сукцессии на сплошных вырубках сосново-березовых лесов центральной части Южного Урала]. *Научные Ведомости* **2013**, *22*, 30–35.
48. Myking, T.; Bohler, F.; Austrheim, G.; Solberg, E.J. Life history strategies of aspen (Populus tremula L.) and browsing effects: A literature review. *Forestry* **2011**, *84*, 61–71. [CrossRef]
49. Runova, E.M.; Soloveva, A.A. The natural regeneration on the logging territories of pine forests adjacent to the Middle Angara region [Естественное возобновление на вырубках сосняков в районе среднего Приангарья]. *Успехи Современного Естествознания* **2017**, *6*, 67–71. Available online: https://natural-sciences.ru/ru/article/view?id=36501 (accessed on 30 November 2023).
50. Ulanova, N.G. Mechanisms of succession of clearcutting vegetation in Southern Taiga spruce forests [Механизмы сукцессий растительности сплошных вырубок в ельниках Южной Тайги]. In *Актуальные проблемы геоботаники. Материалы III Всероссийской школы–конференции 24-28 сентября 2007.* Лекции; КарНЦРАН: Petrosavodsk, Russia, 2007; pp. 199–211.
51. Worrell, R. European aspen (*Populus tremula* L.): A review with particular reference to Scotland I. Distribution, ecology and genetic variation. *For. Int. J. For. Res.* **1995**, *68*, 93–105. [CrossRef]
52. Barker, C.; Davis, M.L.; Ashton, P. Reproductive strategy of a temperate canopy tree Tilia cordata Mill. (Malvaceae) is related to temperature during flowering and density of recent recruits. *Tree Genet. Genomes* **2022**, *18*, 22. [CrossRef]
53. Pigott, C.D. Tilia Cordata Miller. *J. Ecol.* **1991**, *79*, 1147–1207. [CrossRef]
54. Collin, E.; Bilger, I.; Eriksson, G.; Turok, J. The Conservation of Elm Genetic Resources in Europe. In *The Elms: Breeding, Conservation, and Disease Management*; Dunn, C.P., Ed.; Springer: Boston, MA, USA, 2000; pp. 281–293. [CrossRef]
55. Nowak, D.; Rowntree, R. History and Range of Norway Maple. *J. Arboric.* **1990**, *16*, 291–296. [CrossRef]
56. Roussy, A.-M. *The Sexual and Vegetative Propagation of Sugar Maple and its Threat from Norway Maple*; The University of Guelph: Guelph, ON, Canada, 2014.
57. Dormann, C.F.; Bagnara, M.; Boch, S.; Hinderling, J.; Janeiro-Otero, A.; Schäfer, D.; Schall, P.; Hartig, F. Plant species richness increases with light availability, but not variability, in temperate forests understorey. *BMC Ecol.* **2020**, *20*, 43. [CrossRef] [PubMed]
58. Lu, D.; Zhu, J.; Zhang, G.; Sun, Y.; Sun, Y.; Hu, L.; Wang, G.G. Disentangling regeneration by vertical stratification: A 17-year gap-filling process in a temperate secondary forest. *For. Ecol. Manag.* **2023**, *539*, 120994. [CrossRef]
59. Tripathi, S.; Bhadouria, R.; Srivastava, P.; Devi, R.S.; Chaturvedi, R.; Raghubanshi, A.S. Effects of light availability on leaf attributes and seedling growth of four tree species in tropical dry forest. *Ecol. Process.* **2020**, *9*, 2. [CrossRef]

60. Latva-Karjanmaa, T.; Suvanto, L.; Leinonen, K.; Rita, H. Emergence and survival of Populus tremula seedlings under varying moisture conditions. *Can. J. For. Res.* **2003**, *33*, 2081–2088. [CrossRef]
61. Franiel, I.; Kompała-Bąba, A. Reproduction strategies of the silver birch (*Betula pendula* Roth) at post-industrial sites. *Sci. Rep.* **2021**, *11*, 11969. [CrossRef] [PubMed]
62. Karlsson, O. Effects on Natural Seed Regenerated Silver Birch (*Betula pendula* Roth) and Downey Birch (*Betula pubescens* Ehrh) by Mechanical Soil Scarification and Environmental Factors. Master Thesis, Swedish University of Agricultural Sciences, Umea, Sweden, 2022.
63. Tiebel, K. Which factors influence the density of birch (Betula pendula Roth) seeds in soil seed banks in temperate woodlands? *Eur. J. For. Res.* **2021**, *140*, 1441–1455. [CrossRef]
64. Sun, H.; Zhang, J.; Duan, A.G.; He, C. A review of stand basal area growth models. *For. Stud. China* **2007**, *9*, 85–94. [CrossRef]
65. Baranchugov, E.G. Infestation of aspen with core rot and management of healthy aspen stands [Поражаемость осины сердцевидной гнилью и хозяйство на выращивание здоровых осинников]. *Лесное хоз-во.* **1995**, *5*, 26–27.
66. Baranchugov, E.G. Research report for the year 2000 [Научный отчет за 2000 г.]. In *Татарская опытная лесная станция*; All-Russian Research Institute of Silviculture and Mechanization of Forestry: Pushkino, Russia, 2001; p. 49.
67. Nasibullina, A.; van der Maaten-Theunissen, M.; van der Maaten, E.; Fischer, H.; Wagner, S. Thinning effects on growth and occurrence of rotting in aspen stands. *J. For. Sci.* **2023**, *69*, 525–538. [CrossRef]
68. Henderson, J.; Roberts, S.; Grebner, D.; Munn, I. A Graphical Comparison of Loblolly Pine Growth-and-Yield Models. South. *J. Appl. For.* **2013**, *37*, 169–176. [CrossRef]
69. West, P.W.; Ratkowsky, D.A. Models Relating Individual Tree Basal Area Growth Rates to Tree Basal Areas in Even-Aged, Monoculture Forest Stands. *J. For.* **2022**, *9*, 21–38. [CrossRef]
70. Åström, M.; Dynesius, M.; Hylander, K.; Nilsson, C. Slope Aspect Modifies Community Responses to Clear-Cutting in Boreal Forests. *Ecology* **2007**, *88*, 749–758. [CrossRef] [PubMed]
71. Gustafsson, L.; Kouki, J.; Sverdrup-Thygeson, A. Tree retention as a conservation measure in clear-cut forests of northern Europe: A review of ecological consequences. *Scand. J. For. Res.* **2010**, *25*, 295–308. [CrossRef]
72. Gustienė, D.; Varnagirytė-Kabašinskienė, I.; Stakėnas, V. Ground vegetation, forest floor and mineral topsoil in a clear-cutting and reforested Scots pine stands of different ages: A case study. *J. For. Res.* **2022**, *33*, 1247–1257. [CrossRef]
73. Cleland, D.T.; Avers, P.E.; McNab, W.H.; Jensen, M.E.; Bailey, R.G.; King, T.; Russel, W.E. National Hierarchial Framework of Ecological Units. In *Ecosystem Management Applications for Sustainable Forest and Wildlife Resources*; Boyce, M.S., Haney, A., Eds.; Yale University Press: New Haven, CT, USA, 1997; pp. 181–200.
74. Oliver, C.D.; Larson, B.C. *Forest Stand Dynamica*; update edn; Wiley: New York, NY, USA, 1996.
75. Morrissey, R.C.; Jacobs, D.F.; Seifert, J.R.; Kershaw, J.A. Overstory species composition of naturally regenerated clearcuts in an ecological classification framework. *Plant Ecol.* **2010**, *208*, 21–34. [CrossRef]
76. Radoglou, K.; Dobrowolska, D.; Spyroglou, G.; Nicolescu, V. A review on the ecology and eilviculture of limes: (Tilia cordata Mill., Tilia platyphyllos Scop. and Tilia tomentosa Moench.) in Europe. *Die Bodenkultur.* **2009**, *3*, 9–20.
77. Axer, M.; Schlicht, R.; Wagner, S. Modelling potential density of natural regeneration of European oak species (*Quercus robur* L., Quercus petraea (Matt.) Liebl.) depending on the distance to the potential seed source: Methodological approach for modelling dispersal from inventory data at forest enterprise level. *For. Ecol. Manag.* **2021**, *482*, 118802. [CrossRef]
78. Wallraf, A.; Wagner, S. Effects of initial plant density, interspecific competition, tending and age on the survival and quality of oak (*Quercus robur* L.) in young mixed stands in European Russia. *For. Ecol. Manag.* **2019**, *446*, 272–284. [CrossRef]
79. Orman, O.; Wrzesiński, P.; Dobrowolska, D.; Szewczyk, J. Regeneration growth and crown architecture of European beech and silver fir depend on gap characteristics and light gradient in the mixed montane old-growth stands. *For. Ecol. Manag.* **2020**, *482*, 118866. [CrossRef]
80. Martínez-Izquierdo, L.; García, M.M.; Powers, J.S.; Schnitzer, S.A. Lianas suppress seedling growth and survival of 14 tree species in a Panamanian tropical forest. *Ecology* **2016**, *97*, 215–224. [CrossRef]
81. Roberts, M.W.; D'Amato, A.W.; Kern, C.C.; Palik, B.J. Effects of variable retention harvesting on natural tree regeneration in Pinus resinosa (red pine) forests. *For. Ecol. Manag.* **2017**, *385*, 104–115. [CrossRef]
82. Sabo, A.E.; Forrester, J.A.; Burton, J.I.; Jones, P.D.; Mladenoff, D.J.; Kruger, E.L. Ungulate exclusion accentuates increases in woody species richness and abundance with canopy gap creation in a temperate hardwood forest. *For. Ecol. Manag.* **2019**, *433*, 386–395. [CrossRef]
83. Walters, M.B.; Farinosi, E.J.; Willis, J.L. Deer browsing and shrub competition set sapling recruitment height and interact with light to shape recruitment niches for temperate forest tree species. *For. Ecol. Manag.* **2020**, *467*, 118134. [CrossRef]
84. Askeyev, O.; Askeyev, I. *Birdfauna of Tatarstan Republic [Орнитофауна* республики татарстан*]*; Birdfauna of tatarstan republic: Kazan, Russia, 1999.
85. Savin, W.W. Influence of Wild Hoofed Animals on Reforestation in Conditions of Priobsky Water-Protected Pine-Birch Forestry Area [Влияние диких копытных животных на лесовозобновление в условиях Приобского водоохранного сосново-березового лесохозяйственного района]. *ф*ГБОУ ВО «Уральский государственный лесотехнический университет», Ekaterinburg. 2020. Available online: https://dissercat.com/content/vliyanie-dikikh-kopytnykh-zhivotnykh-na-lesovozobnovlenie-v-usloviyakh-priobskogo-vodookhran (accessed on 30 August 2023).

86. Marukhina, A. Wildlife Is a National Asset [Охотничьи ресурсы—это народное достояние]. *KazanFirst*. 2022. Available online: https://kazanfirst.ru/articles/591488 (accessed on 30 August 2023).
87. Afonina, A.; Starlings are dying out, rotan 'devouring' lakes, a disaster with fir: The nature of Tatarstan is transforming right now [Скворцы вымирают, ротан «пожирает» озера, с пихтой катастрофа: природа Татарстана меняется на глазах]. БИЗНЕСONLINE 2022. Available online: https://www.business-gazeta.ru/article/570478 (accessed on 30 August 2023).
88. Axer, M.; Martens, S.; Schlicht, R.; Eisenhauer, D.; Wagner, S. Modelling natural regeneration of Oak in Saxony, Germany: Identifying factors influencing the occurrence and density of regeneration. *IForest Biogeosci. For.* **2023**, *16*, 47–52. [CrossRef]
89. Askeyev, A.; Askeyev, O.; Askeyev, I. Spatial distribution and long-term abundance dynamic of a jay and a three-toed woodpecker in forests of the Republic of Tatarstan [Пространственное распределение и многолетняя динамика численности сойки и трехпалого дятла в лесах республики Татарстан]. *Russ. J. Appl. Ecol.* **2017**, *3*, 9–13.
90. Heydari, M.; Prévosto, B.; Abdi, T.; Mirzaei, J.; Mirab-Balou, M.; Rostami, N.; Khosravi, M.; Pothier, D. Establishment of oak seedlings in historically disturbed sites: Regeneration success as a function of stand structure and soil characteristics. *Ecol. Eng.* **2017**, *107*, 172–182. [CrossRef]
91. Löf, M.; Castro, J.; Engman, M.; Leverkus, A.B.; Madsen, P.; Reque, J.A.; Villalobos, A.; Gardiner, E.S. Tamm Review: Direct seeding to restore oak (*Quercus* spp.) forests and woodlands. *For. Ecol. Manag.* **2019**, *448*, 474–489. [CrossRef]
92. Martínez-Baroja, L.; Rey-Benayas, J.M.; Pérez-Camacho, L.; Villar-Salvador, P. Drivers of oak establishment in Mediterranean old fields from 25-year-old woodland islets planted to assist natural regeneration. *Eur. J. For. Res.* **2022**, *141*, 17–30. [CrossRef]
93. Palma, A.C.; Laurance, S.G.W. A review of the use of direct seeding and seedling plantings in restoration: What do we know and where should we go? *Appl. Veg. Sci.* **2015**, *18*, 561–568. [CrossRef]
94. Zadworny, M.; Jagodziński, A.M.; Łakomy, P.; Ufnalski, K.; Oleksyn, J. The silent shareholder in deterioration of oak growth: Common planting practices affect the long-term response of oaks to periodic drought. *For. Ecol. Manag.* **2014**, *318*, 133–141. [CrossRef]
95. Glebov, V.P.; Verhunov, P.M.; Urmakov, G.N. Izd-vo Chuvashiya: Chuvashia, Russia, 1998; p. 199.
96. Petrov, V.A.; Balasny, V.I. Growth of English oak and linden in forest plantations created by seedling of acorns and planting of seedlings Chuvash Republic cutover areus. *For. Inf.* **2016**, *4*, 66–72.
97. Rahteenko, I.N. *Growth and Interaction of Root Systems of Tree Species [Рост и взаимодействие корневых систем древесных растений]*; AN BSSR: Minsk, Belarus, 1963; p. 254.
98. Kolesnichenko, M.V. *Biochemical Interactions of Trees [Биохимические взаимодействия древесных растений]*; Lesn. prom-st: Moscow, Russia, 1968; p. 150.
99. Woziwoda, B.; Dyderski, M.K.; Kobus, S.; Parzych, A.; Jagodziński, A.M. Natural regeneration and recruitment of native Quercus robur and introduced Q. rubra in European oak-pine mixed forests. *For. Ecol. Manag.* **2019**, *449*, 117473. [CrossRef]
100. Petrov, V.A.; Ilyin, F.S.; Kuznetsova, N.F. Combined Restoration Method Oak in the Chuvash Republic. *Forestry Information.* **2021**, *3*, 35–44. [CrossRef]
101. Saha, S.; Kühne, C.; Kohnle, U.; Bauhus, J. Eignung von Nester- und Truppflanzungen für die Begründung von Eichenbeständen. *AFZ Der Wald* **2013**, *2*, 29–37.
102. 3.03 Trupp-Pflanzung_bf.pdf. Available online: https://www.waldbesitzerportal.de/fileadmin/user_upload/Download/Waldumbau/3.03_Trupp-Pflanzung-bf.pdf (accessed on 6 May 2024).
103. Anreichungskulturen_mb46_bf.pdf. Available online: https://www.lwf.bayern.de/mam/cms04/service/dateien/anreicherungskulturen_mb46_bf.pdf (accessed on 6 May 2024).
104. Glushko, S.G.; Manukova, I.G.; Prokhorenko, N.B. Restoration of oak forest of Middle Volga region. *Вестник Омского Государственного Аграрного Университета* **2017**, *3*, 27. Available online: https://sciup.org/vosstanovlenie-dubrav-srednego-povolzhja-142199376-en (accessed on 20 November 2023).
105. Petrov, V.A.; Ilyin, F.S.; Kuznetsova, N.F. Restoration of Oak Forests on the Basis of Natural Oak Renewal in the Middle Volga Region. *Forestry informaion.* **2022**, *1*, 35–49. [CrossRef]
106. Bartsch, N.; Röhrig, E. *Waldökologie: Einführung für Mitteleuropa*; Springer: Berlin/Heidelberg, Germany, 2016. [CrossRef]
107. Petersen, R.; Schüller, S.; Ammer, C. Early growth of planted pedunculate oak (*Quercus petraea*) in response to varying competition by birch (Betula pendula) over 8 years. *Forstarchiv* **2009**, *80*, 208–214.
108. Wagner, S.; Röker, B. Birkenanflug in Eichenkulturen: Untersuchungen zur Dynamik der Konkurrenz über 5 Vegetationsperioden. *Forst Holz* **2000**, *55*, 18–22.

diversity

MDPI

Article

Dominant Species Composition, Environmental Characteristics and Dynamics of Forests with *Picea jezoensis* Trees in Northeast China

Jichen Duan [1], Zhiyuan Jia [2], Shusen Ge [3], Yutang Li [3], Dongwei Kang [4,*] and Junqing Li [4]

1 College of Forestry, Beijing Forestry University, Beijing 100083, China; djc19970226@bjfu.edu.cn
2 School of Tourism, Xinyang Normal University, Xinyang 464000, China; zyjia_1018@163.com
3 Forest Inventory and Planning Institute of Jilin Province, Changchun 130022, China; shusenge@126.com (S.G.); yutanglee@163.com (Y.L.)
4 School of Ecology and Nature Conservation, Beijing Forestry University, Beijing 100083, China; lijq@bjfu.edu.cn
* Correspondence: kangdw@bjfu.edu.cn

Abstract: To describe the characteristics and dynamics of the *Picea jezoensis* (PJ) community, a survey of 48 forest plots containing PJ trees was conducted in Northeast China. Methods of community grouping, analysis of variance, and linear regression, and indicators of relative basal area (RBA, reflecting the dominance degree of species in the community) and temporary stability (TS, reflecting the stability of forest plant community), were employed. The objectives were to describe the dominant species composition and environment of the PJ community and to quantify the changes in PJ's dominance status. Communities with PJ trees were divided into four groups: PJ as the first dominant species; PJ as the second dominant species; PJ being co-dominant but ranked third or lower; and PJ as a non-dominant species. Among them, the PJ community (i.e., where PJ is the first dominant species) occupied sites at the highest elevation with a mean value of 1408 m, on gentle slopes $\leq 10°$, where the tree species occurring with PJ mainly included *Betula costata* and *Abies nephrolepis*. A linear relationship was found between the TS and RBA. The threshold for PJ shifting between being the first and second dominant species in the community was RBA = 0.387. This study provides a scientific basis for judging the attributes and status of PJ community and the protection and restoration of PJ forests. These findings also provide comparable information for ecological research and conservation of PJ forests in other areas.

Keywords: elevation; community dynamics; dominant species; species composition; temporary stability

1. Introduction

Plant communities are dynamic, characterized by seasonal fluctuations, interannual changes, and succession [1,2]. Community dynamics is an important aspect of ecological studies [3], providing crucial information on changes in community structure and composition over time. Quantitative or attribute changes in species composition, abundance, diversity, and community structure are considered indicators of community dynamics [4–7].

The research method of community dynamics is an important topic in community ecology, and various methods are able to assess different aspects of community dynamics [8]. For example, community characteristics are monitored by field observations [9], and trends in community change can be predicted via model simulations [10]. Stages of community development are often evaluated by comparing the characteristics and relationships of communities with different origins [11]. Community establishment processes are usually determined by analyzing the dominance state of a target species in different communities [12].

Among the various methods for characterizing community dynamics, comparing the composition and status of dominant species determines the community trajectory [12].

Understanding the relationship between the community and local environment is helpful for explaining facets of community performance [13,14], while analyzing stability is critical for quantifying community dynamics and the factors influencing it [15,16]. Integrating these methods enables a better description of community characteristics and stability of a given target species.

Forest ecosystems play an important role in biodiversity conservation, ecosystem services, and climate regulation [2]. Ecological research on forests with different tree species is of great significance. For example, many *Picea* species have attracted much research attention, such as *P. asperata* [17,18], *P. purpurea* [19,20], and *P. crassifolia* [21,22]. *Picea jezoensis* (PJ) is an important evergreen coniferous tree species in Northeast Asia [23,24]. Many ecological studies have focused on PJ, including its regeneration characteristics [25–28], growth–environment relationships [29–32], community and diversity [33,34], and the impacts of climate change [35]. These studies have enriched our understanding of PJ's ecology in places such as Korea, Japan, and Far East Russia. In particular, the stand structure of PJ forests is key research theme at the community level [33,34].

Besides the natural distribution area of PJ mentioned above, China is also an important distribution area of PJ; this species has been the subject of many ecological studies. Most of these focused on radial growth and ecological factors [36–41], community and diversity [42–46], regeneration and growth characteristics [47–50], species–environment relationships [51,52], and the impacts of climate change [53]. That body of research has arguably increased our understanding of the ecological characteristics of PJ in China. Community ecology studies have mainly addressed the species composition and community structure of PJ forests in China [42–45].

Yet, to our best knowledge, few studies have investigated the community dynamics of PJ [33,45–47], leaving several fundamental questions unaddressed, thereby limiting our knowledge of the establishment and changes in the PJ community and hindering the effective protection and restoration of PJ forests. For example, what is the threshold for changes in the dominance position of PJ in the community? Since the PJ community can exhibit regional differences [34], a detailed analysis of the characteristics and dynamics of the PJ community in a specific area is needed. Studying this question provides a scientific basis for the conservation of local PJ forests, as well as the formulation of targeted sustainable forest management strategies.

To answer the above scientific question, this study attempted to propose the hypothesis based on the knowledge accumulation and research experience in forest ecology. Dominant species in forest communities are those with a large number of individuals, extensive crown cover, and high biomass [2,3]. This study presumed that a shift in the dominance status of PJ from being the second dominant species to the first dominant species in a community would occur incrementally. This study hypothesized that this shift would happen when the dominance degree of PJ reached a critical threshold value.

A survey of forests with PJ trees in three counties of Northeast China was conducted to investigate the PJ community's characteristics and dynamics. The objectives were to (1) describe the dominant species composition and environment of the PJ community and (2) determine the threshold for a dominance shift of PJ in the community. By comparing the different forests containing PJ trees, the aim was to comprehensively assess the PJ community and gauge changes in PJ's dominance status to provide empirical information for the conservation of PJ forests, especially in protecting existing PJ forests and restoring degraded PJ forests.

2. Materials and Methods

2.1. Study Area and Data Collection

This field survey was conducted in Antu, Fusong, and Changbai counties of Jilin Province, China. Antu County, in eastern Jilin, encompasses 7444 km^2, and has 86% forest cover, with a wood volume of 98 million m^3, where the annual average temperature is 2.2 °C in the south and 3.6 °C in the north, whose corresponding annual average precipitation is 669.7 and 594.7 mm [54]. Fusong County (6159 km^2) is located in southeastern Jilin, and has 86.7% forest cover, with a wood volume of 4.62 million m^3, where the annual average temperature is 4 °C and the annual average precipitation is 800 mm [55]. Changbai County is also in southeastern Jilin and has an area of 2506 km^2. Its forest cover is 92%, with a wood volume of 23.34 million m^3; the annual average temperature is 3.9 °C, and the annual average precipitation is 605.8 mm [56].

The forests in the study area consist of various tree species. For example, the top seven ranked tree species in terms of relative basal area (RBA) are the following: PJ (RBA = 0.264), *Abies nephrolepis* (AN) (RBA = 0.143), *Larix olgensis* (LO) (RBA = 0.101), *Pinus koraiensis* (PK) (RBA = 0.088), *Populus ussuriensis* (PU) (RBA = 0.086), *Betula costata* (BC) (RBA = 0.082), and *Tilia amurensis* (TA) (RBA = 0.080).

The forest survey was carried out in 2014. Forty-eight forest plots (each 600 m^2) containing PJ trees (23 plots in Antu County, 22 plots in Fusong County, and 3 plots in Changbai County) were surveyed (Figure 1; Table S1). The distance between any two plots was $\geq$4 km. In every plot, the scientific name and stem size of all living trees with a diameter at breast height (DBH) $\geq$ 5 cm were recorded. A total of 2635 trees were measured. The basic information of each plot was also recorded, namely, its elevation, slope aspect, slope degree, and forest origin (Table S1). The elevation of the 48 plots ranged from 840 to 1810 m, and all plots were located at sites with a gently slope ($\leq$10°). All plots were located in natural forests.

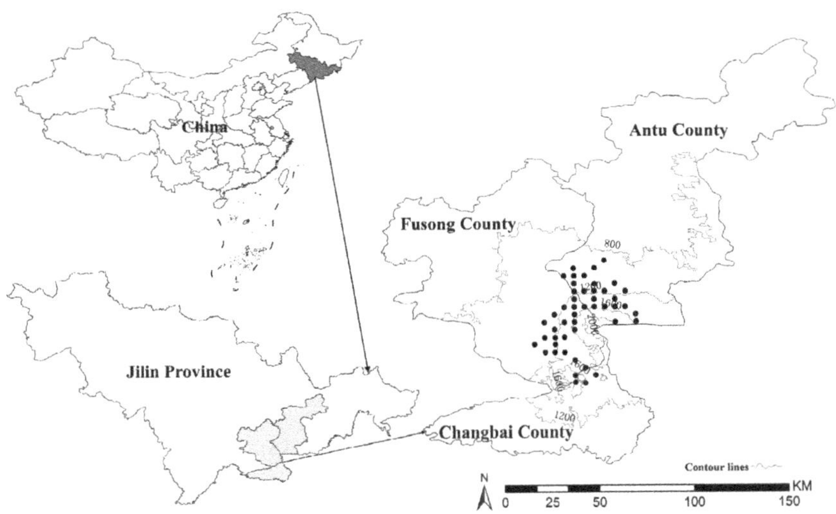

Figure 1. Location of plots containing *Picea jezoensis* trees.

2.2. Data Analysis

Species composition is the most important factor determining the attributes of a community [2]; comparing the species composition of different communities with PJ trees highlights the characteristics of PJ communities. The environment has an impact on the composition and structure of communities [3]; analyzing the relationship between different communities with PJ trees and environmental factors reveals the environmental conditions of PJ community. On the basis of clarifying the PJ community characteristics and its environmental conditions, quantifying the dominance state and change of PJ in the community helps to understand the PJ community dynamics in a simple and convenient way.

2.2.1. Species Composition of Different Forests Containing *P. jezoensis* Trees

This study analyzed the dominant species composition of different forests with PJ trees. The RBA of each tree species per plot was calculated, and the dominant species identified accordingly [12,16]. RBA was calculated by dividing the basal area of a species by the total basal area of all species [57]. Dominant species were defined as those with an RBA > 0.10 [12,16]. In each plot, the first two dominant species and their RBA values were listed, and PJ's rank was determined. The plots were divided into four groups based on the PJ's ranking and dominance: (i) PJ as the first dominant species, (ii) PJ as the second dominant species, (iii) PJ as the third (or less ranked) dominant species, and (iv) PJ as a non-dominant species. Next, the dominant species composition and proportions in the groups were analyzed and compared. To do that, the species name and proportion of the first two dominant species in each group were determined, and then the dominant species that co-appeared among the groups were identified. This study focused on PJ and the top two dominant species and sought to highlight the dominant species composition of the PJ community (i.e., PJ as the first dominant species).

2.2.2. Environmental Characteristics of Different Forests Containing *P. jezoensis* Trees

This study examined the local environment of different forests with PJ trees. Slope is a complex factor affecting the tree species and their growth status [2]. Although there were differences in slope aspect and degree among the sample plots, all were nonetheless situated at gently sloping sites (Table S1). So, we considered that 48 plots had the same slope conditions.

Tree species are usually only distributed within a certain elevational range, and vegetation type and species composition may vary with elevation [2]. Because the elevation span of the 48 plots was about 1000 m, the relationship between forest and elevation cannot be ignored. Hence, the elevational characteristics of different forests with PJ trees were described and compared to highlight the role of elevation for the PJ community.

To determine the distribution pattern of forests with PJ trees vis-à-vis elevation, we calculated the mean elevation of each group and tested whether elevation differed significantly among the four groups using one-way analysis of variance (ANOVA) or a non-parametric test based on the statistical requirements [58]. If so, we further detected differences between any two groups using multiple comparisons; either the least significant difference (LSD) or Games–Howell method was used, based on the homogeneity of variance [59].

2.2.3. Changes to the Dominance Status of *P. jezoensis*

Temporary stability (TS) was employed as a robust metric to assess the dominance dynamics of PJ, providing insights into the critical threshold where shifts in dominance occur. The TS provides a measure of forest community stability [15,16] and was calculated this way:

$$TS = 1 - \frac{d_2}{d_1},$$

where d_1 and d_2 denote the dominance of the first and second ranked species, respectively [15,16]. This study used basal area for dominance. The top two ranked species in dominance have the same basal area when TS = 0, which corresponds to the threshold for a change in the first dominant species.

As seen above, TS conveys the relationship between the top two ranked species in dominance. Therefore, only plots in groups where PJ was the first or second dominant species were retained for this analysis. To determine the feasibility of combining the plots from two groups for a unified analysis, the environmental characteristics of these two groups were evaluated. All plots were located at gently sloping sites (Table S1), so these two groups thus had the same conditions in slope. Then, the *t* test or non-parametric test [58] was used to determine whether elevation differed significantly between the two groups. Furthermore, the dominant species composition of these two groups was also described and compared.

This study analyzed the relationship between the dominance status and degree of dominance of PJ in the community to determine this critical value. TS and the RBA were used to represent the dominant state and the degree of dominance of PJ, respectively. These indicators were calculated for the retained plots. According to its definition [15], if TS has a large value, this indicates that the first dominant species is less likely to be replaced by PJ in plots where it is the second dominant species, and PJ is less likely to be replaced by the second dominant species in plots where PJ is the first dominant species. To integrate these two groups of plots into a unified research system, TS values for plots where PJ was the second dominant species were converted to negative values. The correlation between TS and the RBA of PJ was analyzed to determine whether or not they exhibited a linear trend. If so, to verify whether the linear relationship meets the statistical requirements, ANOVA, the determination coefficient (R^2), significance level, and standardized residuals were used to assess the regression model [60]. The threshold RBA of PJ was calculated by setting TS = 0.

3. Results

3.1. Dominant Species Composition of the *P. jezoensis* Community

Based on the ranking and dominance characteristics of PJ, the 48 plots were divided into four groups. BC and AN mainly constituted the second dominant species in plots where PJ was ranked the first dominant species (13 plots) (Figure 2; Table S2). BC, AN, and PK mainly constituted the first dominant species for plots in which PJ was the second dominant species (14 plots) (Figure 2; Table S3). PU and LO mainly constituted the first dominant species for plots where PJ was a co-dominant but not ranked among the top two dominant species (8 plots) (Figure 2; Table S4). In plots where PJ was not the dominant species (13 plots), LO was the first dominant species in 30.77% of the plots, while PU, QM, and TA had the same proportions (15.38%) (Figure 2; Table S5).

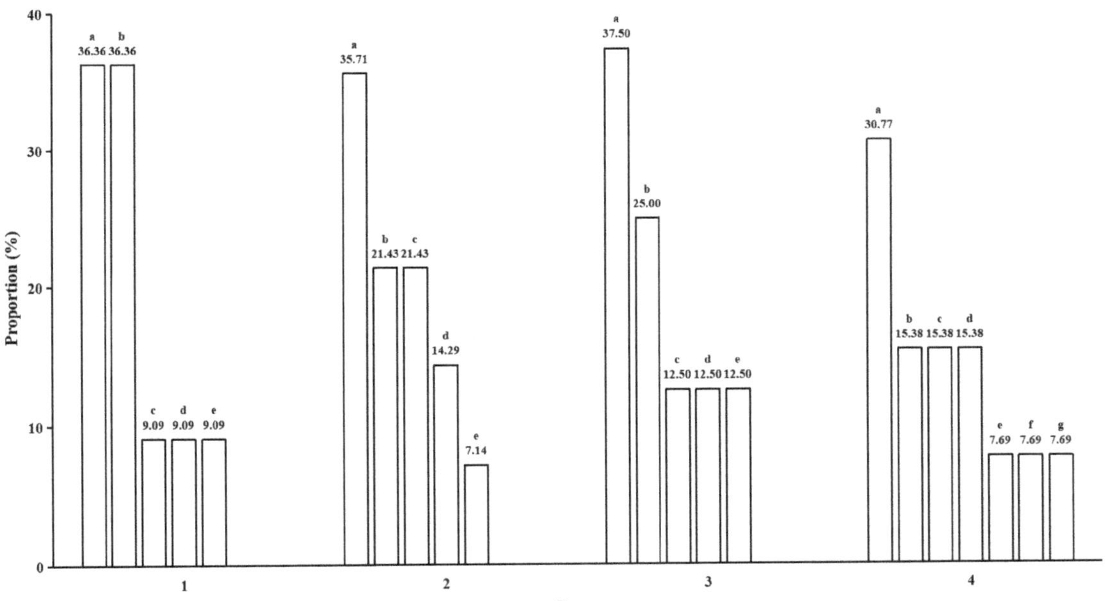

Figure 2. Composition of the first or second dominant species of different forests containing *Picea jezoensis* (PJ) trees (group 1: PJ was the first dominant species; group 2: PJ was the second dominant species; group 3: PJ was a co-dominant but not ranked among the top two dominant species; group 4: PJ was not the dominant species; 1a, 2a, 4f: *Betula costata* (BC); 1b, 2b, 4e: *Abies nephrolepis* (AN); 1c, 2d, 3a, 4b: *Populus ussuriensis* (PU); 1d: *Fraxinus mandshurica* (FM); 1e, 2c, 3d, 4g: *Pinus koraiensis* (PK); 2e, 3b, 4a: *Larix olgensis* (LO); 3c: *Betula platyphylla* (BP); 3e, 4d: *Tilia amurensis* (TA); 4c: *Quercus mongolica* (QM)).

Among the four groups, three species (PK, PU, and LO) co-appeared as the first dominant species in the three groups in which PJ was not the first dominant species. Two species (PK and PU) co-appeared as the second dominant species in the group in which PJ was the first dominant species (Figure 2; Tables S2–S5).

3.2. Distribution of *P. jezoensis* Community Across Elevation

Different forests with PJ trees exhibited a regular distribution with respect to elevation. Forests in which PJ was the first dominant species were distributed at the highest elevation, with a mean value of 1408 (±198) m (Figure 3).

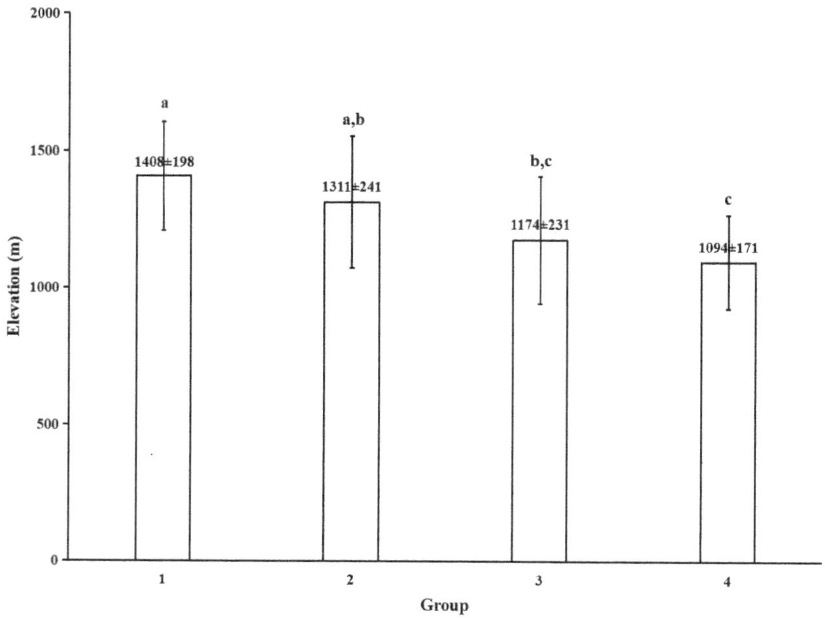

Figure 3. Mean elevation of different forests containing *Picea jezoensis* (PJ) trees (group 1: PJ was the first dominant species; group 2: PJ was the second dominant species; group 3: PJ was a co-dominant but not ranked among the top two dominant species; group 4: PJ was not the dominant species; a, b, c: bars that shared a letter (i.e., a, b, or c) do not differ significantly from one another).

Elevation differed significantly among the four groups (ANOVA, F = 5.549, $p < 0.01$). Based on the multiple comparisons testing, there were no significant differences between adjacent groups (such as groups 1 vs. 2, groups 2 vs. 3, and groups 3 vs. 4) ($p > 0.05$), but there were significant differences between separated groups (such as groups 1 vs. 3, groups 1 vs. 4, and groups 2 vs. 4) ($p < 0.05$; Figure 3; Table 1).

Table 1. Multiple comparisons (using LSD method) of elevation among different forests containing *Picea jezoensis* (PJ) trees (group 1: PJ was the first dominant species; group 2: PJ was the second dominant species; group 3: PJ was a co-dominant but not ranked among the top two dominant species; group 4: PJ was not the dominant species).

Item	Mean Elevation Difference (m)	Standard Error	*p* Value
Group 1–Group 2	97	81	0.239
Group 1–Group 3	234	95	0.017
Group 1–Group 4	314	83	0.000
Group 2–Group 3	138	93	0.148
Group 2–Group 4	217	81	0.010
Group 3–Group 4	80	95	0.405

To sum up, the PJ community occupied sites at high elevation (mean: 1408 m), this being similar to the mean elevation of the community in which PJ was the second dominant species, yet significantly higher than those of communities in which PJ was co-dominant but not among the top two dominant species or not a dominant species.

3.3. Dominance States and Changes of *P. jezoensis* in the Forest Community

Plots in which PJ was not one of the first two dominant species were excluded from the subsequent analysis. A total of 27 plots were retained for this analysis, comprising 13 plots in which PJ was the first dominant species and 14 plots in which PJ was the second dominant species.

In terms of their environmental conditions, both the 13 plots and 14 plots were located at sites with a gentle slope degree of $\leq 10°$. The mean elevation of the 13 plots was 1408 ($\pm$198) m, which exceeded 1311 ($\pm$241) m of the 14 plots, albeit not significantly (*t* test, $t = 1.134$, $df = 25$, $p > 0.05$).

Regarding species composition, the second dominant species of the 13 plots consisted of BC, AN, PU, FM, and PK, of which four (BC, AN, PU, and PK) co-appeared as the first dominant species in the 14 plots (Figure 2; Tables S2 and S3).

The TS was strongly correlated with the RBA of PJ ($r = 0.944$; $p < 0.001$). In the fitted linear model, the R^2 was 0.890, the *p*-value of the ANOVA was < 0.001 ($F = 202.884$), and the standardized residuals ranged from –2.074 to 2.028. Thus, a strong, predictable linear relationship existed between the TS and the RBA of PJ. The regression model's equation was $TS = 2.383 \times RBA - 0.922$. Based on this model, when TS = 0, the RBA of PJ = 0.387 (Table 2; Figure 4).

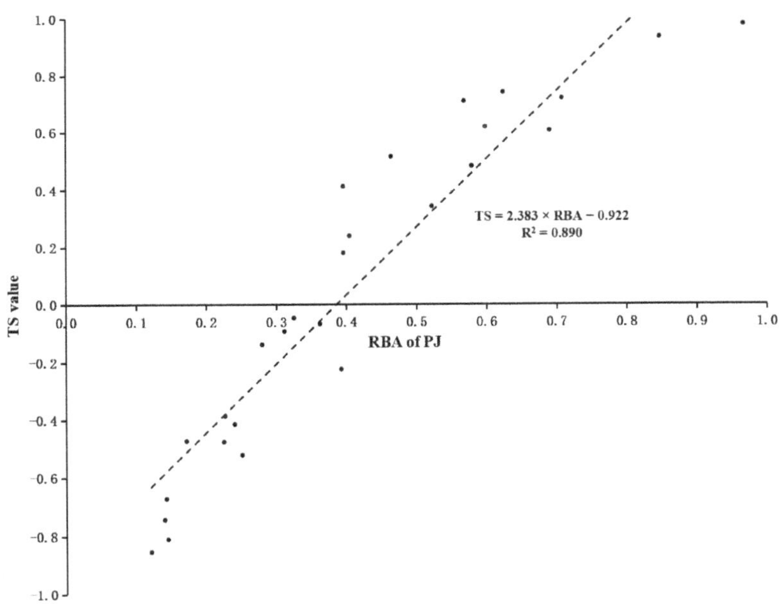

Figure 4. Temporary stability (TS) and relative basal area (RBA) of *Picea jezoensis* (PJ).

Table 2. Linear regression model between temporary stability (TS) and the relative basal area (RBA) of *Picea jezoensis* (PJ).

Model	Unstandardized Coefficient (UC)	Standard Error	t Value	*p* Value	95% Confidence Interval of UC
RBA	2.383	0.167	14.244	0.000	[2.039, 2.728]
Constant	−0.922	0.078	−11.793	0.000	[−1.083, −0.761]

4. Discussion

4.1. *P. jezoensis* Community and Environmental Characteristics

Comparing the dominance of a target species in different communities is capable of revealing the species' community characteristics [12]. Based on the PJ ranking and dominance characteristics, the 48 surveyed plots were divided into four groups. The results show that PJ exhibits different dominance states in different forests, a feature similar to that of a fir species of *Abies faxoniana* [12]. The employed method of community grouping analysis was able to reveal the dominance characteristics of target species in different forests.

Different species have different roles in a given community [3]. From a species composition perspective, evidently BC and AN were the main tree species in the PJ community, while PK and PU were the co-appeared species of PJ. This result is mostly consistent with findings of previous studies [42,53]. Some studies have analyzed the relationship of PJ to other tree species [61,62]. However, the data of the present study did not lend support to such species association and correlation analyses; hence, further studies are needed to reveal the actual relationship between these tree species and PJ. Furthermore, the dominant species differed among forests with PJ trees, indicating that PJ is capable of establishing and growing in a variety of forest types and that suitable PJ habitats are diverse. Alongside PJ's wide distribution range in terms of elevation (Table S1), our results support a previous study that reported PJ has a wide niche breadth [63].

In mountainous conditions, vegetation distribution is usually strongly related to elevation [2]. Although PJ's distribution does range widely across elevation, PJ was not always the first dominant species in every forest community [44]. For example, communities in which PJ is the first dominant species, this being the typical PJ community, prevailed at the highest elevation with a mean value of 1408 m. The first dominant species (PJ) in the PJ community was unlike those of the other three community groups that occupied different elevations. Given that the forest origins of all surveyed plots were natural forest processes, the mean elevation of 1408 m is a pivotal parameter for describing the natural distribution of the PJ community. In this way, the distributional elevation of the PJ community in this study is more specific than previous descriptions for this species [42,44]. Furthermore, many previous studies have analyzed the response of PJ's growth to climate change [37,39,41,49], additional studies should be considered to reveal the potential impact of climate change on the distributional elevation of PJ.

4.2. *P. jezoensis* Community Stability and State Change

Many studies on community stability have been conducted, focusing on the definition of stability [64,65], diversity–stability relationships [66,67], and influential factors [68,69]. Other studies have assessed the PJ community using a subordinate function and spatial structure index [70,71]. However, no empirical studies have yet explored the PJ community's stability. This study, the first to do so, revealed a linear relationship between TS and the RBA of PJ. Importantly, based on this relationship, the critical value when PJ changes from being the second dominant species to the first dominant species was an RBA = 0.387. This key result fills a research gap in quantifying the shift in the dominance status of PJ in the forest community and provides useful information for studying the community dynamics of other forest types.

Under largely consistent slope conditions and similar elevation, this study found that PJ exhibited a different ranking and dominance status and that four species co-appeared as the dominants in forests where PJ was the first and second dominant species. We conjecture that the PJ ranking and dominance state are reversible. Considering the PJ community occupied the highest elevation of 1408 m, this could be the most suitable elevation for the growth of PJ trees. Therefore, the higher the elevation, the more likely a site is to be conducive to improving PJ's ranking and dominance status. This phenomenon requires further observation and verification, however.

Many indicators are used to gauge forest community stability [69–73]. As a recently proposed indicator, TS has been adopted to analyze the community stability of forests in Northeast and Southwest China [15,16], but its scope of application should still be determined. In this study, our results confirm that TS can be used to analyze and detect the shift in dominance status from the second to the first dominant species. However, this indicator could be improved to accommodate a wider range of community dynamics.

4.3. Limitations and Implications

This study described the characteristics and dynamics of the PJ community. Nonetheless, it has some limitations. First, this work is based on a single survey; long-term continuous observations should therefore be considered. Second, the sample size was 48 plots; using a larger sample size would provide even more comprehensive information. Third, this study described the characteristics of the PJ community from the perspective of trees; incorporating the understory vegetation would facilitate a more detailed description. Fourth, this study depicted the PJ community mainly in relation to elevation and slope. We suggest that future studies incorporate other plausible influencing factors, such as soil [70], wind [74], and insects [75]. Fifth, due to differences in environment and tree species, the linear model and critical value obtained in this study may not necessarily have broad applicability, so it should be used cautiously in other areas and with other tree species.

By assessing the dominant species composition and environmental characteristics of the PJ community, this study offers an approach to studying the community ecology of a targeted plant species. We analyzed the relationship between the dominance status and the degree of dominance to obtain the critical value when PJ changed from being the second dominant species to the first dominant species, thus providing a quantitative method to objectively analyze changes in a plant community. PJ is not only distributed in China but also occurs in regions of Northeast Asia. Furthermore, many other tree species are growing in forests with PJ trees. Accordingly, the research ideas and methods from this study can provide a valuable reference for studying the PJ forests outside of China, as well as the community ecology of other tree species.

The forests of Northeast China play a fundamental role in timber reserves, biodiversity conservation, mitigating climate change, and ecological functions [76], so protecting these forests is of great significance. This study provides a scientific basis for the conservation of PJ forests. For instance, in this study, PK and PU co-appeared as the first dominant species in the three groups where PJ was not the first dominant species, and they also served as the second dominant species in the group where PJ was the first dominant species. That is, the status of PJ in communities dominated by these two species changed from no advantage to a slight advantage to a strong advantage and to replacing these species. The establishment of the PJ community consisted of replacing the first dominant species with PJ, providing potential paths for restoring PJ forests.

Furthermore, the derived elevation distribution provides location information for the protection or restoration site selection of PJ forest tracts. The biotic (such as species composition) and abiotic (such as slope and elevation) factors reported on here can provide environmental information for restoring PJ forest. The critical value threshold is helpful for judging the attributes and status of forest communities. Moreover, this study's findings also provide comparable information for ecological research and conservation of PJ forests in other areas in its natural range across Northeast Asia.

5. Conclusions

Based on its ranking and dominance characteristics, *Picea jezoensis* (PJ) exhibited differing states of dominance in different forests surveyed. By comparing the forests containing PJ trees, we find that the PJ community occupies gentle sites at high elevations, and that tree species co-occurring with PJ mainly included *Betula costata* and *Abies nephrolepis*. The linear relationship between temporary stability (TS) and the relative basal area (RBA) of PJ explained its dominance status shift in the local community. To facilitate greater comprehensive understanding, studying the dynamics of more PJ populations in broader areas during long-term surveys, while also considering more soil, geographical, and biotic factors, is strongly suggested.

Supplementary Materials: The following supporting information can be downloaded at: https://www.mdpi.com/article/10.3390/d16120731/s1, Table S1: Location of plots in this study; Table S2: First two dominant species and relative basal area (RBA) values in plots where *Picea jezoensis* (PJ) was the first dominant species; Table S3: First two dominant species and relative basal area (RBA) values in plots where *Picea jezoensis* (PJ) was the second dominant species; Table S4: First two dominant species and relative basal area (RBA) values in plots where *Picea jezoensis* (PJ) was a co-dominant but not ranked among the top two dominant species; Table S5: First two dominant species and relative basal area (RBA) values in plots where *Picea jezoensis* (PJ) was a non-dominant species.

Author Contributions: Conceptualization, D.K.; methodology, D.K.; formal analysis, J.D., Z.J. and D.K.; investigation, S.G. and Y.L.; writing—original draft preparation, J.D.; writing—review and editing, D.K.; supervision, J.L.; funding acquisition, J.D. and D.K. All authors have read and agreed to the published version of the manuscript.

Funding: This research was funded by the Training Program of Innovation and Entrepreneurship for Undergraduates of Beijing Forestry University (grant number X202410022007), and the technical consulting project of Beijing Forestry University (grant number 2023HXFWBH-KDW-01).

Institutional Review Board Statement: Not applicable.

Data Availability Statement: The original contributions presented in the study are included in the article/Supplementary Materials, further inquiries can be directed to the corresponding author.

Conflicts of Interest: The authors declare no conflicts of interest.

References

1. Shang, Y. *General Ecology*, 3rd ed.; Peking University Press: Beijing, China, 2010; pp. 338–362.
2. Li, J.; Niu, S.; Liu, Y. *Forest Ecology*, 3rd ed.; Higher Education Press: Beijing, China, 2017; pp. 17–32, 95–118, 208–254, 312–351.
3. Niu, C.; Lou, A.; Sun, R.; Li, Q. *Foundations in Ecology*, 4th ed.; Higher Education Press: Beijing, China, 2023; pp. 72–100, 160–201.
4. He, J.N.; Ning, C.W.; Zhang, W.T.; ÜmÜt, H.; Shen, Z.H. Altitudinal patterns of population structure dynamics of *Picea schrenkiana* in the northern slope of the eastern Tianshan Mountains. *Acta Ecol. Sin.* **2024**, *44*, 5389–5398.
5. Li, J.Y.; Chen, X.X.; Li, M.W.; Zhang, N.N.; Zhou, Z.Q.; Li, X.M.; Ding, J.L.; He, Q.H.; Shi, F.S. Community composition and structure under different restoration pathways in subalpine forest, Sichuan Province, China. *Acta Ecol. Sin.* **2024**, *44*, 3471–3482.
6. Niu, Y.D.; Cai, T.J. Changes in species diversity and influencing factors in secondary forest succession in northern Da Hinggan Mountains. *Chin. J. Plant Ecol.* **2024**, *48*, 349–363.
7. Tian, Z.; Wang, Y.; Duan, W.; Wang, Z.; Yang, W.; Chen, L. Study on spatial structure characteristics of broadleaved *Pinus koraiensis* forest at different succession stages in the Xiaoxing'an Mountains. *J. Southwest For. Univ. (Nat. Sci.)* **2024**, *44*, 130–137.
8. Sun, Z.; Zhou, D. *Ecology Research Methods*; Science Press: Beijing, China, 2010; pp. 120–131.
9. Li, J.; Li, X.Y.; Cao, J.; Tan, L.Z.; Zhao, X.H. Community structure and dynamics of secondary coniferous and broad-leaved mixed forest in Changbai Mountains. *Acta Ecol. Sin.* **2020**, *40*, 1195–1206.
10. Zhang, W.W.; Yin, D.H.; Lei, Y.T.; Han, X.T.; Peng, J.Y.; Zhao, Z. Structural dynamics and succession simulation of the dominant populations in three types of *Quercus acutissima* community in the southern Loess Plateau. *Acta Ecol. Sin.* **2024**, *44*, 2572–2581.
11. Ouyang, Z.Y.; Li, Z.X.; Liu, J.G.; An, L.; Zhang, H.M.; Tan, Y.C.; Zhou, S.Q. The recovery processes of giant panda habitat in Wolong Nature Reserve, Sichuan, China. *Acta Ecol. Sin.* **2002**, *22*, 1840–1849.
12. Kang, D.; Lv, J.; Li, S.; Chen, X.; Wang, X.; Li, J. Community change characteristics of *Abies faxoniana*: A case study in the Wanglang Nature Reserve. *Ecol. Indic.* **2019**, *107*, 105594. [CrossRef]
13. Lu, J.Y.; Zhang, C.Y.; Zhao, X.H. Relationship between forest strata structure and topography of confierous and broad-leaved mixed forest in Jiaohe City, Jilin Province. *Acta Ecol. Sin.* **2021**, *41*, 6613–6620.
14. Tong, Y.W.; Qu, L.L.; Fu, Q.X.; Chen, Y.B.; Xiang, X.Y.; Zhu, W.D.; Qi, G.; Dai, L.M. Species diversity of forest plant communities on the southern slope of the Dabie Mountains and its relationship with altitude factors. *Acta Ecol. Sin.* **2024**, *44*, 5307–5317.
15. Jia, Z.; Ge, S.; Li, Y.; Kang, D.; Li, J. Relationship between tree richness and temporary stability of plant communities: A case study of a forest in Northeast China. *Forests* **2021**, *12*, 1756. [CrossRef]
16. Kang, D.; Li, J. Temporary stability characteristics and influencing factors of *Abies faxoniana*-dominated communities in the Wanglang Nature Reserve. *Forests* **2022**, *13*, 1186. [CrossRef]
17. Liu, Q.; Yin, H.J.; Wu, Y. A study on the structure of *Picea asperata* comunity of Miyaluo subalpine in Western Sichuan, China. *J. Mt. Sci.* **2003**, *21*, 695–701.
18. Yang, X.; Li, J.; Fan, N.; Wang, Y.; Zhang, Z. Effects of gap size on natural regeneration in *Picea asperata* forests of Northern China. *Forests* **2023**, *14*, 2102. [CrossRef]
19. Zhao, Z.; Kang, D.; Guo, W.; Zhao, L.; Cui, L.; Li, J. Climate sensitivity of purple cone spruce (*Picea purpurea*) across an altitudinal gradient on the eastern Tibetan Plateau. *Dendrochronologia* **2019**, *56*, 125586. [CrossRef]

20. Zhao, Y.; Cao, J.; Li, B.; Qi, R.; Liu, T.; Chen, X.; Gao, B.; Zhao, Y.; Liu, J.; Cao, X. Niche of woody plant populations of *Picea purpurea* community in Dayugou forest area, Taohe Nature Reserve, Gansu Province. *Acta Ecol. Sin.* **2022**, *42*, 1865–1875.
21. He, Z.; Zhao, W.; Liu, H.; Zhang, Z. Successional process of *Picea crassifolia* forest after logging disturbance in semiarid mountains: A case study in the Qilian Mountains, northwestern China. *For. Ecol. Manag.* **2010**, *260*, 396–402. [CrossRef]
22. Zhao, W.J.; Liu, X.D.; Jing, W.M.; Xu, L.H.; Niu, Y.; Qi, P.; Zhao, Y.H. Spatial heterogeneity of community structure of *Picea crassifolia* forest in Qilian Mountains, China. *Chin. J. Appl. Ecol.* **2015**, *26*, 2591–2599.
23. Aizawa, M.; Yoshimaru, H.; Saito, H.; Katsuki, T.; Kawahara, T.; Kitamura, K.; Shi, F.; Sabirov, R.; Kaji, M. Range-wide genetic structure in a north-east Asian spruce (*Picea jezoensis*) determined using nuclear microsatellite markers. *J. Biogeogr.* **2009**, *36*, 996–1007. [CrossRef]
24. Thomas, P.; Zhang, D.; Katsuki, T.; Rushforth, K. *Picea jezoensis*. The IUCN Red List of Threatened Species 2013: e.T42325A2972665. Available online: https://www.iucnredlist.org/species/42325/2972665 (accessed on 16 May 2024).
25. Nakagawa, M.; Kurahashi, A.; Hogetsu, T. The regeneration characteristics of *Picea jezoensis* and *Abies sachalinensis* on cut stumps in the sub-boreal forests of Hokkaido Tokyo University Forest. *For. Ecol. Manag.* **2003**, *180*, 353–359. [CrossRef]
26. Iijima, H.; Shibuya, M. Evaluation of suitable conditions for natural regeneration of *Picea jezoensis* on fallen logs. *J. For. Res.* **2010**, *15*, 46–54. [CrossRef]
27. Yano, K.; Shibuya, M. Site preference and occurrence patterns of *Picea jezoensis* and *Abies sachalinensis* on decayed logs in natural coniferous forests in Hokkaido, northern Japan. *J. For. Res.* **2010**, *15*, 108–114. [CrossRef]
28. Han, A.R.; Kim, H.J.; Jung, J.B.; Park, P.S. Seed germination and initial seedling survival of the subalpine tree species, *Picea jezoensis*, on different forest floor substrates under elevated temperature. *For. Ecol. Manag.* **2018**, *429*, 579–588. [CrossRef]
29. Han, A.R.; Lee, S.K.; Suh, G.U.; Park, Y.; Park, P.S. Wind and topography influence the crown growth of *Picea jezoensis* in a subalpine forest on Mt. Deogyu, Korea. *Agric. For. Meteorol.* **2012**, *166–167*, 207–214. [CrossRef]
30. Jang, W.; Keyes, C.R.; Running, S.W.; Lim, J.H.; Park, P.S. Climate growth relationships of relict *Picea jezoensis* at Mt. Gyebang, South Korea. *For. Sci. Technol.* **2015**, *11*, 19–26.
31. Ukhvatkina, O.; Omelko, A.; Zhmerenetsky, A. Local topography has significant impact on dendroclimatic response of *Picea jezoensis* and determines variation of factors limiting its radial growth in the Southern Sikhote-Alin. *Forests* **2023**, *14*, 2050. [CrossRef]
32. Ukhvatkina, O.N.; Omelko, A.M.; Zhmerenetsky, A.A. Changes in the dendroclimatic response of the *Picea jezoensis* (Siebold & Zucc.) Carriere along altitudinal gradient in the southern Sikhote-Alin. *Contemp. Probl. Ecol.* **2023**, *16*, 745–757.
33. Jang, W.; Park, P.S. Stand structure and maintenance of *Picea jezoensis* in a northern temperate forest, South Korea. *J. Plant Biol.* **2010**, *53*, 180–189. [CrossRef]
34. Park, B.J.; Byeon, J.G.; Heo, T.I.; Cheon, K.; Yang, J.C.; Oh, S.H. Comparison of species composition among *Picea jezoensis* (Siebold & Zucc.) carrière forests in Northeast Asia (from China to South Korea). *J. Asia-Pac. Biodivers.* **2023**, *16*, 272e281.
35. Park, C.; Byeon, J.; Park, J.; Lee, S.; Park, B.; Kim, J.; Woo, H. Impact of climate change on species distribution of *Picea jezoensis* Carrière in Baekdudaegan using ensemble modeling. *Sens. Mater.* **2022**, *34*, 4629–4637. [CrossRef]
36. Yu, D.; Wang, Q.; Wang, G.G.; Dai, L. Dendroclimatic response of *Picea jezoensis* along an altitudinal gradient in Changbai Mountains. *Sci. China Ser. E Technol. Sci.* **2006**, *49*, 150–159. [CrossRef]
37. Li, G.Q.; Bai, F.; Sang, W.G. Different responses of radial growth to climate warming in *Pinus koraiensis* and *Picea jezoensis* var. *komarovii* at their upper elevational limits in Changbai Mountain, China. *Chin. J. Plant Ecol.* **2011**, *35*, 500–511.
38. Wang, S.L.; Zhao, Y.; Gai, X.R.; Wang, X.Y.; Yu, D.P.; Zhou, W.M.; Zhou, L.; Dai, L.M. Response of radial growth of *Picea jezoensis* var. *komarovii* to climate factors along an altitudinal gradient on Changbai Mountain, Northeast China. *Chin. J. Ecol.* **2017**, *36*, 3131–3137.
39. Yu, J.; Liu, Q.J.; Zhou, G.; Meng, S.W.; Zhou, H.; Xu, Z.Z.; Shi, J.N.; Du, W.X. Response of radial growth of *Pinus koraiensis* and *Picea jezoensis* to climate change in Xiaoxing'anling Mountains, Northeast China. *Chin. J. Appl. Ecol.* **2017**, *28*, 3451–3460.
40. Zhu, L.; Cooper, D.J.; Yang, J.; Zhang, X.; Wang, X. Rapid warming induces the contrasting growth of Yezo spruce (*Picea jezoensis* var. *microsperma*) at two elevation gradient sites of northeast China. *Dendrochronologia* **2018**, *50*, 52–63.
41. Gong, X.; Xie, R.; Yang, H. Response of radial growth of three common tree species to climate change in a spruce-fir mixed stand in Changbai Mountain of northeastern China. *J. Beijing For. Univ.* **2023**, *45*, 1–10.
42. Chen, L. Preliminary study on the structure of *Picea jezoensis* forest on the southwest slope of Changbai Mountain. *Ser. Plant Ecol. Geobot.* **1963**, *1*, 69–80.
43. Shao, B.; Deng, K.M. The composition and important value of plant species in subalpine spruce-fir forest on the northern slope of Changbai Mts. *J. Nat. Resour.* **2000**, *15*, 66–73.
44. Jiang, P.; Zhao, G.; Ye, J.; Cui, G.; Deng, H. Structure of forest communities on the northern slope of Changbai Mountain and its variation along elevation gradients. *Chin. J. Ecol.* **2003**, *22*, 28–32.
45. Lei, X.; Lu, Y.; Peng, C.; Zhang, X.; Chang, J.; Hong, L. Growth and structure development of semi-natural larch-spruce-fir (*Larix olgensis–Picea jezoensis–Abies nephrolepis*) forests in northeast China: 12-year results after thinning. *For. Ecol. Manag.* **2007**, *240*, 165–177. [CrossRef]
46. Chen, K.Y.; Zhang, H.R.; Zhang, B.; He, Y.J. Population structure and dynamic characteristics of typical constructive species in natural secondary forest on the northern slope of Changbai Mountain. *Acta Ecol. Sin.* **2021**, *41*, 5142–5152.
47. Deng, K.M.; Shao, B.; Li, F. Analysis for diameter height structures and growth of Spruce-fir forest on northern slope of Changbai Mountains. *Resour. Sci.* **1999**, *21*, 77–84.

48. Yu, Z.L.; Yu, G.R.; Wang, Q.F.; Zhao, S.D.; Yan, X.D. Canopy gap characteristics and their influences on the regeneration of broad-leaved *Pinus Koriensis* forests in Changbai Mountain. *Resour. Sci.* **2001**, *23*, 64–68.
49. Qin, X. Growth response of main tree species to climate change and disturbance in an ecotone on Changbai Mountains. *J. Northeast. For. Univ.* **2015**, *43*, 42–45.
50. He, D.N.; Yang, H.; Wen, J.; Xie, R. Density and spatial distribution of seedlings and saplings in different gap sizes of a spruce-fir mixed stand in Changbai Mountains, China. *Chin. J. Appl. Ecol.* **2020**, *31*, 1916–1922.
51. Bu, R.C.; Chang, Y.; Hu, Y.M.; Li, X.Z.; He, H.S. Sensitivity of coniferous trees to environmental factors at different scales in the Small Xing'an Mountains, China. *J. Plant Ecol.* **2008**, *32*, 80–87.
52. Liu, Q.; Bi, L.; Song, G.; Wang, Q.; Jin, G. Species-habitat associations in an old-growth temperate forest in northeastern China. *BMC Ecol.* **2018**, *18*, 20. [CrossRef] [PubMed]
53. Zhang, P.Y.; Liang, Y.; Ma, T.X.; Liu, B.; Wu, M.M. The migration of main tree species lags behind climate change in Changbai Mountains. *Chin. J. Ecol.* **2022**, *41*, 1674–1682.
54. Antu County Government. 2022. Available online: http://www.antu.gov.cn/xq/dldm/202204/t20220414_377913.html (accessed on 24 May 2024).
55. Fusong County Government. 2024. Available online: http://www.fusong.gov.cn/lyfs/fsgk/zrdl/201706/t20170612_96794.html (accessed on 24 May 2024).
56. Changbai County Government. 2023. Available online: http://www.changbai.gov.cn/cbgk/zrzy/201801/t20180108_257152.html (accessed on 24 May 2024).
57. Fang, J.; Wang, X.; Shen, Z.; Tang, Z.; He, J.; Yu, D.; Jiang, Y.; Wang, Z.; Zheng, C.; Zhu, J.; et al. Methods and protocols for plant community inventory. *Biodivers. Sci.* **2009**, *17*, 533–548.
58. Yan, H.; Xu, Y.; Zhao, N.; Yang, S.; Wang, T. *Medical Statistics*, 3rd ed.; People's Medical Publishing House: Beijing, China, 2015; pp. 444–457.
59. Xie, L.H.; Shang, T. *SPSS Statistical Analysis and Data Mining*; Publishing House of Electronics Industry: Beijing, China, 2012; pp. 163–188.
60. Backhaus, K.; Erichson, B.; Plinke, W.; Wang, X.; Weiber, R. *Multivariate Statistical Analysis: Using SPSS Tools*; Truth and Wisdom Press: Shanghai, China; Shanghai People's Publishing House: Shanghai, China, 2009; pp. 35–80.
61. Ni, R.Q.; Tang, J.Y.; Cheng, Y.X.; Zhang, C.Y.; Chen, B.B.; He, H.J.; Xia, F.C. Spatial distribution patterns and associations of main tree species in spruce-fir forest in Changbai Mountains, northeastern China. *J. Beijing For. Univ.* **2013**, *35*, 28–35.
62. Qiang, Y.Q.; Fan, C.Y.; Zhang, C.Y. Species diversity maintenance mechanism of dark coniferous forests community in Changbai Mountain. *Acta Ecol. Sin.* **2023**, *43*, 1884–1891.
63. Liu, W.; Cao, W. Niche characteristics of main plant species in spruce-fir forests in Changbai Mountains. *Chin. J. Ecol.* **2011**, *30*, 1766–1774.
64. Pimm, S.L. The complexity and stability of ecosystems. *Nature* **1984**, *307*, 321–326. [CrossRef]
65. Ives, A.R. Diversity and stability in ecological communities. In *Theoretical Ecology: Principles and Applications*, 3rd ed.; May, R.; Mclean, A.; Tao, Y.; Wang, B., Translators; Higher Education Press: Beijing, China, 2010; pp. 116–129.
66. McCann, K.S. The diversity–stability debate. *Nature* **2000**, *405*, 228–233. [CrossRef] [PubMed]
67. Zhang, J.H.; Huang, Y.M. Biodiversity and stability mechanisms: Understanding and future research. *Acta Ecol. Sin.* **2016**, *36*, 3859–3870.
68. Guo, Q.Q.; Zhang, W.H.; Cao, X.P. Establishment of an evaluation model of the forest community stability based on fuzzy synthetic evaluation: A case study of main forest communities in Huanglong Mountains. *Sci. Silvae Sin.* **2009**, *45*, 19–24.
69. Song, W.; Zhang, H. Research status and hot spot analysis of forest ecosystem stability based on CNKI. *For. Sci. Technol.* **2022**, *3*, 48–52.
70. Zhang, M.T.; Zhang, Q.; Kang, X.G.; Yang, Y.J.; Xu, G.; Zhang, L.X. Community stability for spruce-fir forest at different succession stages in Changbai Mountains, Northeast China. *Chin. J. Appl. Ecol.* **2015**, *26*, 1609–1616.
71. Zhou, M.L.; Zhang, Q.; Kang, X.G.; Guo, W.W.; Ji, L.; Li, C.F. Study on the stability of forest communities based on the spatial structure index. *Plant Sci. J.* **2016**, *34*, 724–733.
72. Zheng, Y.R. Comparison of methods for studying stability of forest community. *Sci. Silvae Sin.* **2000**, *36*, 28–32.
73. Zhang, M.X.; Wang, D.X.; Peng, S.L.; Huang, Y.K.; Zhang, G.G. Community stability analysis for the oak-pine mixed forest in Qinling Mountains. *Acta Ecol. Sin.* **2015**, *35*, 2564–2573.
74. Wang, H.Y.; Zhang, Y.J.; Jin, Y.H.; Xu, J.W.; Tao, Y.; He, H.S.; Gao, X.; Bai, Y.Y.; Hu, R.; Han, Y.Y. Vegetation damage, post-disaster change degree and driving factors in the cold temperate wind disaster area of Changbai Mountain, China. *Chin. J. Appl. Ecol.* **2019**, *30*, 1580–1588.
75. Yu, D.J.; Zheng, Y.G.; Sun, C.H.; Zhao, C.; Jin, Y.H.; Zhang, L.J.; Li, J.G.; Liu, L.J. Investigation and factors analysis of *Dendrolimus superans* outbreaks in Changbai Mountain National Nature Reserve. *For. Res.* **2022**, *35*, 103–111.
76. National Forestry and Grassland Administration. *China Forest Resources Report*; China Forestry Publishing House: Beijing, China, 2019; pp. 78–241.

diversity

MDPI

Article

Early Vegetation Recovery After the 2008–2009 Explosive Eruption of the Chaitén Volcano, Chile

Ricardo Moreno-Gonzalez [1,2,*], Iván A. Díaz [1], Duncan A. Christie [1,3,4] and Antonio Lara [1,3]

1 Instituto de Conservación, Biodiversidad y Territorio, Universidad Austral de Chile, Valdivia 5090000, Chile; ivan.diaz@docentes.uach.cl (I.A.D.); duncanchristieb@gmail.com (D.A.C.); antoniolara@uach.cl (A.L.)
2 Calahuala Cooperative, Valdivia 5090000, Chile
3 Center for Climate and Resilience Research (CR)2, Valdivia 5090000, Chile
4 Cape Horn International Center for Global Change Studies and Biocultural Conservation (CHIC), O'Higgins 310, Puerto Williams 6350000, Chile
* Correspondence: ricardo.moreno@uach.cl

Abstract: In May 2008, Chaitén volcano entered an eruptive process, leading to one of the world's largest eruptions in recent decades. The magnitude of tephra ejected by the eruption left different types of disturbances and caused diverse forms of environmental damage that were heterogeneously distributed across the surrounding area. We went to the field to assess the early vegetation responses a year after the eruption in September 2009. We evaluated the lateral-blast disturbance zone. We distributed a set of plots in three disturbed sites and one in an undisturbed site. In each of these sites, in a rectangular plot of 1000 m^2, we marked all standing trees, recording whether they were alive, resprouting, or dead. Additionally, in each site of 80 small plots (~4 m^2), we tallied the regenerated plants, their coverage, and the log volume. We described whether the plant regeneration was occurring on a mineral or organic substrate (i.e., ash or leaf litter, respectively). In the blast zone, the eruption created a gradient of disturbance. Close to the crater, we found high levels of devastation marked by no surviving species, scarcely standing-dead trees and logs, and no tree regeneration. At the other extreme end of the disturbance zone, the trees with damaged crowns were resprouting, small plants were regrowing, and seedlings were more dispersed. The main form of regeneration was the resprouting of trunks or buried roots; additionally, a few seedlings were observed in the small plots and elsewhere in disturbed areas. The results suggest that the early stages of succession are shaped by life history traits like dispersion syndrome and regeneration strategy (i.e., vegetative), as was found after other volcanic eruptions. Likewise, the distribution of biological legacies, which is related to disturbance intensity, can cause certain species traits to thrive. For instance, in the blow-down zone, surviving species were chiefly those dispersed by the wind, while in the standing-dead zone, survivors were those dispersed by frugivorous birds. Additionally, we suggest that disturbance intensity variations are related to the elevation gradient. The varying intensities of disturbance further contribute to these ecological dynamics. The early succession in the blast zone of Chaitén volcano is influenced by the interaction between species-specific life history, altitudinal gradient, and biological legacies. Further studies are required to observe the current successional patterns that occur directly in the blast zone and compare these results with those obtained following other volcanic disturbances.

Keywords: 2008 Chaitén volcano; volcanic disturbance; vegetation dynamics; *Nothofagus* dynamics; South American temperate rainforest; early succession

1. Introduction

Volcanic eruptions are large-scale disturbances that modify the composition, structure, and ecosystem processes [1–3]. As volcanoes enter the eruptive phase, a complex mechanism of events is involved in multiple types of disturbances, such as explosive blasts, block and ash flows, thermal and toxic chemical waves, landslides, glowing avalanche deposits, debris flows, lava flows, and the air fall of volcanic tephra [1,4]. As a consequence, eruptions can have effects at very large spatial and temporal scales [5,6], as well as direct effects on the surrounding forest through burning, burying, and/or blowing down trees or other plants [4]. However, since this kind of large disturbance is infrequent [2] and has seldom been monitored [7], our understanding of vegetation responses remains limited. The effect of volcanism has been studied by comparing different stands in chronosequences, assuming that different forest patches reflect different successional stages after the disturbance, e.g., [8–11]. Despite the importance of volcanic eruptions, few studies have analyzed the effect of eruptions shortly after the event [4,7]. Notwithstanding this, important contributions have been made to research on plant community dynamics. These include studies conducted after the 1996 eruption of Mount Koma in northern Japan [12], the 1907 Ksudach volcano eruption in the Kamchatka Peninsula [13], the eruption of Mount Oyama in Miyake-Jima Island, Japan [14], and many others in the Northern Hemisphere. While in the Southern Hemisphere, related studies are largely concentrated in New Zealand [15,16].

Forest recovery is considerably slow following disturbances that heavily impact the soils as well as the aboveground vegetation [17]. Although volcanoes are considered catastrophic disturbance events, it is possible for diverse biological legacies to remain after the event. As observed after the eruptions of Mount St Helens, the persistence and heterogeneous distribution of these biological legacies can aid the recolonization of plants [2,18]. For example, the life history traits of the pre-disturbed community could mean that particular survivors are perhaps able to recolonize and expand under the new conditions presented by the disturbed environment [2,18]. Therefore, the arrangement and amount of such biological legacies could influence the rate and pathways of new colonizers. The recovery rate is negatively related to the distance from the undisturbed area [19].

One of the most active tectonic and volcanic areas in the world corresponds to the subduction area of the Nazca and the South American plate, e.g., [20], which also contains the longest and the second-highest mountain range in the world, the Andes. The volcanoes in the Chilean Andes are quite active. The Southern and Austral Volcanic Zones comprise 74 volcanoes that have been active since the post-glacial period (~17 ka) up to the present, and at least 21 have had one large explosive eruption (Volcanic Eruptive Index $\geq$ 5) [21]. The South American temperate rainforest (SATR) is located in the south of the Andes [22]. This cordillera plays a critical role in the SATR dynamics, where the current successional models after volcanic disturbances suggest that the *Nothofagus* are dominant pioneer trees that benefit from such disturbances [23]. The SATR represents a biogeographical island that is rich in endemic species and dominated by evergreen, broad-leaved species [24–26]. South of 40° S, extensive areas of old-growth forest dominate the mountainous landscape. These forests are not directly disturbed by modern human activities and represent one of the last pristine areas with pre-industrial biogeochemical conditions [27].

In May 2008, the last eruption of the Chaitén Volcano (42°50′ S) started near the small town of Chaitén in southern Chile (Figure 1, left panel). This eruption was the largest rhyolite eruption since the great eruption of Katmai Volcano in 1912 and the first rhyolite eruption for which some attributes of its dynamics and impacts were monitored [28]. The eruption consisted of an approximately 2-week-long explosive phase that generated as much as 1 km^3 bulk volume tephra (~0.3 km^3 dense rock equivalent), followed by an approximately 20-month-long effusive phase during which 0.8 km^3 of high-silica rhyolite lava erupted, forming a new dome within the volcano's caldera [28]. Chaiten Volcano was surrounded by extensive old-growth forests undisturbed by humans, providing a unique opportunity to evaluate the effects of volcanism over an undisturbed landscape.

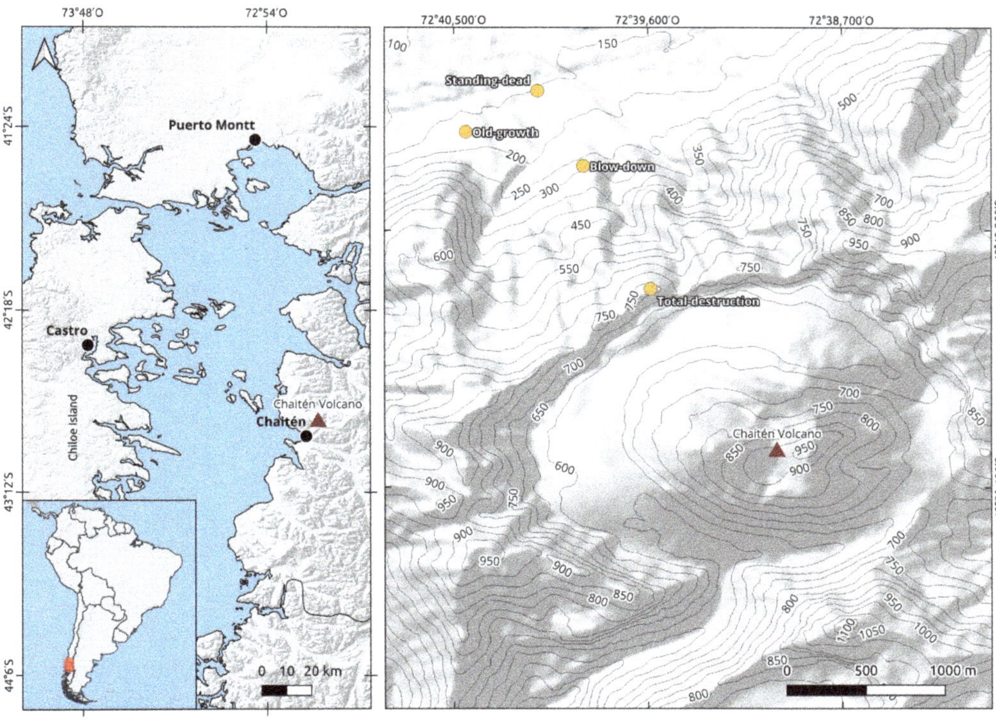

Figure 1. Map of the study site. (**Left panel**) indicates the position of Chaiten Volcano (red triangle) in relation to the main cities (black dots). (**Right panel**) shows a close-up of the volcano's blast zone, the plots' distribution along the disturbance/elevation gradient (total-destruction plot, blow-down plot, and standing-dead plot), and the undisturbed sector (old-growth plot).

In this study, we characterized the early establishment of vegetation after the eruption of Chaitén Volcano along a disturbance gradient one year after the eruption started, from near the crater to the closest old-growth forest with vegetation completely alive. We predict that plant species diversity and the number of living trees will decrease close to the crater, and pioneering *Nothofagus* species are colonizing the areas affected by the eruption. Our objectives included the creation of a baseline for further long-term monitoring of the vegetation that is recovering and to identify and analyze the importance of biological legacies for forest recovery.

2. Material and Methods

2.1. Study Site

Chaitén Volcano is located on the west side of the Andean Range in southern Chile, at 42°50′ S and 72°39′ W (Figure 1), and is part of the so-called Southern Volcanic Zones. Chaitén Volcano is a dome of ~1100 m above sea level, located 10 km NW of the town of Chaitén, populated by approximately 5000 inhabitants before the eruption (Figure 1). The climate of the area corresponds to a humid temperate with a strong oceanic influence [29]. The landscape is characterized by sharp mountains shaped by glaciers and valleys originating from fluvial and glacio-fluvial outwash [1]. Vegetation within the influence area of Chaiten Volcano is dominated by a dense old-growth temperate rainforest of northern Patagonian and Valdivian types [1,29–31]. These forests' compositions are dominated by broad-leaved evergreen trees such as *Nothofagus dombeyi* (Mirb.) Oerst. (Nothofagaceae), *Laureliopsis philippiana* (Looser) Schodde, *Gevuina avellana* (Molina), *Amomyrtus luma* (Molina) D. Legrand and Kausel, *Luma apiculata* (DC.) Burret, *Drymis winteri* J.R. Forst. and G. Forst., *Eucryphia cordifolia* Cav., and *Weinmannia trichosperma* Cav. (both Cunoniaceae). These species are normally arranged in multi-layered canopy strata and old, large emergent trees (up to c. 40 m height) and abundant snags and logs. The understory is densely covered by ferns, such as *Lophosoria quadripinnata* (J.F.Gmel.) C.Chr. and *Blechnum magellanicum* (Desv.) Mett., seedlings, saplings, and bamboo tickets (*Chusquea* spp.). Specific plant composition and structure previous to the disturbance are not available because it was an area largely unexplored.

2.2. Study Design

We carried out the fieldwork in September 2009. We first conducted a general inspection of the study area as suggested by del Moral [32]. Later, we selected four sites with different degrees of volcanic impact (Figure 1), following similar classifications of previous studies [32]. The first site corresponds to a forest with low volcanic impact that only received a few millimeters of tephra deposition. This site is called "Old-growth forest". It is located 5 km from the center of the new dome at an elevation of 600 m a.s.l. (Figure 1, right panel). The second site is characterized by the presence of abundant dead-standing trees and higher tephra deposition. This site is called the "Standing-dead" plot and is located 3.5 km from the center of the new dome at 160 m a.s.l. The third site was more affected by the eruption and is characterized by high tephra deposition, where most trees are broken or blown over by the eruption. This site is called "Blow down" and is located 2.3 km from the center of the new dome at an elevation of 300 m a.s.l. Finally, the fourth and most affected site is on the border of the crater at 1.5 km from the center of the new dome and at 750 m a.s.l. This site presented the higher tephra deposition, and all trees were blown down by the eruption and heavily buried under volcanic ash. This site is called "Total destruction".

In each one of these four areas was installed a 50 $\times$ 20 m permanent plot, where we measured the diameter at the breast height (dbh) of all trees $\geq$ 5 cm in diameter. Every tree was identified at the species level when possible and labeled with a numbered aluminum tag. We registered if the stems were alive, resprouting after the eruption, or snags (standing-dead stems). To estimate the presence and abundance of tree regeneration, we traced four transects 120 m long. The transects were distributed parallel, separated 20 m from each other. In each transect, we mounted five circular plots of 1.5 m radius (4.71 m^2) every 30 m. On these plots, we counted all seedlings and saplings present and recorded the substrate where they were rooted (forest soil, logs, or volcanic tephra). We registered other non-tree plant species in the circular plots, estimating their abundance as the relative cover, and also recorded if the substrate where they were rooted was organic or mineral (e.g., forest soil, logs, or volcanic tephra). We assessed the volume of all exposed logs within the plots. Assuming a cylindrical form of each log, we calculated the volume of logs present inside the plot and standardized its measurements to m^3 of logs per m^2 of plot surface.

2.3. Data Analysis

The species richness of tree seedlings, herbs, and shrubs was compared among sites using rarefaction analysis. Rarefaction curves allow for comparing species richness among environments, avoiding the bias in the number of species detected due to the increased probability of species detection in areas where individuals were more abundant, by plotting the number of species detected in the function of the number of individuals observed [33,34]. This procedure includes Monte Carlo simulations delivering an average value of species richness with a confidence level of 95% (Colwell 2006) [35]. Rarefaction analyses were conducted using the free software EstimateS Win 8 [35]). Non-vascular species were excluded from the rarefaction analyses since not all of them were classified up to the species level, and one genus could include several species. Plant composition among sites was compared using Sørensen's similarity index based on the presence/absence of plant species. Rarefaction and similitude analyses were both carried out with the software EstimateS Win 8 as well. In attempting to define if the plant composition of one site was a subset of the composition of another site, we conducted a nested analysis in NestCalc v.1 [36,37].

3. Results

3.1. General Description of the Effects of the Eruption

The eruption of Chaitén Volcano has an impressive effect on the forest and landscape (Figure 2). The eruption in the study area produced a gradient of destruction (Figure 2B,C). Around the crater, all the trees were blown down and heavily buried by tephra. Other trees were partially buried and/or broken in half, with the trunks covered with the impact marks of many little stones—this resembled damage caused by gunfire (Figure 2A,D). Moving away and down from the crater, trees were blown down and densely covered by ash (Figure 2E), while further away, trees were standing deadwood ('snag'), presenting progressively finer branches (Figure 2F). Many small patches of organic litter were exposed and sparsely distributed and protected from the direct effect of the eruption by root discs and large trunks or were exposed by small ravines that flush away the volcanic ash. These small patches looked like clumps of organic soil, with roots of ferns and other small plants that were not buried by the tephra. These clumps were covered by plants that survived in the area affected by the eruption, and their frequency increased when moving away from the crater toward the undisturbed forests (Figure 2G).

In the surrounding area, we did not observe evidence of fire caused by the eruption, but the trunk and bark of several trees looked charred within the influence area. Trees were dead and dry from the eruptive explosions, but some trees in the blow-down and standing-dead areas maintained bark on the opposite side of the crater, perhaps with possibilities to resprout hereafter. Tephra buried the fallen trees in the whole study area, but principally in the blow-down and total-destruction areas. Here, the tephra was very compacted, seeming like pavement that covered the fallen forests. Heterogeneous tephra deposition exposed some trunks, clumps, and remnants with organic forest soil that was likely from the original forest. Shrubs and epiphytes were spread everywhere in the standing-dead area. Ferns, bamboo, and tree seedlings were growing in remnants of exposed organic soil, usually below or at one side of trunks or unearthed roots. Surviving trees increased when moving away and down from the crater.

3.2. Species Composition

We found 10 species of trees in the area and 50 species of smaller plants, including seedlings, herbs, and ferns (Table 1). The rarefaction analysis for tree species showed that the standing-dead plot held one more species than the old-growth forest plot (Figure 3). This same analysis showed that the blow-down plot held very few individuals, but the initial slope and shape of the curve were similar to the other two curves, indicating that all curves showed a similar number of tree species. The rarefaction analysis of small vascular plants showed that the old-growth forest site had the richest species composition while decreasing in the standing-dead and blow-down plots (Figure 3). The total-destruction plot showed practically no species.

Figure 2. Photographic records of the study site. (**A**) showing impressive impacts close to the crater. Comparison of the volcanic dome and vegetation cover before (**B**) and during the 2008–2009 eruption (**C**). Red dot indicates the approximate position of the researcher (Dr. Díaz) in panel (**A**). Images (**D–G**) represent a disturbance condition where plots were established: (**D**) Total-destruction, (**E**) blow-down, (**F**) standing-dead, and (**G**) old-growth plots.

Table 1. List of plant species present in Chaitén Volcano and their frequency in the big plot and in 20 plots of 4.71 m^2 in the different studied areas.

Families	Species	Old-Growth	Standing-Dead	Blow-Down	Near Crater
SEEDLINGS					
Cunoneaceae	*Weinmannia trichosperma*		2		
	Caldcluvia paniculata	4	4		
Eucryphiaceae	*Eucryphia cordifolia*	5			
Flacourtiaceae	*Azara lanceolata*			1	
Monimiacae	*Laureliopsis philippiana*	6			
Podocarpaceae	*Podocarpus saligna*	1			
Winteraceae	*Drimys winteri*	3			
Proteaceae	*Embothrium coccineum*		1	4	
	Lomatia ferruginea	5			
	Gevuina avellana	1			
Myrtaceae	*Myrceugenia parviflora*	3			
	Myrceugenia planipes	12			
	Amomyrtus luma	16	4		
Araliacae	*Raukaua laetevirens*	3			
HERBS AND SMALL SHRUBS					
Gesneraceae	*Mitraria coccinea*	8			
	Asteranthera ovata	1	1		
	Campsidium valdivianum	5			
Poaceae	*Chusquea uliginosa*	7	1	1	
Apocinaceae	*Elytropus chilensis*	6	1		
Fitolaceae	*Ercilla syncarpellata*	3			
Saxifragaceae	*Ribes magellanicum*		1		
Bromeliaceae	*Greigia landbeckii*	4			
Griselineaceae	*Griselinia racemosa*	1			
	Griselinia ruscifolia	4	1	1	
Guneraceae	*Gunnera tinctorea*			1	
Rubiaceae	*Nertera granadensis*	1			
Hydrangeaceae	*Hydrangea serratifolia*	12			
Luzuriagaceae	*Luzuriaga poliphylla*	8	1	1	
	Luzuriaga radicans	14			
FERNS					
Dicksoniaceae	*Lophosoria quadripinnata*	6	4	3	
Hymenophylaceae	*Hymenophyllum tortuosum*			2	
	Hymenophyllum magellanico	2			
	Hymenoglossum cruentum	6			
	Hymenophyllum caudiculatum	3			
	Hymenophyllum cuneatum	2			
	Hymenophyllum dentatum	5			
	Hymenophyllum dicranotrichium	12			
	Hymenophyllum krauseanum	5			
	Hymenophyllum pectinatum	3			
	Hymenophyllum plicatum	3			
	Serpillopsis caespitosa	5			
Aspleniaceae	*Asplenium daeroides*	4			
	Pleurosorus papaverifolius	4		3	
Dryopteridaceae	*Megalastrum spectabile*	7			
Blechnaceae	*Blechnum mochaenum*	1			
	Blechnum arcuatum	1			

Table 1. *Cont.*

Families	Species	Old-Growth	Standing-Dead	Blow-Down	Near Crater
	Blechnum chilense	2	1	1	
	Blechnum magellanicum	1			
Dennstaedtiaceae	*Hypolepis rugosula*			1	
Total general		**201**	**22**	**20**	**0**

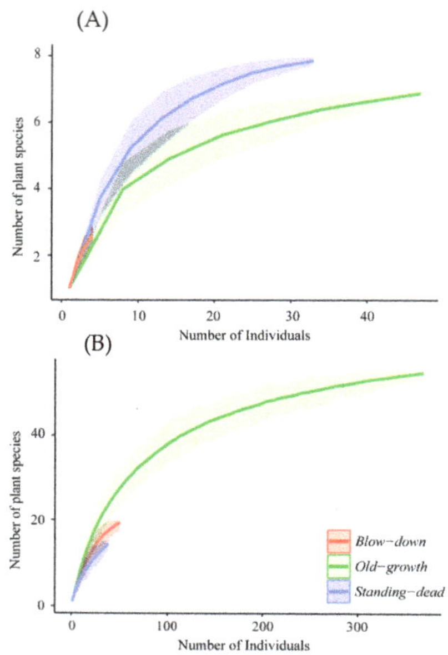

Figure 3. Rarefaction analysis shows the relationship between the number of tree seedling species (**A**) and small vascular plant species (**B**) as a function of the number of individuals. In both panels, the shaded area indicates one standard error. All sites affected by the eruption of Chaitén old-growth, standing-dead, and blow-down plots are compared, except for the total-destruction plot, which was not included because of the absence of plants.

Plant composition among old-growth and standing-dead plots showed some similarities (34%) but was less related to the blow-down (20%) and the total-destruction plots (0%). The standing-dead and blow-down plots showed the highest similarities in species composition (Table 2). In the disturbed plots, the recorded plant species were a subgroup of the species present in the old-growth forest plot (Nestedness calculator T = 13.85°). The most frequent plant was the fern *L. quadripinnata*, which survived the direct impact of the eruption in small clumps of organic matter that were protected by trunks or the root disc of trees and were not completely covered by tephra. Bamboo and ferns were sprouting from roots in the remnants of exposed forest soil.

Table 2. Sørensen similarity index for the plant composition among the different sites in the Chaitén influence area.

Site	Old-Growth	Standing-Dead	Blow-Down	Total-Destruction
Old-growth	1			
Standing-dead	0.342	1		
Blow-down	0.2	0.526	1	
Total-destruction	0	0	0.1	1

3.3. Forest Structure

The forest structure of the old-growth site showed individuals of all sizes, with several large emergent trees (Figure 4). Most trees in the old-growth plot were alive, and the snags most probably represent trees that died before the eruption, while in the other plots, most trees were killed by the eruption. The standing-dead plot showed a similar dbh distribution to the old-growth plot, with a higher frequency of small dbh. The blow-down plot showed a very small number of standing trees, concentrated in the intermediate size class, while the total-destruction plot practically had no individual standing trees (Figure 4).

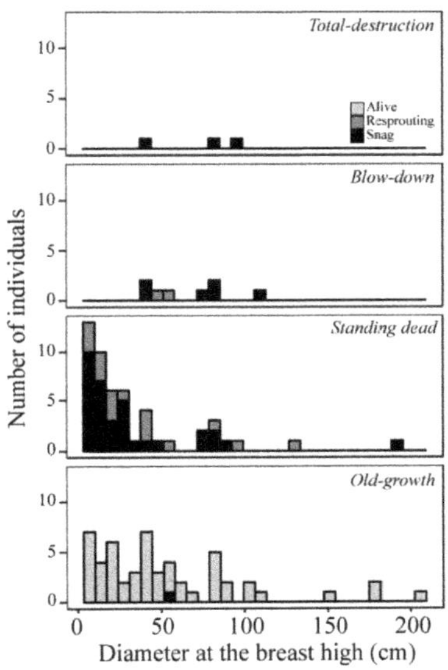

Figure 4. Diameter at the breast height (dbh) distribution for living and standing-dead trees in the study sites. Each panel represents a plot condition influenced by the eruption of Chaitén Volcano.

The basal area of the old-growth plot was dominated by *L. philippiana,* followed by *E. cordifolia* and *C. paniculata* (Table 3). In the standing-dead and blow-down plots *W. trichosperma* and *N. dombeyi* were the dominant components of the basal area, but also with a high number of unidentified trees (Table 3). We were not able to identify the species of the standing-dead trees in the total-destruction plot. The volume of fallen trees increased in the blow-down and standing-dead plots, but it was much lower in the old-growth and the total-destruction plots (Figure 5). In the total-destruction plot, most logs were heavily buried by a dense tephra layer.

Table 3. Total basal area (m^2) of tree species in the studied plots: Total-destruction, blow-down, standing-dead, and old-growth. Trees within plots were classified as alive, resprouting after the eruption, or snag when corresponding.

	Total-Destruction	Blow-Down		Standing-Dead		Old-Growth	
	Snag	Snag	Resprouting	Snag	Resprouting	Snag	Alive
Amomyrtus luma				0.02	0.34		0.47
Amomyrtus meli					0.06		
Caldcluvia paniculata							2.54
Drimys winterii				0.14			
Lomatia ferruginea				0.04	0.01		0.02
Myrceugenia planipes							0.84

Table 3. *Cont.*

	Total-Destruction	Blow-Down		Standing-Dead		Old-Growth	
	Snag	Snag	Resprouting	Snag	Resprouting	Snag	Alive
Eucryphia cordifolia		0.55		0.59	0.61	0.26	6.37
Laureliopsis philippiana		0.25		0.87			9.64
Nothofagus dombeyi				3.46			
Weinmannia trichosperma		0.95	0.41		2.38		0.71
Not identified	1.41	0.94		1.45			
Total	**1.41**	**2.69**	**0.41**	**6.57**	**3.39**	**0.26**	**20.59**

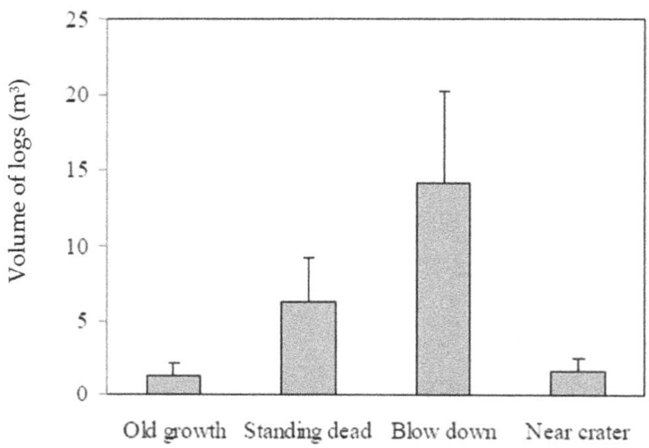

Figure 5. Volume of fallen dead trees (logs) in the study sites influenced by the eruption of Chaitén Volcano in the study sites.

3.4. Seedlings and Small Plants

We found a total of 90 seedlings from five trees and shrub species in the disturbed plots, while we registered 174 seedlings of 12 species in the old-growth plots. The standing-dead plots had 83 seedlings from five species, while the blow-down plots presented 7 seedlings from two species. We did not find any seedlings in the total-destruction plot. In the old-growth plots, seedlings were largely dominated by the shade-tolerant *A. luma, M. parviflora,* and *M. planipes*. In the standing-dead plots, seedlings were dominated by *C. paniculata* (48% of the total), but also present were the shade-tolerant tree *A. luma* (24%), and the pioneer tree species *W. trichosperma* and *E. coccineum* (18% and 8% respectively). In the blow-down plot, *E. coccineum,* a pioneer species, and *A. lanceolata* were the only recorded species.

We found different mechanisms of forest regeneration in the plots. In the old-growth plot, most regeneration was from seeds, while in the standing-dead plots, most regeneration came from resprouts and a few tree seedlings from seeds (Table 4). In the old-growth plots, most of the tree seedlings corresponded to species with fleshy fruits dispersed by birds, while in the standing-dead plots, most seeds were wind-dispersed. In the blow-down plots, all seedlings came from seeds (Table 4), but half of the seedlings corresponded to species dispersed by birds (e.g., *Ribes magellanicum*), and the other half were species dispersed by wind (e.g., *E. coccineum*). Second, in all plots, most seedlings were growing on organic substrata (Table 4), such as organic soil on the ground, root discs, and logs, while few seedlings were growing directly either on the tephra or on rocks. In the blow-down plot, we found 7 seedlings of *E. coccineum* growing just in remnants of organic litter; in the standing-dead plot, 11 seedlings were growing directly in the tephra; while the other 72 seedlings were resprouting directly from the standing tree's trunks. In the old-growth plots, we found just one species growing in an exposed rock while the rest of the seedlings were growing in organic substrates (Table 4).

Table 4. Number of seedlings per regeneration type and substratum type in the plots.

	Regeneration Type		Substratum Type	
	Resprouting	**Seed**	**Mineral**	**Organic**
Total-destruction	-	-	-	-
Blow-down	-	7	-	7
Standing-dead	74	9	11	72
Old-growth	31	127	1	157

Other small plant species in the plots studied (Table 1) showed differences along the disturbance gradient (Table 2). For example, the old-growth plot has species in all categories of plant cover, with few species covering over 75% of the ground and many species covering between 1–5% (Figure 6). In contrast, standing-dead and blow-down plots were covered by fewer species and concentrated in the lower ranges of coverage (Figure 6). In the total-destruction plot, we only found a non-vascular plant (*Marchantia* sp.) with a cover between 5–25% (Figure 6) growing on a log exposed to the surface. The dominant understory species in the old-growth forest were the bamboo *Chusquea* spp. and the fern *L. quadripinnata*. These species were also present in the standing-dead and blow-down plots, although they were scattered and with a low percent of coverage in the disturbed plots.

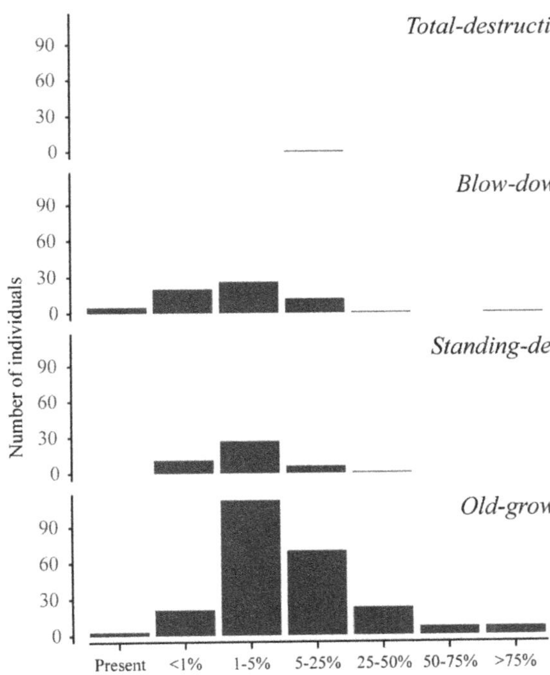

Figure 6. Number of individuals or colonies per relative coverage range at the ground level of the study areas affected by Chaitén Volcano, including vascular and non-vascular species. Panels correspond to the plots distributed in the different disturbance zones.

4. Discussion

4.1. The Relevance of Biological Legacies

Few studies have been conducted shortly after an eruption in the southern Andes; however, successional models predict that disturbed areas in low- and mid-altitude (<1000 m a.s.l.) forests would be dominated by *Nothofagus* species due to shade intolerance, e.g., [38]. In the present study, one year after the eruption, we did not register the *Nothofagus* species despite their abundance in the surrounding undisturbed areas. It is worth noting that the *Nothofagus* species have a cyclic seed production [39,40], which could have resulted in low seed dispersal in the preceding year, although due to the short-term and limited spatial extension of this study, we were unable to confirm this. However, our results suggest that regeneration in the blast zone occurs mainly due to the in situ development of plants associated with biological legacies such as logs, organic soil clumps, and individuals that survive the eruption. Other areas disturbed by Chaiten Volcano presented similar types and abundance of biological structures as well [1].

Similar patterns have been documented in other volcanic regions, too. For instance, studies on Mount St. Helens in the United States revealed that surviving vegetation and organic matter played pivotal roles in ecosystem recovery following the 1980 eruption. Franklin and collaborators [41] emphasized that biological legacies (including surviving plants and organic debris) provided critical refugia and sources for recolonization. This aligns with findings from the Chaitén and Calbuco volcanos in Chile [42], where the presence of biological legacies facilitated resprouting and regeneration.

However, the significance of biological legacies can vary depending on the eruption's intensity and the ecosystem's resilience. In some cases, severe eruptions may obliterate all organic matter and the biological legacies it provides, necessitating primary succession processes. For example, the 2008 eruption of Kasatochi Island in Alaska resulted in the complete removal of pre-existing vegetation and soil, leading to a primary succession dominated by colonizing species, e.g., [43]. This contrasts with Chaitén, where the preservation of biological legacies enabled, in some areas, secondary succession through resprouting. As volcanic eruptions are highly variable, this study, like others, underlines [44,45] the importance of identifying the disturbance type and its internal variability to understand successional processes.

4.2. Resprouting and Seed Dispersal Mechanism

The initial establishment of tree species, besides the disturbance gradient, may be influenced by the interaction between the life history of the species and the altitudinal gradient. Life history attributes, such as seed dispersal, flowering phenology, growth form, resprouting ability, or light-temperature stress resistance, are all mechanisms that influence the recovery response of post-disturbance vegetation [46–50]. In the disturbed plots, the principal regeneration mechanism was the resprouting capability of species (Tables 1 and 4). This is a frequent strategy in many plant species in response to disturbances of different types and intensities [51,52].

Seed dispersion syndrome is another important mechanism that influences colonization after disturbance. Valdivian temperate rainforests have one of the highest proportions of plants dispersed by frugivorous birds when compared to other temperate rainforests [53,54]. Birds consume a high proportion of fruits and may disperse seeds to specific places, such as to a perch [55,56]. Seed availability might increase in sites where remnant snags, shrubs, or perching trees are available, e.g., [57]. Among the wind-dispersed seeds, we only found seedlings of *E. coccineum* in disturbed plots. While this tree has adapted to stressful conditions, the seed is able to disperse for long distances [58].

The importance of animal-mediated seed dispersal in post-eruption recovery has been observed in other contexts than in Chaiten. On Surtsey Island, Iceland, birds contributed to the spread of plant species by transporting seeds, enhancing plant colonization rates [59]. Similarly, studies on Mount St. Helens reported that birds and mammals facilitated seed dispersal, aiding vegetation recovery [60]. Evidence from past volcanic disturbance (Sollipulli-Alpehue, ~3 ka BP) suggests that in semi-arid conditions, zoochory might play an important role, as well, in the spread of *Ephedra* after burying a huge area into the steppe [44].

On the contrary, in other volcanic landscapes, wind-dispersed species played a more prominent role in early succession. For instance, following the eruption of Mount Usu in Japan, wind-dispersed herbaceous species rapidly colonized the volcanic deposits, initiating succession processes [61]. This suggests that the relative importance of seed dispersal mechanisms can vary based on species traits and environmental conditions.

4.3. Environmental Gradients and Recovery

For tree species, the standing-dead plot had more species than the old-growth plot, while the reverse was true for non-tree species, revealing the effect of the eruption on the small plant species. In the blow-down and total-destruction plots, i.e., from mid to high disturbance intensity at mid to high altitudes and without/or sparse biological legacies, we found low species richness and coverage and sparse tree regeneration. This mosaic of legacies in the landscape offers opportunities for the recolonization of many different species in the affected area. The different species compositions would promote different pathways to early succession with different communities but also to different recovery rates. We propose that in low- to mid-elevation areas of Chaiten Volcano, the vegetation recovery will be faster, dominated by species that are shade-tolerant to semi-tolerant and dispersed by birds, aided by corresponding biological legacies. However, we can expect that at higher altitudes, vegetation recovery would be slower (due, in part, to lower temperatures) with recolonization made by shade-intolerant plants whose seed is dispersed by winds.

Environmental gradients, such as altitude and disturbance intensity, are well-known to influence successional trajectories. Research on Mount St. Helens demonstrated that elevation and disturbance severely affected species composition and recovery rates, with lower elevations supporting more rapid vegetation establishment [60]. Similarly, studies in Hawaii's volcanic landscapes revealed that soil development and nutrient availability along altitudinal gradients significantly impacted plant succession [62]. However, the interplay between environmental gradients and succession is complex and context-dependent. In some cases, higher elevations may experience faster recovery due to favorable microclimatic conditions or proximity to seed sources. For example, research on the 1977–1978 eruptions of Mount Usu indicated that certain high-altitude areas exhibited rapid colonization by pioneer species, which were facilitated by wind dispersal and suitable substrates [61]. Our study further highlights the need to consider local factors when assessing successional dynamics across environmental gradients.

5. Conclusions

The 2008 eruption of Chaitén Volcano provided a unique opportunity to study ecological succession in response to volcanic disturbances. The primary findings from this study highlight the critical role of biological legacies and resprouting in early vegetation recovery in the Andean temperate forests.

This study observed that areas near the crater exhibited high devastation, with no surviving species or tree regeneration after the first year. In contrast, less affected zones permitted the resprouting of damaged trees and the regeneration of small plants and seedlings.

This pattern underscores the importance of biological legacies—such as surviving plant material and organic substrates—in facilitating ecosystem recovery. In the blast zone of the volcano, resprouting from trunks and buried roots emerged as the dominant regeneration strategy, while seedling establishment was limited.

This study highlights the role of frugivorous birds in seed dispersal, facilitated by remnant structures such as snags and perches. This interaction may be promoting the establishment of fleshy-fruited species in disturbed areas. In contrast, wind-dispersed species were less prevalent, except for certain (unexpected) pioneer species that had adapted to stressful conditions.

Vegetation recovery was faster in low-altitude, less-disturbed areas dominated by shade- and semi-tolerant species, while higher altitudes and severely disturbed zones exhibited slower recovery—colonization by shade-intolerant and wind-dispersed species was more important under this condition, although scarce.

Author Contributions: Conceptualization, R.M.-G., I.A.D., D.A.C. and A.L.; methodology, R.M.-G., I.A.D. and D.A.C.; formal analysis, R.M.-G. and I.A.D.; resources, I.A.D., D.A.C. and A.L.; data curation, R.M-G.; writing—original draft preparation, R.M.-G. and I.A.D.; writing—review and editing, R.M.-G., I.A.D., D.A.C. and A.L.; visualization, R.M.-G.; project administration, I.A.D., D.A.C. and A.L.; funding acquisition, I.A.D., D.A.C. and A.L. All authors have read and agreed to the published version of the manuscript.

Funding: The fieldwork was financed by CONICYT PDA-24. A.L. and D.C. acknowledge support from grant CONICYT/FONDAP/1523A0002, and D.C. from CHIC ANID/BASAL FB210018. R.M-G. acknowledges the scholarship DAAD/Becas Chile 2014 during the mid stages of this manuscript and Fundación Mar Adentro during the last stages.

Institutional Review Board Statement: Not applicable.

Data Availability Statement: Data will be available upon request.

Acknowledgments: Our thanks to Julio Gerding for preparing the map and, Pierre Gilbert and Rafael Coopman for their contribution during the fieldwork. Special thanks go to George Swan for revising the text and its English edition.

Conflicts of Interest: The authors declare no conflicts of interest.

References

1. Swanson, F.J.; Jones, J.A.; Crisafulli, C.M.; Lara, A. Effects of volcanic and hydrologic processes on forest vegetation: Chaitén Volcano, Chile. *Andean Geol.* **2013**, *40*, 359–391. [CrossRef]
2. Turner, M.G.; Baker, W.L.; Peterson, C.J.; Peet, R.K. Factors influencing succession: Lessons from large, infrequent natural disturbances. *Ecosystems* **1998**, *1*, 511–523. [CrossRef]
3. Turner, M.G.; Dale, V.H.; Everham, E. Fires, hurricanes, and volcanoes: Comparing large disturbances. *Bioscience* **1997**, *47*, 758–768. [CrossRef]
4. Dale, V.H.; Swanson, F.J.; Crisafulli, C.M. Ecological perspectives on management of the mount St. Helens landscape. In *Ecological Responses to the 1980 Eruption of Mount St Helens*; Dale, V.H., Swanson, F.J., Crisafulli, C.M., Eds.; Springer: New York, NY, USA, 2005; pp. 277–286. ISBN 978-0-387-23868-5.
5. Robock, A.; Oppenheimer, C. Volcanoes and Earth's Atmosphere. *Geophys. Monogr.* **2003**, 139. [CrossRef]
6. Peet, R.K. Community structure and ecosystem function. In *Plant Succession: Theory and Prediction*; Glenn-Lewin, D., Peet, R.K., Veblen, T.T., Eds.; Chapman & Hall: London, UK, 1992; pp. 103–151.
7. Crisafulli, C.M.; Dale, V.H. *Ecological Responses at Mount St. Helens: Revisited 35 Years After the 1980 Eruption*; Springer: Berlin/Heidelberg, Germany, 2018; ISBN 9781493974511.
8. Franklin, J.F.; Swanson, M.E. Long-term ecological research in the forests of the United States: Key lessons for its application in Chile and around the world. *Rev. Chil. Hist. Nat.* **2010**, *83*, 185–194. [CrossRef]
9. Drury, W.H.; Nisbet, I.C. Succession. *J. Arnolod Arbor.* **1973**, *54*, 331–368. [CrossRef]
10. Pickett, S.T.A. Space-for-time substitution as an alternative to long term studies. In *Long-Term Studies in Ecology*; Likens, G.E., Ed.; Springer: New York, NY, USA, 1988; pp. 110–135.

11. Walker, L.R.; Wardle, D.A.; Bardgett, R.D.; Clarkson, B.D. The use of chronosequences in studies of ecological succession and soil development. *J. Ecol.* **2010**, *98*, 725–736. [CrossRef]
12. Tsuyuzaki, S.; Hase, A. Plant community dynamics on the volcano Mount Koma, northern Japan, after the 1996 Eruption. *Folia Geobot.* **2005**, *40*, 319–330. [CrossRef]
13. Grishin, S.Y.; del Moral, R.; Krestov, P.V.; Verkholat, V. Succession following the catastrophic eruption of Ksudach (Kamchattka, 1907). *Vegetatio* **1996**, *127*, 129–153. [CrossRef]
14. Peng, X.; Tamura, K.; Asano, M.; Takano, A.; Kawagoe, M.; Kamijo, T. Changes in soil physical and chemical properties during vegetation succession on miyake-jima island. *Forests* **2021**, *12*, 1435. [CrossRef]
15. Wyse, S.; Wilmshurst, J.; Burns, B.; Perry, G. New Zealand forest dynamics: A review of past and present vegetation responses to disturbance, and development of conceptual forest models. *N. Z. J. Ecol.* **2018**, *42*, 87–106. [CrossRef]
16. Clarkson, B.D. A review of vegetation development following recent (<450 years) volcanic disturbance in North Island, New Zealand. *N. Z. J. Ecol.* **1990**, *14*, 59–71.
17. Robin, L.C. Tropical forest recovery: Legacies of human impact and natural disturbances. *Perspect. Plant Ecol. Evol. Syst.* **2003**, *6*, 51–71. [CrossRef]
18. Foster, D.R.; Knight, D.H.; Franklin, J.F. Landscape patterns and legacies resulting from large, infrequent forest disturbances. *Ecosystems* **1998**, *1*, 497–510. [CrossRef]
19. del Moral, R.; Grishin, S.Y. Volcanic disturbances and ecosystem recovery. In *Ecosystems of Disturbed Ground*; Walker, L.R., Ed.; Elsevier: Amsterdam, The Netherlands, 1999; p. 868.
20. Stern, C.R. Active Andean volcanism: Its geologic and tectonic setting. *Rev. Geológica Chile* **2004**, *31*, 161–206. [CrossRef]
21. Fontijn, K.; Lachowycz, S.M.; Rawson, H.; Pyle, D.M.; Mather, T.A.; Naranjo, J.A.; Moreno-roa, H. Late Quaternary tephrostratigraphy of southern Chile and Argentina. *Quat. Sci. Rev.* **2014**, *89*, 70–84. [CrossRef]
22. Armesto, J.; Rozzi, R.; Smith-Ramirez, C.; Arroyo, M.T. Conservation Targets in South american Temperate Forests. *Science* **1998**, *282*, 1271–1272. [CrossRef]
23. Veblen, T.T.; González, M.E.; Stewart, G.H.; Kitzberger, T.; Brunet, J. Tectonic ecology of the temperate forests of South America and New Zealand Tectonic ecology of the temperate forests of South America and New Zealand. *N. Z. J. Bot.* **2016**, *54*, 223–246. [CrossRef]
24. Armesto, J.J.; Arroyo, M.T.K.; Villagrán, C. *Ecología de los Bosques Nativos de Chile*; Editorial Universitaria: Santiago, Chile, 1996.
25. Arroyo, M.T.K.; Marquet, P.A.; Marticorena, J.A.S.C.; Cavieres, L.; Squeo, F.; Rozzi, R. Chilean winter rainfall-Valdivian forests. In *Hotspots Revisted: Earth's Biologically Wealthiest and Most Threatened Ecosystems*; Mittermeier, R.A., Gil, P.R., Hoffmann, M., Pilgrim, J., Brooks, T., Mittermeier, C.G., Lamoreux, J., da Fonseca, G.A.B., Eds.; CEMEX: Monterrey, México, 2004; pp. 99–103.
26. Villagrán, C.; Hinojosa, F. Esquema Biogeográfico de Chile. In *Regionalización Biogeográfica en Iberoámerica y tópicos afines*; Llorente Bousquets, J., Morrone, J.J., Eds.; Jiménez Editores: México, Mexico, 2005; pp. 551–577.
27. Perakis, S.S.; Hedin, L. Nitrogen loss from unpolluted South American forest mainly via dissolved organic compounds. *Nature* **2002**, *415*, 416–419. [CrossRef]
28. Major, J.J.; Lara, L.E. Overview of Chaitén Volcano, Chile, and Its 2008–2009 eruption. *Andean Geol.* **2013**, *40*, 196–215. [CrossRef]
29. Luebert, F.; Pliscoff, P. *Sinopsis Bioclimática y Vegetacional de Chile*; Editorial Universitaria: Stgo, Chile, 2006; ISBN 9561118327.
30. Veblen, T.T.; Schlegel, F.M.; Oltremari, J. Temperate broad-leaved evergreen forest of South America. In *Temperate Broad-Leaved Evergreen Forest*; Ovington, J.D., Ed.; Elsevier B.V.: Amsterdam, The Netherlands, 1983; pp. 5–31.
31. Veblen, T.T.; Donoso, C.; Kitzberger, T.; Rebertus, A.J. Ecology of Southern Chilean and Argentinean Nothofagus Forests Geology and Physiography Climate Distribution, Habitats and Forest Types Nothofagus pumilio Reproductive Biology Flowering, Pollination and S. In *Ecology and Biogeography of Nothofagus Forests*; Veblen, T.T., Hill, R., Read, J., Eds.; Yale University Press: New Haven, CT, USA, 1996.
32. del Moral, R. Life returns to Mount St. Helens. *Nat. Hist. N. Y.* **1981**, *90*, 36.
33. Colwell, R.K.; Coddington, J.A. Estimating terrestrial biodiversity through extrapolation. *Phil. Trans. R. Soc. Lond. B* **1994**, *345*, 101–118. [CrossRef]
34. Colwell, R.K.; Chao, A.; Gotelli, N.J.; Lin, S.; Mao, C.; Chazdon, R.L.; Longnio, J.T. Models and estimators linking individual-based and sample-based rarefaction, extrapolation and comparison of assemblages. *J. Plant Ecol.* **2012**, *5*, 3–21. [CrossRef]
35. Colwell, R.K. EstimateS: Statistical Estimation of Species Richness and Shared Species from Samples. 2006. Available online: http://purl.oclc.org/estimates (accessed on 21 October 2024).
36. Atmar, W.; Patterson, B.D. The measure of order and disorder in the distribution of species in fragmented habitat. *Oecologia* **1993**, *96*, 373–382. [CrossRef]
37. Atmar, W.; Patterson, B.D. *The Nestedness Temperature Calculator: A Visual Basic Program, Including 294 Presence-Absence Matrices*; AICS Research, Inc.: University Park, NM, USA; The Field Museum: Chicago, IL, USA, 1995.
38. Veblen, T.T.; Ashton, D.H. Catastrophic influences on the vegetation of the Valdivian Andes, Chile. *Vegetatio* **1978**, *36*, 149–167. [CrossRef]

39. Donoso, C. *Bosques templados de Chile y Argentina. Variación, estructura y dinámica*; Editorial Universitaria: Santiago, Chile, 1993.
40. Pickett, S.T.A.; White, P.S. *The Ecology of Natural Disturbance and Patch Dynamics Preface*; Academic Press: Orlando, FL, USA, 1985; ISBN 9780080504957.
41. Franklin, J.F.; MacMahon, J.A.; Swanson, F.J.; Sedell, J.R. Ecosystem responses to the eruption of Mount St. Helens. *Natl. Geaographic Res.* **1985**, *1*, 198–216.
42. Hintz, L.; Fischer, D.; Ferrari, N.; Crisafulli, C.M. Vegetation dynamics under residual large trees following a volcanic eruption in a Valdivian temperate rainforest. *Plant Ecol.* **2021**, *222*, 915–931. [CrossRef]
43. Talbot, S.S.; Talbot, S.L.; Walker, L.R. Post-eruption legacy effects and their implications for long-term recovery of the vegetation on Kasatochi Island, Alaska. *Arctic, Antarct. Alp. Res.* **2010**, *42*, 285–296. [CrossRef]
44. Moreno-Gonzalez, R. Vegetation responses to past volcanic disturbances at the *Araucaria araucana* forest-steppe ecotone in northern Patagonia. *Ecol. Evol.* **2022**, *12*, e9362. [CrossRef]
45. Swanson, F.J.; Crisafulli, C.M. Volcano Ecology: State of the Field and Contributions of Mount St. Helens Research. In *Ecological Responses at Mount St. Helens: Revisited 35 Years After the 1980 Eruption*; Springer: New York, NY, USA, 2018; pp. 305–323.
46. Eriksson, O. Funtional roles of remnants plant population in communities and ecosystem. *Glob. Ecol. Biogeogr.* **2000**, *9*, 443–449. [CrossRef]
47. McIntyre, S.; Lavorel, S.; Tremont, M. Plant Life-History Attributes: Their relationship to disturbance response in herbaceous vegetation. *J. Ecol.* **1995**, *83*, 31–44. [CrossRef]
48. McLachlan, S.M.; Bazely, D.R. Recovery Patterns of Understory Herbs and Their Use as Indicators of Deciduous Forest RegenerationRecuperación de Patrones del Sotobosque Herbáceo y su Uso como Indicadores de Regeneración del Bosque Deciduo. *Conserv. Biol.* **2001**, *15*, 98–110. [CrossRef]
49. Pausas, J.G.; Lavorel, S. A hierchical deductive approach for functional types in disturbed ecosystem. *J. Veg. Sci.* **2003**, *14*, 409–416. [CrossRef]
50. Tilman, D. Community invasibility, recruitment limitation, and grassland biodiversity. *Ecology* **1997**, *78*, 81–92. [CrossRef]
51. Bellingham, P.; Sparrow, A. Resprouting as a life-history strategy in woody plant communities. *Oikos* **2000**, *89*, 409–416. [CrossRef]
52. Bond, W.J.; Midgley, J.J. Ecology of sprouting in woody plants: The persistence niche. *Trends Ecol. Evol.* **2001**, *16*, 45–51. [CrossRef]
53. Armesto, J.; Rozzi, R.; Miranda, P.; Sabag, C. Plant/frugivore interactions in south american temperate forests. *Rev. Chil. Hist. Nat.* **1987**, *60*, 321–336.
54. Willson, M.F.; De Santo, T.L.; Sabag, C.E.; Armesto, J.J. Avian communities in temperate rainforest of North and South America. In *High-Latitude Rainforests and Associated Ecosystems of the West Coast of the Americas*; Lawford, R., Alaback, P.B., Fuentes, E., Eds.; Springer: New York, NY, USA, 1996; pp. 228–247.
55. Armesto, J.J.; Díaz, I.A.; Papic, C.; Willson, M.F. Seed rains of fleshy and dry propagules in different habitats in the temperate rainforest of Chiloé Island, Chile. *Austral Ecol.* **2001**, *26*, 311–320. [CrossRef]
56. Lindenmayer, D.B.; Franklin, J.F. *Conserving Forest Biodiversity: A Comprehensive Multiscaled Approach*; Island Press: Washigton, DC, USA, 2002.
57. Bustamante-Sánchez, M.A.; Armesto, J.J.; Halpern, C.B. Biotic and abiotic controls on tree colonization in three early successional communities of Chiloé Island, Chile. *J. Ecol.* **2011**, *99*, 288–299. [CrossRef]
58. Donoso, C. *Las especies arbóreas de los bosques templados de Chile y Argentina. Autoecología*; 2nd ed.; Marisa Cuneo Ediciones: Valdivia, Chile, 2006; ISBN 9789567173273.
59. Magnússon, B.; Magnússon, S.H.; Fridriksson, S. Developments in plant colonization and succession on Surtsey during 1999–2008. *Surtsey Res.* **2009**, *12*, 57–76. [CrossRef]
60. Dale, V.H.; Crisafulli, C.M.; Swanson, F.J. 25 Years of ecological change at Mount St. Helens. *Science* **2005**, *308*, 961–962. [CrossRef] [PubMed]
61. Tsuyuzaki, S.; Moral, R. del Species attributes in early primary succession on volcanoes. *J. Veg. Sci.* **1995**, *6*, 517–522. [CrossRef]
62. Vitousek, P.M.; Walker, L.R.; Whiteaker, L.D.; Matson, P.A. Nutrient limitations to plant growth during primary succession in Hawaii Volcanoes National Park. *Biogeochemistry* **1993**, *23*, 197–215. [CrossRef]

diversity

MDPI

Article

Allelopathic Properties of the Species Comprising Communities of Invasive *Impatiens* spp. and Antioxidant System of Invaders' Populations

Eugenija Kupcinskiene [1], Ruta Budreviciute [1], Vaida Jasionyte [1], Laura Simanaviciute [1], Lina Jociene [1], Edvina Krokaite-Kudakiene [1], Tomas Rekasius [2,3] and Vitas Marozas [4,*]

1 Department of Biology, Faculty of Natural Sciences, Vytautas Magnus University, K. Donelaicio Str. 58, 44248 Kaunas, Lithuania; e.kupcinskiene@gmail.com (E.K.); ruta.budreviciute@stud.vdu.lt (R.B.); vaida.jasionyte@stud.vdu.lt (V.J.); laura.simana@gmail.com (L.S.); lina.jociene@vdu.lt (L.J.); edvina.krokaite@vdu.lt (E.K.-K.)

2 Department of Mathematics and Statistics, Faculty of Informatics, Vytautas Magnus University, K. Donelaicio Str. 58, 44248 Kaunas, Lithuania; tomas.rekasius@vgtu.lt

3 Department of Mathematical Statistics, Faculty of Fundamental Sciences, Vilnius Gediminas Technical University, Sauletekio Ave. 11, 10223 Vilnius, Lithuania

4 Department of Environment and Ecology, Faculty of Forest Sciences and Ecology, Agriculture Academy, Vytautas Magnus University, K. Donelaicio Str. 58, 44248 Kaunas, Lithuania

* Correspondence: vitas.marozas@vdu.lt

Abstract: Globalization has greatly expanded the opportunities for plant species to enter new areas through a wide range of pathways. Elucidating the pathways of spread of alien species and the characteristics of organisms that make them invasive is one of the most pressing problems in ecological sciences. Once established, alien species may have serious implications for communities and *vice versa*. Allelopathy has been proposed as one of the possible invasion mechanisms of exotic plants. *Impatiens parviflora* and *Impatiens glandulifera* are among the widely spread invasive plant species in the Baltic region. The aim of the study was to evaluate the allelopathic effect of invasive Lithuanian *Impatiens* spp. and their neighboring plants (11 pristine species) using parameters of germination and seedling growth of biotest species *Lepidium sativum*, and to expand this study by determining the content of phenolic compounds and the radical scavenging activity in the leaf extracts of Lithuanian *Impatiens* spp. populations (20 of each *Impatiens* species). Leaf extracts of all species examined had an inhibitory effect on *Lepidium sativum* germination and morphology of the seedlings. In our study, at all leaf extract concentrations, significantly higher allelopathic potential on radicle growth was characteristic of *I. glandulifera* compared to native species *Alnus glutinosa*, *Calystegia sepium*, and *Urtica dioica*. At all leaf extract concentrations, *I. parviflora* showed significantly higher allelopathic potential on radicle growth compared to native species *Urtica dioica*. *Impatiens glandulifera* had a higher juglone index than *Impatiens parviflora*. *I. glandulifera* also had the highest juglone index compared to all neighboring species studied. The differences between the populations in the content of phenolic compounds, DPPH and ABTS radical scavenging activity were 2.3, 2.2 and 2.7 times for *I. glandulifera* and 2.6, 5.2 and 2.7 times for *I. parviflora*. The mean values of total phenolic content, DPPH and ABTS radical scavenging activity of *I. glandulifera* populations were 2.1, 2.7 and 3.3 times higher than those of *I. parviflora* populations, respectively. In general, our results about allelopathic potential and phenolics content, as well as free radical scavenging ability, confirm the supreme competitive ability of *I. glandulifera* compared to both *I. parviflora* and to the native co-occurring species.

Diversity **2025**, *17*, 20

https://doi.org/10.3390/d17010020

Keywords: alien plants; DPPH; ABTS; phenolic compounds; small balsam; Himalayan balsam; biotest; garden cress; juglone index

1. Introduction

Plant invasions are the man-mediated spread of organisms outside their natural range. Invasive alien species often disrupt native ecosystems by outcompeting indigenous flora, altering habitat structures, changing nutrient cycling processes, exacerbating soil degradation and water scarcity, as a consequence leading to profound changes in ecosystem functions and community dynamics [1,2]. The need for invasive species management and eradication strategies is arising from their potential to reduce biodiversity. Finally, invasions are causing economic losses. Therefore, the management of invasive plant species is a critical environmental necessity. Successful eradication of an invasive species requires comprehensive knowledge of the invasion mechanism. [3]. For developing effective eradication practices, understanding of the invaders' ability to thrive in a variety of environmental conditions is required.

In the second part of the 20th century, the role of numerous features of the invasive plant species and their habitat properties were investigated. The combined findings led to the development of invasion hypotheses, which can be broadly categorized into those focusing on species traits, community structure, and environmental factors. Starting from the "Baker's rule" and "Tens rule" [4,5], many hypotheses regarding plant invasion emerged, significantly advancing understanding of the mechanisms and factors that contribute to the invasiveness of certain plant species [6–9]. Analyzing the mechanisms of invasion led to very important insights about competitive advantage of invaders due to chemical peculiarities. The "novel weapon hypothesis" posited that some invasive plants may have stronger allelopathic impacts on other species in introduced ranges than in native ranges because the former may not be as adapted to specific biochemical signatures of invaders compared to species in the native range [10–13].

Allelopathy refers to the biochemical interactions between plants, where one plant species releases allelochemicals that can inhibit or promote the growth of neighboring plants [14–16] and can be regarded as a natural strategy protecting plants against environmental "enemies" and competing plants. The "novel weapon hypothesis" led to a large body of work on the allelopathy of invasive plants, which continues to this day [17–21]. Analyses on the allelopathic potential of over 500 known invasive plant species have shown that allelopathy occurred in more than half of the examined plants [22]. Recent research on allelopathy indicates that allelopathy influences the succession of boreal forests, playing a vital role in ecosystem regeneration under soil erosion; thus, allelopathic interactions affect species composition and biodiversity, particularly in the face of climate change [23,24]. Allelochemicals may influence soil microbial communities, which in turn affect plant performance [25]. In some cases, it was detected that the allelopathic effects of native species may shape the composition of plant communities by suppressing invasive species, thereby maintaining biodiversity [26–28]. Analysis of several native species and several non-native species indicated that it is important to compare the allelopathic effects of invasive and native species to correctly estimate the phytotoxic effect of invasive species on their invasiveness [29]. Despite the understanding that some invasive plants may rely on allelopathy mechanisms to gain competitive superiority over native species, the significance of interference competition for invasion ecology remains uncertain [30].

The phenomenon of allelopathy has been extensively studied in relation to chemical compounds. These compounds include a diverse range of secondary metabolites, such as phenolics, alkaloids, glycosides, and terpenoids [31,32]. Terpenoids can act as allelochemicals that either suppress or stimulate the growth of competing plant species [33,34]. Terpenoids are among the key allelochemicals produced by some invasive species, like *Solidago canadensis* [19]. Phenolic compounds play a crucial role in plant adaptation and defense, and are known to be involved in various ecological interactions, including allelopathy [31,35]. The influence of allelopathy is most often mediated through secondary metabolites with antioxidant properties. ABTS$^{•+}$ (2,2′-azino-bis(3-ethylbenzothiazoline-6-sulfonic acid) radical scavenging assay is specifically advantageous as it can evaluate both lipophilic and hydrophilic antioxidants, making it suitable for biological samples, including those derived from allelopathic plants [36]. The DPPH$^{•}$ (2,2-diphenyl-2-picrylhydrazyl hydrate) assay is also widely used to assess the free radical scavenging ability of antioxidants in plant extracts and is sensitive to a wide range of antioxidants, including phenolic compounds [37]. Phenolic compounds are known to have strong antioxidant properties, since phenolic compounds often correlate with antioxidant activity and can support and validate the results of DPPH$^{•}$ and ABTS$^{•+}$ assays [38]. To study allelopathy, researchers employ various bioassay methods that assess the effects of these compounds on plant growth, particularly focusing on seed germination and seedling development [39].

Impatiens glandulifera (Himalayan balsam) [40–42] and *Impatiens parviflora* (small balsam) [43,44], representatives of family *Balsaminaceae*, are among the plants originally native to Asia that nowadays are spreading across other continents. *Impatiens* spp. have emerged as a significant concern in Europe due to their invasive nature. These two annuals have shown a remarkable ability to adapt and thrive in various habitats across the continent. *Impatiens* spp. grows in riparian zones along rivers on humid soils and in wet woodlands, often dominating in such sites [45–48]. As an invasive alien species of concern to the European Union, *Impatiens glandulifera* has been included in the 2017 update to Regulation (EU) No. 1143/2014 [49]. *Impatiens parviflora* is presently the most common alien invader in European woodlands [50]. The presence of invasive *Impatiens* spp. can significantly reduce species richness and diversity in invaded communities [51–55]. Among the possible reasons for such poverty might be competitive success, at least partially based on biochemical peculiarities of these species. In various countries of Central Europe, much research has been done analyzing the ecophysiology of populations of invasive *Impatiens* spp. [45,46,53,56], investigating allelopathic features among members of communities comprising *Impatiens* [17,57–59], and identifying allelopathic compounds [35,60–63].

In recent years, more and more attention has been paid to allelopathy research in Lithuania, in particular to the effects of some invasive plants, such as *Erigeron annuus* [64], *Solidago* spp. [65–67], and *Heracleum* spp. [68], on the germination and growth of seeds of biotest species, and to the identification of allelopathy-related compounds; however, research on a similar level on the invasive *Impatiens* is still lacking.

Chemicals that influence the growth of neighboring plants play a significant role in shaping plant communities and ecological interactions, and are specific for each site [18]. That is why plant population research is vital for elucidating the complex interactions between plants, their environments, and other organisms [69]. Invasive plant population research in Lithuania has gained significant attention in respect of genetic diversity of invaders in various country sites, evaluating *Erigeron annuus* [64], *Bidens frondosus* [70], *Bunias orientalis* [71], *Impatiens glandulifera* [72–74], and *Impatiens parviflora* [74–77]; meanwhile, population-based analyses of the biochemical diversity of invasive plants have not received sufficient attention, until now.

The aim of the study was to evaluate the allelopathic effect of invasive *Impatiens* spp. and their neighboring species in Lithuania using the parameters of germination and seedling growth of the biotest species *Lepidium sativum*, and to expand this study by determining the content of phenolic compounds and the radical scavenging activity of DPPH• and ABTS•+ in leaf extracts of Lithuanian *Impatiens* spp. populations. By examining the interactions between these invasive species and their neighbors, this study seeks to contribute to the broader understanding of plant invasions and the ecological implications of allelopathy.

2. Materials and Methods

2.1. Sampling Sites and Plant Material

Based on community data [50,76–80], the following species, as frequent neighbors of *Impatiens glandulifera* Royle (abbreviation for the figures—Imp gla), and/or *I. parviflora* DC. (Imp par), were selected: *Urtica dioica* L. (Urt dio), *Geum urbanum* L. (Geu urb), *Aegopodium podagraria* L. (Aeg pod), *Stellaria media* (L.) Vill. (Ste med), *Alliaria petiolata* (M.Bieb.) Cavara & Grande (All pet), *Galium aparine* L. (Gal apa), *Rubus caesius* L. (Rub cae), *Geranium robertianum* L. (Ger rob), *Epilobium hirsutum* L. (Epi hir), *Calystegia sepium* (L.) R. Br. (Cal sep), *Phragmites australis* (Cav.) Steud. (Phr aus), *Alnus glutinosa* (L.) Gaertn. (Aln glu); also, some other alien species: *Erigeron annuus* (L.) Pers (Eri ann), *Helianthus tuberosus* L. (Hel tub), and cultivated tree *Juglans regia* L. (Jug reg). Sampling was carried out during the period July 25–August 5. After plant collection, materials were placed into cool bags and transported to the laboratory on the same day. More detailed procedures of sampling of various species were described earlier [81–83].

For biochemical analyses, 20 populations of *Impatiens glandulifera* and 20 populations of *Impatiens parviflora* DC.) were selected in order to cover all the territory of Lithuania (Figure 1, Table 1).

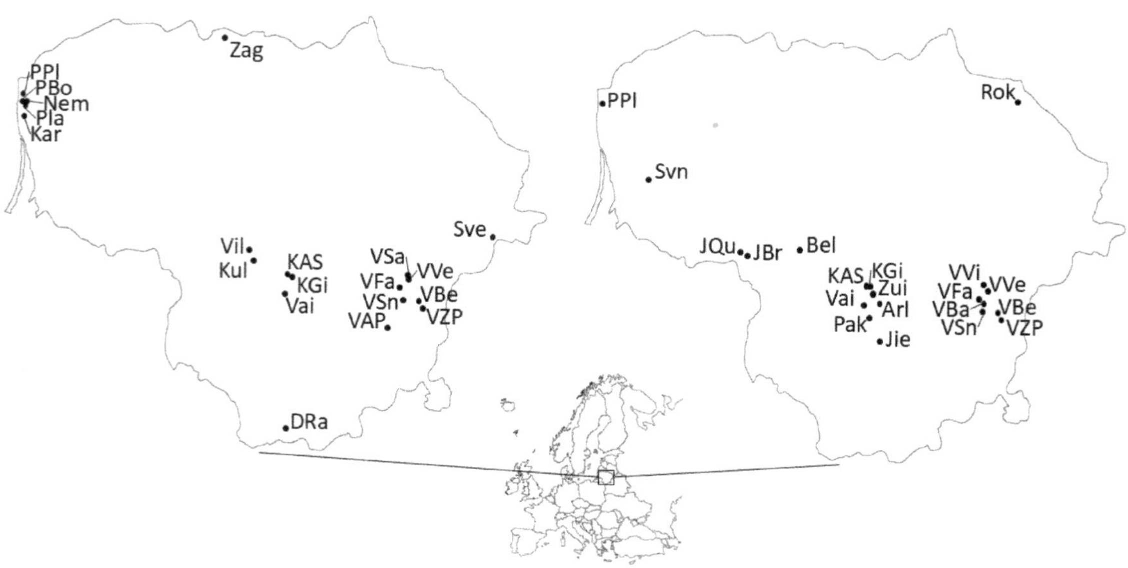

Figure 1. Location of Lithuanian populations (names and name abbreviations) of *I. parviflora* (left) and *I. glandulifera* (right). Note: Rok—Rokiškis; VVi—Vilnius-Visoriai; VSn—Vilnius-Šnipiškės; VFa—Vilnius-Fabijoniškės; VBa—Vilnius-Baltupiai; VBe—Vilnius-Belmontas; VVe—Vilnius-Verkiai; VSa—Vilnius-Santariškės; VAP—Vilnius-A. Paneriai; VZP—Vilnius-Ž. Pavilnys; Svn—Švenčionys; DRa—Druskininkai-Ratnyčia; KAS—Kaunas-A. Šančiai; Vai—Vaišvydava; KGi—Kaunas-Giraitė; Zui—Kaunas-Zuikinė; Jie—Jieznas; Pak—Pakuonis; Arl—Arlaviškės; Kul—Kulautuva; Vil—Vilkija; Bel—Belvederis; JBr—Jurbarkas-Bridge; JQu—Jurbarkas-Quarry; Sve—Švėkšna; Kar—Karklė; Nem—Nemirsėta; PBo—Palanga-Botanical garden; PPl—Palanga; Pla—Plazė; Zag—Žagarė.

Table 1. Geographical location of sampling sites of Lithuanian populations of *Impatiens glandulifera* and *I. parviflora*.

Population		Latitude (N)	Longitude (E)
	Impatiens glandulifera		
Rokiškis	Rok	55°58′	25°35′
Vilnius-Visoriai	VVi	54°45′	25°15′
Vilnius-Šnipiškės	VSn	54°41′	25°17′
Vilnius-Fabijoniškės	VFa	54°44′	25°14′
Vilnius-Baltupiai	VBa	54°44′	25°17′
Vilnius-Belmontas	VBe	54°41′	25°21′
Vilnius-Verkiai	VVe	54°44′	25°18′
Vilnius-Ž. Pavilnys	VZP	54°40′	25°21′
Kaunas-A. Šančiai	KAS	54°52′	23°56′
Kaunas-Girionys	KGi	54°56′	23°52′
Kaunas-Zuikinė	Zui	54°51′	24°06′
Vaišvydava	Vai	54°50′	24°01′
Arlaviškės	Arl	54°48′	24°11′
Jieznas	Jie	54°35′	24°10′
Pakuonis	Pak	54°44′	24°03′
Belvederis	Bel	55°04′	23°23′
Jurbarkas-Bridge	JBr	55°04′	22°46′
Jurbarkas-Quarry	JQu	55°05′	22°41′
Palanga	PPl	55°55′	21°03′
Švėkšna	Sve	55°30′	21°37′
	Impatiens parviflora		
Vilnius-Verkiai	VVe	54°44′	25°18′
Vilnius-A. Paneriai	VAP	54°37′	25°09′
Vilnius-Santariškės	VSa	54°45′	25°16′
Vilnius-Ž. Pavilnys	VZP	54°40′	25°21′
Vilnius-Šnipiškės	VSn	54°42′	25°15′
Vilnius-Fabijoniškės	VFa	54°43′	25°15′
Vilnius-Belmontas	VBe	54°41′	25°21′
Švenčionys	Svn	55°08′	26°09′
Druskininkai-Ratnyčia	DRa	53°59′	24°23′
Kaunas-A. Šančiai	KAS	54°53′	23°56′
Kaunas-Vaišvydava	KVa	54°50′	24°02′
Kaunas-Giraitė	KGi	54°56′	23°52′
Kulautuva	Kul	54°56′	23°39′
Vilkija	Vil	55°02′	23°35′
Karklė	Kar	55°48′	21°04′
Nemirsėta	Nem	55°53′	21°03′
Palanga-Botanical Garden	PBo	55°54′	21°03′
Palanga	Pal	55°54′	21°04′
Plazė	Pla	55°52′	21°03′
Žagarė	Zag	56°21′	23°15′

Sites with *Impatiens* were named by three letters reflecting their geographical location. Geographical coordinates of the examination are ranged from 53°59′ to 56°21′ latitude (N), from 20°59′ to 26°09′ longitude (E), with annual rainfall of 722–936 mm, and mean annual temperature of 6.0–6.8 °C [73,75]. Sampling of populations was carried out as it is described earlier [72,74].

2.2. Preparation of Plant Leaf Extracts for Toxicity Assessment

One of the most widely utilized bioassay methods is the seed germination test, which evaluates the impact of allelochemicals on the germination rates of seeds from sensitive plant species [84]. The effects of leaf extracts of *Impatiens* species and adjacent plant species on germination, as well as on radicle and hypocotyl lengths of garden cress (*Lepidium sativum* L.), were investigated. The selection of *Lepidium sativum* was based on our former experience using it as a test object [85] and related to the other researchers [86,87].

Dried leaves were placed in distilled water (dry mass to distilled water, ratio 1:10 mass/volume (hereinafter referred to as 100 mg/mL)) and left stirring at room temperature for 24 h. The ratio of plant dry mass to water is comparable to the natural amount of litter to rainwater in European semi-natural grasslands and, hence, results in a realistic concentration of extracted compounds [20,28,88].

The extraction was performed overnight in an orbital shaker PSU-20i (BioSan, Riga, Latvia) at room temperature (18 °C). From the resulting aqueous extract, aqueous extracts of other concentrations (1:20, 1:40, hereinafter referred to respectively, as 50 mg/mL, 25 mg/mL) were prepared by dilution. The aqueous extracts prepared in each variant were then transferred in 7 mL aliquots into glass Petri plates lined with 4 filter papers (Whatman #1). One hundred seeds of the test plant, *Lepidium sativum* (*Brassicaceae* family), were sown in each plate. Target seeds were treated with leaf leachate only once at the beginning of the experiment. The seeds were germinated at 25 °C in a thermostat (Venticell, MMM Medcenter Einrichtungen GmbH, Gräfelfing, Germany) in the dark for seven days, each day moistening with 1 mL of distilled water. The tests were carried out in triplicate. For radicle and hypocotyl data collection, treatments with the highest concentration of the leaf extracts were replicated nine times. Seven days after sowing, the number of germinated seeds of garden cress was counted and the lengths of the hypocotyls and radicles of the seedlings were measured. The data obtained were compared with those obtained by soaking the seeds in distilled water.

To study allelopathic potential, Juglone index (Ij/x) was used to compare the effects of treatment with 1 mM juglone and substance extracted of plant species with unknown effect [89,90]. The effect of two other concentrations of juglone—0.1 mM and 0.1 mM—was also examined. The juglone index (Ij/x) was calculated as $Ij/x = (Gj + Rj + Hj)/(Gx + Rx + Hx)$; "*Impatiens glandulifera* index" was calculated as $I_{Ig}/x = (G_{Ig} + R_{Ig} + H_{Ig})/(Gx + Rx + Hx)$; and "*Impatiens parviflora* index" as $I_{Ip}/x = (G_{Ip} + R_{Ip} + H_{Ip})/(Gx + Rx + Hx)$, where—G—germination, R—radicle length (mm), S—hypocotyl length (mm) under treatment of juglone (j), extract of *Impatiens glandulifera* leaves (Ig), extract of *Impatiens parviflora* leaves (Ip), or extract of the leaves of other plant species (x) with unknown allelopathic potential; G, R, and H are means of three replicates.

2.3. Preparation of Impatiens spp. Leaf Extracts for Determination of Concentration of the Total Phenol, DPPT$^{\bullet}$, and ABTS$^{\bullet+}$

Among various solvents used for the preparation of plant extracts for assessment of antioxidant activity and phenols, we have selected ethanol [91–93]. One g of the ground dry leaves of *Impatiens glandulifera* and *I. parviflora* were extracted in a shaker PSU-20i (BioSan, Latvia) with 10 mL of 100% ethanol at room temperature overnight. The resulting suspension was filtered through Whatman #1 filter paper for analysis. The extraction was carried out 3 times. After extraction, the 30 mL filtrates were evaporated to dryness using a rotary evaporator (Heidolph G3, Wood Dale, IL, USA) at 40 °C. Dry extracts were stored at +4 °C for 24 h until analysis.

2.4. Determination of Concentration of the Total Phenols

The phenolic compounds were determined spectrophotometrically by the Folin–Ciocâlteu reagent (FCR) or Folin's phenol reagent (1927) method [94,95]. For the determination of phenolic compounds, the test extracts were dissolved in 100% ethanol [91,92]. Mixture, containing 0.05 mL of extract, 3.55 mL of 7.5% Na_2CO_3 solution and 2 M Folin–Ciocâlteu reagent (Sigma-Aldrich, Steinheim, Germany) was prepared. The reaction was carried out at room temperature for 60 min, the reaction mixtures were centrifuged before measurement, and the optical density was measured at $\lambda = 765$ nm in disposable PMMA macro cuvettes. Similar reactions were carried out with different concentration gallic acid dissolved in ethanol, and the absorbance was measured. The measurements were repeated 4 times. The results were expressed as gallic acid equivalent mg/g d. m. (GRE mg/g).

2.5. Determination of the Radical Scavenging Activity of the DPPH$^•$

The radical scavenging capacity of DPPH$^•$ (2,2-diphenyl-2-picrylhydrazyl hydrate, hereinafter referred to as DPPH) was assessed spectrophotometrically according to the modified method [96]. An ethanolic solution of 2×10^{-8} M DPPH$^•$ was prepared. The plant extracts were also dissolved in 100% ethanol (concentration: 5 mg extract/1 mL ethanol). 2.95 mL of the ethanolic DPPH$^•$ solution was mixed with 0.05 mL of the extract solution in 1 cm optical glass cuvettes, and after 15 min the absorbance at 517 nm was measured. The analyses were carried out in 3 replicates. The results are expressed as gallic acid equivalent in mg/g dry mass (GRE mg/g d. m.). For this purpose, a calibration curve was constructed from 8 different concentrations of ethanolic solutions of gallic acid in analogy with the reaction with DPPH$^•$.

2.6. Determination of the Radical Scavenging Activity of the ABTS$^{•+}$ Cation

The radical binding capacity of the ABTS$^{•+}$ (2,2′-azino-bis(3-ethylbenzothiazoline-6-sulfonic acid)) cation was determined spectrophotometrically. The ABTS$^{•+}$ cation radical was produced by reacting ABTS$^{•+}$ with potassium persulphate. Before measurement, the ABTS$^{•+}$ solution was diluted with phosphate buffer to determine an absorbance of 0.800 ± 0.030 at 734 nm. A quantity of 3.45 mL of ABTS$^{•+}$ was mixed with 0.05 mL of the extract dissolved in ethanol. Reaction time was 15 min. Absorbance was measured at $\lambda = 734$ nm in disposable PMMA macro cuvettes. Solutions of different concentrations of Trolox (6-hydroxy-2,5,7,8-tetramethylchroman-2-carboxylic acid) dissolved in ethanol were used to construct the calibration curve. The radical-cation binding activity of ABTS$^{•+}$ was expressed as mg/g (TE mg/g) of dry extract equivalent of the standard antioxidant Trolox [97].

2.7. Statistical Analyses

Descriptive statistics were performed, and figures were drawn using the R statistical package (version 4.4.2) [98]. Pearson's correlation coefficient was used to determine the relationship between content of total phenols, DPPH$^•$, ABTS$^{•+}$ of populations of *Impatiens glandulifera* and *I. parviflora*, employing the R package "correlation" (version 0.8.3) [99]. Dunnett's test was used to compare effects of different species extract on germination, as well as radicle and hypocotyl parameters of *Lepidium sativum*. The R package PMCMRplus was used for this (version 1.9.12) [100].

3. Results

3.1. *Lepidium sativum* Test on Invasive *Impatiens* spp. and Their Neighbours

In our study, the allelopathic impact of leaf extracts of various plant species on *Lepidium sativum* was analyzed, providing absolute values of seed germination and seedling growth (Figures 2–4) and employing juglone index.

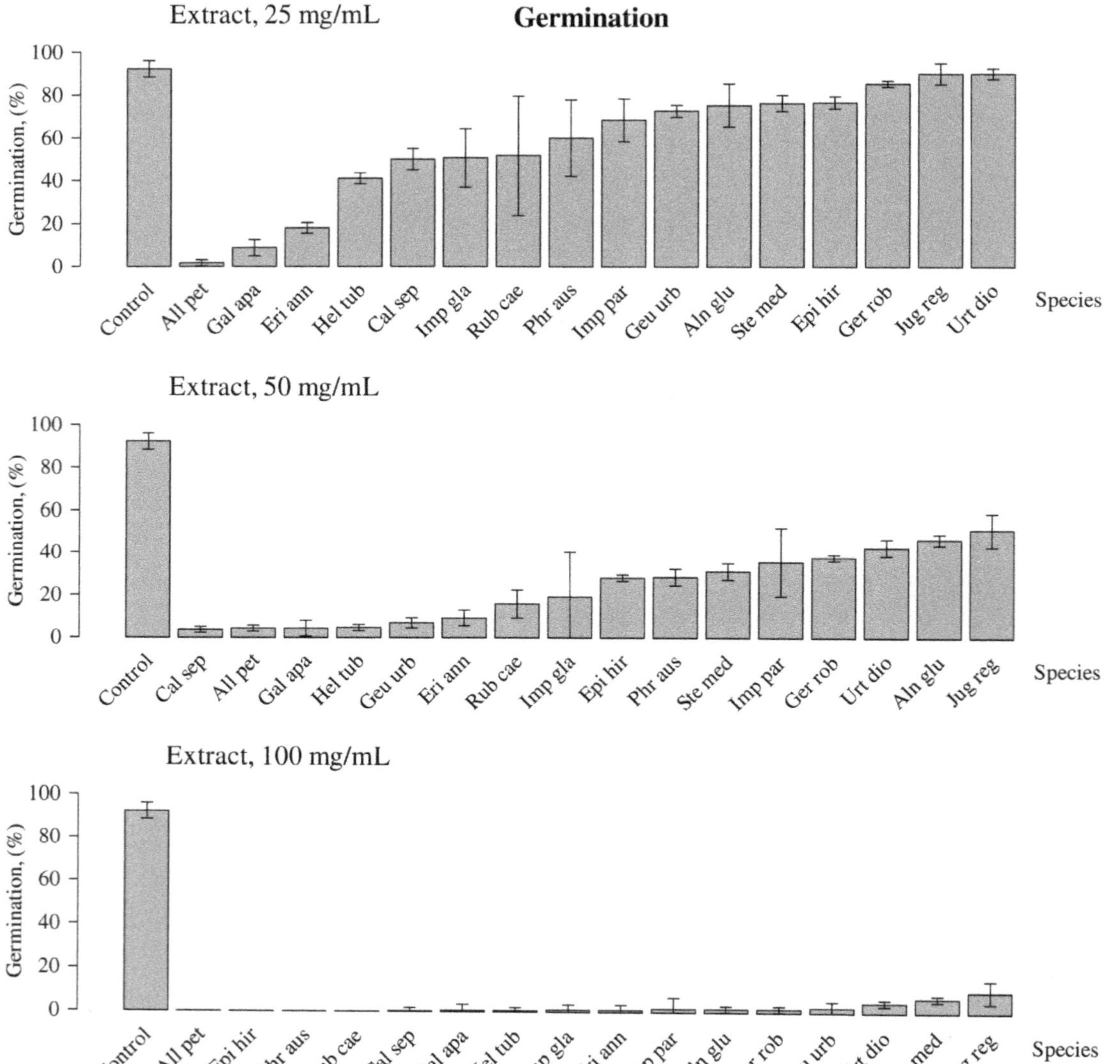

Figure 2. Germination (%) of *Lepidium sativum* seeds in the different concentration water extracts (25, 50, and 100 mg/mL) of the leaves of selected plant species. Columns indicate mean values, and whiskers indicate intervals of confidence.

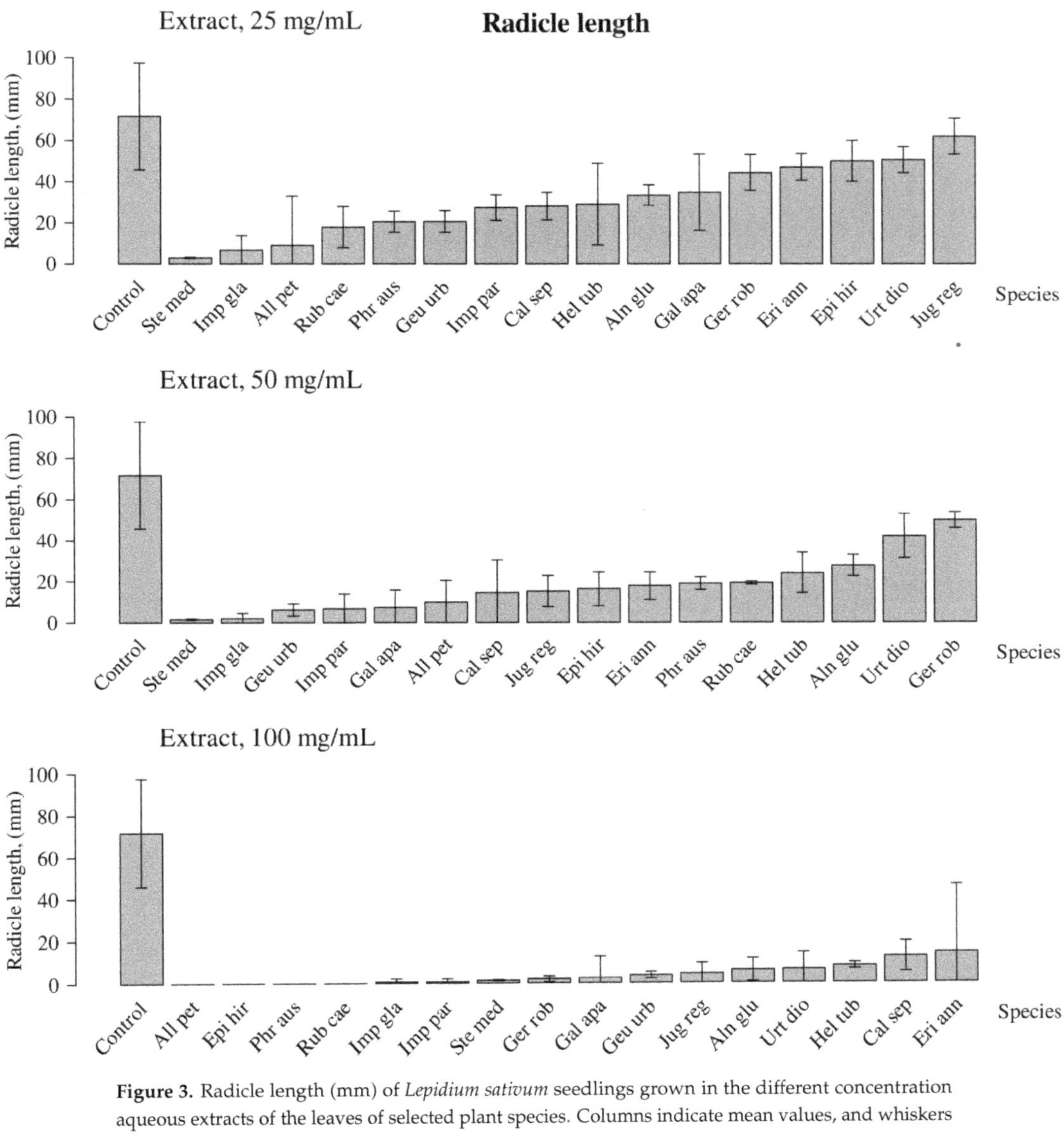

Figure 3. Radicle length (mm) of *Lepidium sativum* seedlings grown in the different concentration aqueous extracts of the leaves of selected plant species. Columns indicate mean values, and whiskers indicate intervals of confidence.

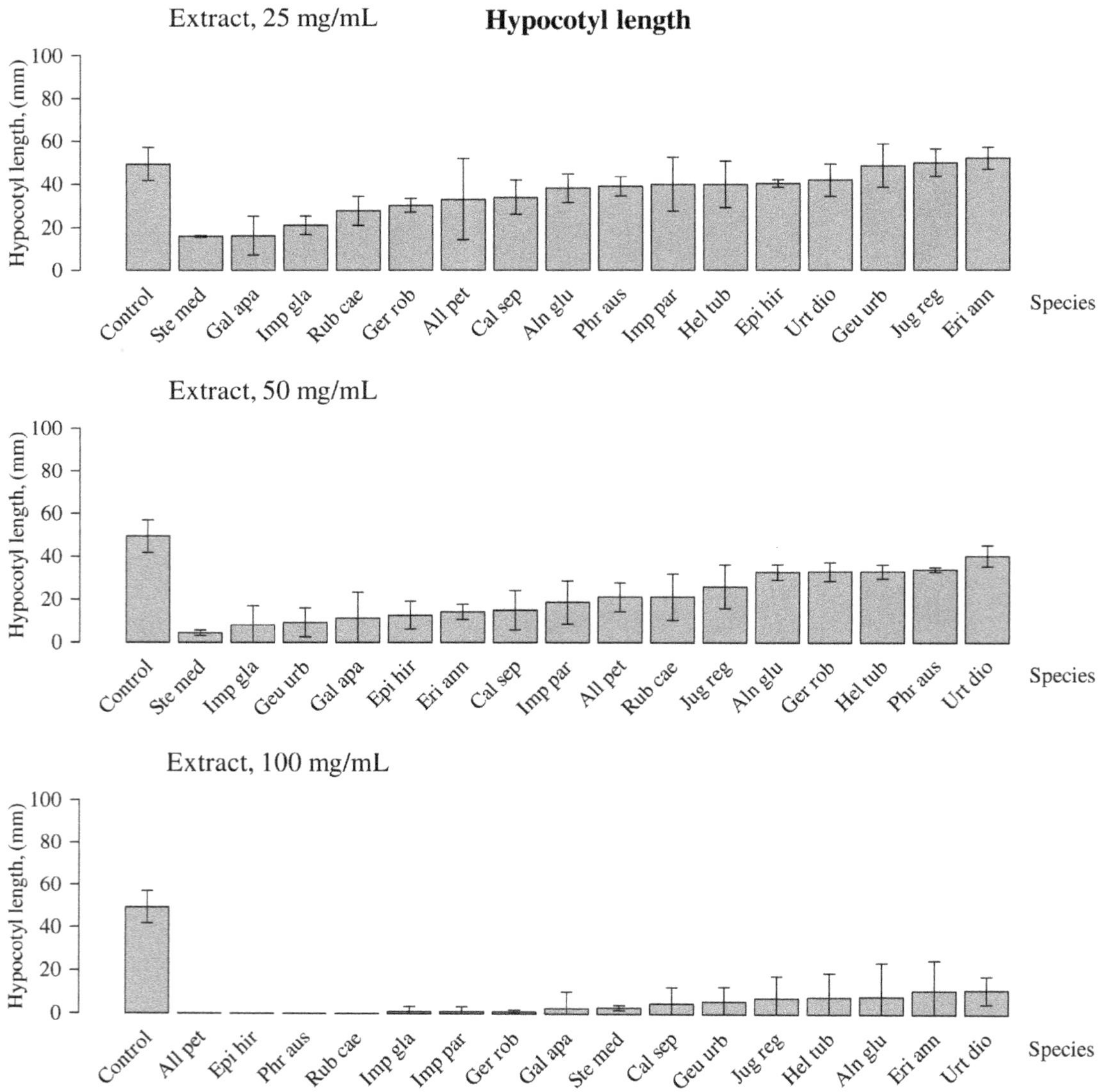

Figure 4. Hypocotyl length (mm) of *Lepidium sativum* seedlings grown in different concentration aqueous extracts of the leaves of selected plant species. Columns indicate mean values, and whiskers indicate intervals of confidence.

Analysis of the effect of different concentrations of aqueous extracts of the leaves of *I. glandulifera, I. parviflora,* and neighboring species on the germination of *Lepidium sativum* seeds showed that the number of germinated seeds increased with decreasing concentrations of aqueous extracts for all plant species (Figure 2).

The lowest concentration (25 mg/mL) of aqueous extracts of *Alliaria petiolata, Galium aparine,* and *Erigeron annuus* had the strongest inhibition effect on seed germination (2–19%); other species effect was moderate (45–97%). The results show that increasing concentrations of aqueous extracts of *Impatiens glandulifera, I. parviflora,* and neighboring species inhibited germination of *Lepidium sativum* seeds. Regarding the effect of aqueous extracts of *Impatiens glandulifera* and *I. parviflora* on the germination of *Lepidium sativum* seeds at all concentrations tested, the allelopathic effect of *Impatiens glandulifera* (1–55%) was stronger compared to *I. parviflora* (2–74%). The intermediate concentration (50 mg/mL) of aqueous extracts of *Calystegia sepium*, *Alliaria petiolata*, *Galium aparine*, *Helianthus tuberosus*, *Geum urbanum*, and *Erigeron annuus* had the strongest inhibition on germination of *L. sativum* seeds (4–10%); other species had similar inhibition effect (17–55%). The highest concentration (100 mg/mL) of aqueous extracts of *Alliaria petiolata*, *Epilobium hirsutum*, *Phragmites australis*, and *Rubus caesius* completely inhibited the germination of *Lepidium sativum* seeds (0%); aqueous extracts of other plant species also had a strong inhibitory effect on seed germination (1–3%); the least inhibition on seed germination was observed with aqueous extracts of *Urtica dioica*, *Stellaria media*, and *Juglans regia* (5–11%).

Analysis of the effect of different concentrations of aqueous extracts of leaves of *I. glandulifera*, *I. parviflora*, and neighboring species on the radicle length of *Lepidium sativum* revealed that radicle length increased as the concentration of the aqueous extracts decreased (Figure 3). At the lowest concentration (25 mg/mL), aqueous extracts of *Stellaria media*, *Impatiens glandulifera*, and *Alliaria petiolata* exhibited the greatest inhibitory effect on radicle length of *Lepidium sativum* (3–9 mm); other species demonstrated a moderate effect (17.87–44 mm), while *Erigeron annuus*, *Epilobium hirsutum*, *Urtica dioica*, and *Juglans regia* exhibited the weakest effect (47–62 mm). Significant difference ($p < 0.05$) from *I. glandulifera* was found for *Juglans regia*, *Epilobium hirsutum*, and *Urtica dioica*. At the intermediate concentration (50 mg/mL), aqueous extracts of *Stellaria media* and *Impatiens glandulifera* had the strongest inhibitory effect on radicle length of *Lepidium sativum* (1.6–2 mm); extract from other species showed a moderate impact (6.2–27.61 mm), while *Urtica dioica* and *Geranium robertianum* had the least inhibitory effect (49.3–42 mm). Significant difference ($p < 0.05$) from *I. glandulifera* was found for *Alnus glutinosa*, *Geranium robertianum*, *Urtica dioica*, and *Helianthus tuberosus*. Significant difference ($p < 0.05$) from *Impatiens parviflora* was found for *Geranium robertianum* and *Urtica dioica*. At the highest concentration (100 mg/mL), aqueous extracts of *Alliaria petiolata*, *Epilobium hirsutum*, *Phragmites australis*, and *Rubus caesius* completely inhibited radicle growth. Extracts of other species at this concentration had a strong effect on the length of *L. sativum* radicle (0.5–6.30 mm), except for *Calystegia sepium* and *Erigeron annuus*, which exhibited weaker effect (8–14.2 mm). *Stellaria media*, *Alliaria petiolata*, and *Impatiens glandulifera* had the strongest effect on the seedling radicle growth.

At the lowest concentration (25 mg/mL) (Figure 4), aqueous extracts of *Stellaria media*, *Galium aparine*, and *Impatiens glandulifera* exhibited the strongest effect on the hypocotyl length (16–21 mm; extracts from other species produced a moderate impact on hypocotyl length (27.73–42 mm), while *Geum urbanum*, *Juglans regia*, and *Erigeron annus* exhibited less pronounced effect (46.6–52 mm). Significant difference ($p < 0.05$) from *I. glandulifera* was found for *Juglans regia*, *Geum urbanum*, and *Erigeron annuus*. At the intermediate concentration (50 mg/mL), aqueous extracts of *Impatiens glandulifera*, *Geum urbanum*, and *Stellaria media* exhibited the strongest inhibitory effects on hypocotyl length (4.3–9.2 mm); other species demonstrated a moderate inhibitory effect (11.32–26 mm), while *Alnus glutinosa*, *Geranium robertianum*, *Helianthus tuberosus*, *Phragmites australis*, and *Urtica dioica* exhibited less pronounced effect (32.77–40.3 mm). Significant difference ($p < 0.05$) from *I. glandulifera* was found for *Phragmites australis* and *Urtica dioica*. At the highest concentration (100 mg/mL), aqueous extracts of *Alliaria petiolata*, *Epilobium hirsutum*, *Phragmites australis*, and *Rubus caesius* completely inhibited *L. sativum* seed germination, resulting in no hypocotyl growth. In contrast, extracts of *Impatiens glandulifera*, *Impatiens parviflora*, *Geranium robertianum*, *Galium aparine*, and *Stellaria media* had the greatest effect on the hypocotyl length of *L. sativum* (0.75–3 mm). Extracts from other species exhibited less pronounced effect (5–11 mm). The results indicate that the hypocotyl length of *Lepidium sativum* increased with decreasing extract concentration as in the case of the radicle length, discussed above. Extracts of *Stellaria media*, *Alliaria petiolata*, and *Impatiens glandulifera* had the strongest effect on the seedling hypocotyl growth.

In most cases, germination, radicle and hypocotyl length of *Lepidium sativum* grown in the aqueous leaf extracts of various concentration of plant species differed significantly ($p < 0.05$; $p < 0.01$; $p < 0.001$) compared to control (H_2O) (Table 2), with some were for *Urtica dioica* and *Juglans regia*. For the highest concentration of the extracts of various plant species, in most cases the effect on seedlings did not differ significantly from *Impatiens glandulifera*; generally, few exceptions were documented for germinations, with more exceptions for radicle and hypocotyl growth. Lower concentration extracts of the neighboring plants have caused significant difference in *Lepidium sativum* growth compared to *Impatiens glandulifera*; exceptions for both lower concentration extracts were *Rubus caesius*, *Phragmites australis* (according to germination), *Geum urbanum*, *Stellaria media* (according to the radicle growth), *Callystegia sepium*, *Galium aparine*, and *Stellaria media* (according to the hypocotyl growth). For the highest concentration of the extracts of various plant species, in some cases the effect on seedlings did not differ significantly from *Impatiens parviflora*, although more exceptions (compared to *Impatiens glandulifera*) were documented for germinations, in addition to radicle and hypocotyl growth. In the case of radicle of *Galium aparine* and *Geum urbanum*, as well as hypocotyls of *Alliaria petiolata*, *Callystegia sepium*, and *Epilobium hirsutum*, there was no significant difference when compared to the *Impatiens parviflora* (25 and 50 mg/mL extract cases). Effects of *Impatiens glandulifera* and *Impatiens parviflora* differed significantly in most parameters, recorded for 25 and 50 mg/mL extract cases.

Applying the juglone index, we have made comparisons of two types: (1) the effect of the extracts was compared with the effect of the juglone solution; (2) the effect of the *I. glandulifera*- and *I. parviflora*-neighboring plants was compared with the effect of *I. glandulifera* and *I. parviflora* by calculating the "*Impatiens glandulifera* index" and the "*Impatiens parviflora* index", analogous to the juglone index (Table 3). Since three concentrations of juglone were tested (the authors who proposed this index also used several concentrations of juglone [89]), the actual effects of *I. glandulifera* and *I. parviflora* and neighboring species were compared with three juglone indices: juglone-low index (juglone concentration 0.01 mM), juglone-intermediate index (juglone concentration 0.1 mM) and juglone-high index (juglone concentration 1.0 mM). Similarly, three "*Impatiens glandulifera* indexes" and three "*Impatiens parviflora* indexes" had to be introduced, as the effects of three concentrations of leaf extracts of each plant species were used in the experiments: "*Impatiens glandulifera*-low index" (25 mg/mL), "*Impatiens glandulifera*-intermediate index" (50 mg/mL), and "*Impatiens glandulifera*-high index" (100 mg/mL), and similarly for the three *Impatiens parviflora* indexes.

The juglone-low index (0.01 mM) for aqueous extracts of all species tested was greater than one, indicating strong inhibitory effect, with *I. glandulifera* showing the highest index (96.50). An exception was observed for *Juglans regia*, where the effect was weaker (0.96) (Table 3). The juglone-intermediate index (0.1 mM) for extracts of all species was also greater than one, with *I. glandulifera* exhibiting the highest value (85.50). Exceptions were indicated for *Juglans regia* and *Urtica dioica* at concentrations of 25 mL/mL, where the index was 0.85 and 0.94, respectively. The juglone-high index (0.1 mM) was greater than one for most species, with *I. glandulifera* exhibiting the highest value (69.00). However, values below one were recorded for *Juglans regia*, *Alnus glutinosa*, *Epilobium hirsutum*, *Geranium robertianum*, *Geum urbanum*, and *Urtica dioica* indicating lower inhibitory effect for these species.

Table 2. Significance of differences ($p < 0.05$ *, $p < 0.01$ **, $p < 0.001$ ***) in germination, radicle and hypocotyl length of *Lepidium sativum* germinated in the aqueous leaf extracts of various concentration of plant species comprising communities of *Impatiens glandulifera* and *I. parviflora*, compared to control (H_2O), *I. glandulifera*, and *I. parviflora*.

Concentration of Extract	Significance ($p < 0.05$ *, $p < 0.01$ **, $p < 0.001$ ***) of Differences Compared to								
	Control (H_2O)			*Impatiens glandulifera*			*Impatiens parviflora*		
	25	50	100	25	50	100	25	50	100
Species					**Germination**				
Impatiens glandulifera	***	***	***	-	-	-	**	***	ns
Impatiens parviflora	***	***	***	**	***	ns	-	-	-
Juglans regia	ns	ns	ns	***	***	**	***	***	*
Alliaria petiolata	***	***	***	***	***	ns	***	***	ns
Alnus glutinosa	***	**	**	***	***	ns	*	**	ns
Calystegia sepium	***	***	***	ns	***	ns	**	***	ns
Epilobium hirsutum	***	***	***	***	ns	ns	***	**	ns
Galium aparine	***	***	***	***	***	ns	***	***	ns
Geranium robertianum	*	***	**	***	***	ns	***	ns	ns
Geum urbanum	***	***	*	***	**	ns	ns	***	ns
Phragmites australis	***	***	***	ns	ns	ns	ns	**	ns
Rubus caesius	***	***	***	ns	ns	ns	**	***	ns
Stelaria media	***	***	ns	***	**	*	**	ns	*
Urtica dioica	ns	***	ns	***	***	ns	***	ns	ns
Erigeon annuus	***	***	***	**	ns	ns	***	***	ns
Helianthus tuberosus	***	***	***	*	***	ns	***	***	ns
					Radicle				
Impatiens glandulifera	***	***	***	-	-	-	***	ns	ns
Impatiens parviflora	***	***	***	***	ns	ns	-	-	-
Juglans regia	ns	***	**	***	***	*	***	*	*
Alliaria petiolata	***	***	***	ns	*	ns	***	ns	ns
Alnus glutinosa	***	*	*	***	***	**	ns	***	**
Calystegia sepium	***	***	ns	***	***	***	ns	*	***
Epilobium hirsutum	ns	***	***	***	***	ns	***	**	ns
Galium aparine	***	***	***	***	ns	ns	ns	ns	ns
Geranium robertianum	***	ns	***	***	***	ns	**	***	ns
Geum urbanum	***	***	**	ns	ns	*	ns	ns	*
Phragmites australis	***	***	***	ns	***	ns	ns	***	ns
Rubus caesius	***	***	***	ns	***	ns	*	***	ns
Stelaria media	***	***	***	ns	ns	ns	***	ns	ns
Urtica dioica	ns	ns	*	***	***	**	***	***	**
Erigeon annuus	**	***	ns	***	***	**	***	***	**
Helianthus tuberosus	***	**	ns	***	***	***	ns	***	***
					Hypocotyl				
Impatiens glandulifera	***	***	***	-	-	-	***	**	ns
Impatiens parviflora	*	***	***	***	**	ns	-	-	-
Juglans regia	ns	***	*	***	***	**	*	ns	**
Alliaria petiolata	***	***	***	ns	***	ns	ns	ns	ns
Alnus glutinosa	***	***	*	**	***	**	ns	**	**
Calystegia sepium	***	***	**	ns	ns	ns	ns	ns	ns
Epilobium hirsutum	*	***	***	***	ns	ns	ns	ns	ns
Galium aparine	***	***	***	ns	ns	ns	***	*	ns
Geranium robertianum	***	**	***	ns	***	ns	*	**	ns
Geum urbanum	ns	***	**	***	ns	*	ns	**	*
Phragmites australis	**	*	***	***	***	ns	ns	***	ns
Rubus caesius	***	***	***	ns	***	ns	**	ns	ns
Stelaria media	***	***	***	ns	ns	ns	***	**	ns
Urtica dioica	ns	ns	ns	***	***	***	ns	***	***
Erigeon annuus	ns	***	ns	***	ns	***	*	ns	***
Helianthus tuberosus	**	**	ns	***	***	**	ns	**	**

Notes: ns—not significant.

Table 3. Juglone index [89], "*Impatiens glandulifera* index", "*Impatiens parviflora* index", determined for *Lepidium sativum* by treating its seeds with leaf extracts of various plant species (full names of the species in respect to abbreviations are provided in Section 2).

Species Abbr.	Leaf Extract	Juglone Index			"*Impatiens glandulifera* Index" Concentrations			"*Impatiens parviflora* Index"		
		Juglone, mM			Leaf Extract, mg/mL			Leaf Extract, mg/mL		
		0.01	0.1	1	25	50	100	25	50	100
Control (H_2O)		0.90	0.80	0.65	**0.37**	**0.14**	**0.01**	**0.64**	**0.29**	**0.01**
Juglone, 0.01 mM		-	-	-	**0.40**	**0.15**	**0.01**	**0.70**	**0.32**	**0.02**
Juglone, 0.1 mM		-	-	-	**0.46**	**0.17**	**0.01**	**0.80**	**0.36**	**0.02**
Juglone, 1.0 mM		-	-	-	**0.57**	**0.21**	**0.01**	**0.99**	**0.44**	**0.02**
Imp gla	25	2.46	2.18	1.76	-	-	-	1.74	**0.78**	**0.04**
Imp gla	50	6.58	5.83	4.70	-	-	-	4.64	2.08	**0.10**
Imp gla	100	96.50	85.50	69.00	-	-	-	68.00	30.50	1.50
Imp par	25	1.42	1.26	1.02	**0.57**	**0.21**	**0.01**	-	-	-
Imp par	50	3.16	2.80	2.26	1.28	**0.48**	**0.03**	-	-	-
Imp par	100	57.90	51.30	41.40	23.40	8.70	**0.60**	-	-	-
Jug reg	25	**0.96**	**0.85**	**0.68**	**0.39**	**0.14**	**0.01**	**0.67**	**0.30**	**0.01**
Jug reg	50	2.11	1.87	1.51	**0.85**	**0.32**	**0.02**	1.48	**0.67**	**0.03**
Jug reg	100	8.94	7.92	6.39	3.61	1.34	**0.09**	6.30	2.82	**0.14**
All pet	25	4.42	3.92	3.16	1.79	**0.66**	**0.05**	3.11	1.40	**0.07**
All pet	50	5.46	4.84	3.90	2.21	**0.82**	**0.06**	3.85	1.73	**0.08**
All pet	100	NA	NA	NA	NA	NA	**NA**	NA	NA	**NA**
Aln glu	25	1.31	1.16	**0.94**	**0.53**	**0.20**	**0.01**	**0.93**	**0.41**	**0.02**
Aln glu	50	1.81	1.61	1.30	**0.73**	**0.27**	**0.02**	1.28	**0.57**	**0.03**
Aln glu	100	12.06	10.69	8.63	4.88	1.81	**0.13**	8.50	3.81	**0.19**
Cal sep	25	1.72	1.53	1.23	**0.70**	**0.26**	**0.02**	1.21	**0.54**	**0.03**
Cal sep	50	5.85	5.18	4.18	2.36	**0.88**	**0.06**	4.12	1.85	**0.09**
Cal sep	100	17.55	15.55	12.55	7.09	2.64	**0.18**	12.36	5.55	**0.27**
Epi hir	25	1.16	1.02	**0.83**	**0.47**	**0.17**	**0.01**	**0.81**	**0.37**	**0.02**
Epi hir	50	3.39	3.00	2.42	1.37	**0.51**	**0.04**	2.39	1.07	**0.05**
Epi hir	100	NA	NA	NA	NA	NA	**NA**	NA	NA	**NA**
Gal apa	25	3.24	2.87	2.32	1.31	**0.49**	**0.03**	2.28	1.02	**0.05**
Gal apa	50	8.42	7.46	6.02	3.40	1.27	**0.09**	5.93	2.66	**0.13**
Gal apa	100	34.06	30.18	24.35	13.76	5.12	**0.35**	24.00	10.76	**0.53**
Ger rob	25	1.21	1.07	**0.86**	**0.49**	**0.18**	**0.01**	**0.85**	**0.38**	**0.02**
Ger rob	50	1.61	1.43	1.15	**0.65**	**0.24**	**0.02**	1.13	**0.51**	**0.03**
Ger rob	100	43.86	38.86	31.36	17.73	6.59	**0.45**	30.91	13.86	**0.68**
Geu urb	25	1.36	1.21	**0.97**	**0.55**	**0.20**	**0.01**	**0.96**	**0.43**	**0.02**
Geu urb	50	8.50	7.53	6.08	3.44	1.28	**0.09**	5.99	2.69	**0.13**
Geu urb	100	15.82	14.02	11.31	6.39	2.38	**0.16**	11.15	5.00	**0.25**
Phr aus	25	1.61	1.43	1.15	**0.65**	**0.24**	**0.02**	1.14	**0.51**	**0.03**
Phr aus	50	2.38	2.11	1.70	**0.96**	**0.36**	**0.02**	1.67	**0.75**	**0.04**
Phr aus	100	NA	NA	NA	NA	NA	**NA**	NA	NA	**NA**
Rub cae	25	1.98	1.76	1.42	**0.80**	**0.30**	**0.02**	1.40	**0.63**	**0.03**
Rub cae	50	3.44	3.05	2.46	1.39	**0.52**	**0.04**	2.42	1.09	**0.05**
Rub cae	100	NA	NA	NA	NA	NA	**NA**	NA	NA	**NA**
Ste med	25	2.03	1.79	1.45	**0.82**	**0.30**	**0.02**	1.43	**0.64**	**0.03**
Ste med	50	5.19	4.60	3.71	2.10	**0.78**	**0.05**	3.66	1.64	**0.08**
Ste med	100	17.71	15.69	12.66	7.16	2.66	**0.18**	12.48	5.60	**0.28**
Urt dio	25	1.06	**0.94**	**0.76**	**0.43**	**0.16**	**0.01**	**0.74**	**0.33**	**0.02**
Urt dio	50	1.55	1.37	1.10	**0.62**	**0.23**	**0.02**	1.09	**0.49**	**0.02**
Urt dio	100	8.69	7.70	6.22	3.51	1.31	**0.09**	6.13	2.75	**0.14**
Eri ann	25	1.65	1.46	1.18	**0.67**	**0.25**	**0.02**	1.16	**0.52**	**0.03**
Eri ann	50	4.70	4.16	3.36	1.90	**0.71**	**0.05**	3.31	1.48	**0.07**
Eri ann	100	7.37	6.53	5.27	2.98	1.11	**0.08**	5.19	2.33	**0.11**
Hel tub	25	1.75	1.55	1.25	**0.71**	**0.26**	**0.02**	1.23	**0.55**	**0.03**
Hel tub	50	3.13	2.78	2.24	1.27	**0.47**	**0.03**	2.21	**0.99**	**0.05**
Hel tub	100	11.35	10.06	8.12	4.59	1.71	**0.12**	8.00	3.59	**0.18**

Notes: NA—non applicable; numbers in bold denote a weaker effect of the plant species compared to juglone, *Impatiens glandulifera*, and *Impatiens parviflora*, respectively; and vice versa, numbers in ordinary black denote a stronger effect.

The "*Impatiens glandulifera*-low index" (25 mg/mL) at species extract concentration of 25 mg/mL was greater than one for *Galium aparine* and *Alliaria petiolata* (1.31–1.79). Other species and juglone exhibited lower values (0.39–0.82). The index at species extract concentration of 50 mg/mL was greater than one for all species, with *Geum urbanum* exhibiting the highest value (3.44). Exceptions were indicated for juglone, *Juglans regia*, *Alnus glutinosa*, *Geranium robertianum*, *Phragmites australis*, and *Urtica dioica* (0.46–0.96). The index at species extract concentration of 100 mg/mL was also greater than one for all species, except for juglone (0.57), with *Impatiens parviflora* exhibiting the highest value (23.40).

"*Impatiens glandulifera*-intermediate index" (50 mg/mL) at species extract concentration of 25 and 50 mg/mL was lower than one for all tested species, ranging from 0.14 to 0.66. However, an extract concentration of 100 mg/mL was greater than one for all species, except for juglone (0.21), with *Impatiens parviflora* exhibiting the highest value (8.7). "*Impatiens glandulifera*-high index" (100 mg/mL) remained below one across all species at extract concentrations of 25, 50, and 100 mg/mL, with values ranging from 0.01 to 0.6. The results suggest that the index for *Impatiens glandulifera* extracts was concentration-dependent, being greater than one for most species at 50 and 100 mg/mL, with *Impatiens parviflora* showing the highest value at 100 mg/mL. At 25 mg/mL, only *Galium aparine* and *Alliaria petiolata* had indices greater than one. The "*Impatiens glandulifera*-high index" remained below one at all concentrations, indicating weaker inhibition.

The "*Impatiens parviflora*-low index" (25 mg/mL) at species extract concentration of 25 mg/mL was greater than one for *Impatiens glandulifera*, *Alliaria petiolata*, *Calystegia sepium*, *Galium aparine*, *Phragmites australis*, *Rubus caesius*, *Stellaria media*, *Erigeron annuus*, and *Helianthus tuberosus*, ranging from 1.14 to 3.11, with *Alliaria petiolata* having the highest value. Other species and juglone exhibited lower values (0.70–0.96). The index at species extract concentration of 50 mg/mL was greater than one for all species, except for juglone (0.8); *Geum urbanum* exhibited the highest value (5.99). The index at species extract concentration of 100 mg/mL was also greater than one for all species, except for juglone (0.99); *Impatiens glandulifera* exhibited the highest value (68). "*Impatiens parviflora*-intermediate index" (50 mg/mL) at species extract concentration of 25 mg/mL was greater than one for *Galium aparine* and *Alliaria petiolata* (1.02–1.40). Other species and juglone exhibited lower values (0.30–0.78). The index at species extract concentration of 50 mg/mL was greater than one for *Impatiens glandulifera*, *Alliaria petiolata*, *Calystegia sepium*, *Epilobium hirsutum*, *Galium aparine*, *Geum urbanum*, *Rubus caesius*, *Stellaria media*, and *Erigeron annuus*; *Geum urbanum* exhibited the highest value (2.69). An extract concentration of 100 mg/mL was greater than one for all species, except for juglone (0.44), with *Impatiens glandulifera* exhibiting the highest value (30.5). "*Impatiens parviflora*-high index" (100 mg/mL) remained below one across all species at extract concentrations of 25, 50, and 100 mg/mL, with values ranging from 0.02 to 0.68. Exception was indicated for *Impatiens glandulifera*, where the value was 1.5.

3.2. Total Phenol Content and Radical Scavenging Activities of *Impatiens glandulifera* and *I. parviflora*

The total phenolic concentration of the leaves of 20 *Impatiens glandulifera* populations ranged from 64.45 ± 4.46 to 149.06 ± 10.33 GAE mg/g d. m., min and max values differing significantly ($p < 0.05$) 2.3 times, with a mean value of 92.97 ± 8.76 GAE mg/g d. m. (Figure 5). The lowest phenolic content was observed in Kaunas-Zuikinė (Central Lithuania) population, while the highest was recorded in Belvederis population (Central Lithuania). DPPH$^{\bullet}$ radical scavenging activity of *I. glandulifera* populations varied from 5.2 ± 0.39 to 11.3 ± 0.18 GAE mg/g d. m., min and max values differing significantly ($p < 0.05$) 2.2 times, with an average activity of 8.82 ± 0.91 GAE mg/g d. m. The lowest DPPH$^{\bullet}$ activity was detected in Kaunas-Zuikinė (Central Lithuania) population, whereas the highest activity was observed in Rokiškis population (Eastern Lithuania). ABTS radical scavenging activity ranged from 221 ± 7 to 590 ± 6 μmol/g d. m., min and max values differing significantly ($p < 0.05$) 2.7 times, with an average value of 351.7 ± 38 μmol/g d. m. The lowest ABTS$^{\bullet+}$ activity was observed in Kaunas-Zuikinė population (Central Lithuania), while the highest was recorded in Belvederis population (Central Lithuania), indicating variation in antioxidant properties among populations. Correlation analysis between all the measured indices revealed that *I. glandulifera* populations exhibited a significant positive correlation between total phenols and DPPH$^{\bullet}$ ($r = 0.59$, $p = 0.06$), between phenols and $ABTS^{\bullet +}$ (r = 0.91, $p < 0.001$), and between $DPPH^{\bullet}$ and $ATBS^{\bullet +}$ (r = 0.79, $p < 0.001$).

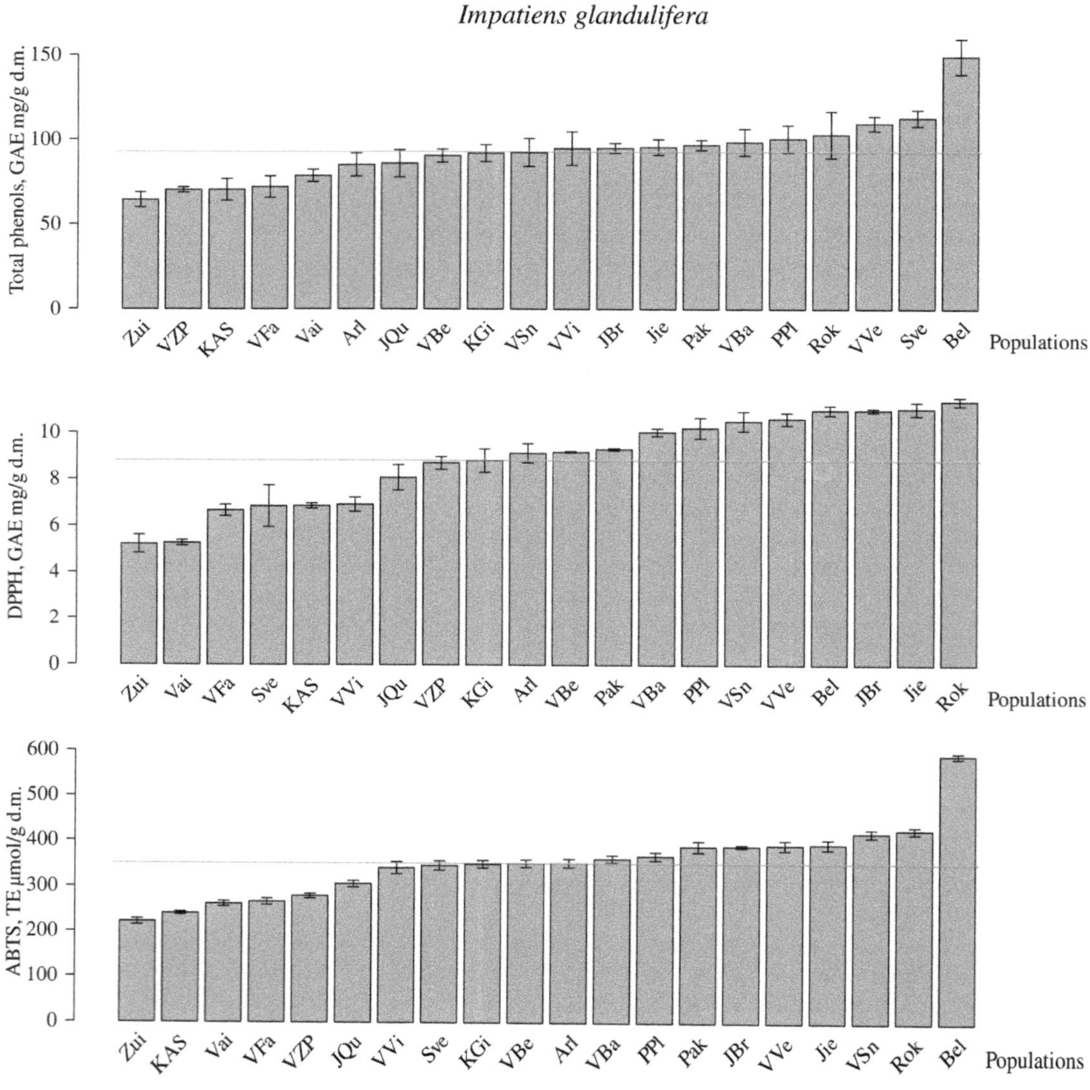

Figure 5. Content of the total phenols (GAE mg/g d. m.), 2,2-Diphenyl-1-picrylhydrazyl (DPPH•) radical scavenging activity (GAE mg/g d. m.), and 2,20-azinobis-(3-ethyl-benzothiazoline-6-sulfonic acid) $ABTS^{\bullet +}$ radical scavenging activity (TE µmol/g d. m.) in the ethanolic extracts of the leaves of *Impatiens glandulifera* populations. Columns indicate mean values, and whiskers indicate intervals of confidence.

The total phenolic concentration of 20 *Impatiens parviflora* populations ranged from 24.44 ± 0.71 to 64.72 ± 1.42 GAE mg/g d. m., min and max values differing significantly ($p < 0.05$) 2.6 times, with a mean value of 44.09 ± 4.49 GAE mg/g d. m. (Figure 6). The lowest phenolic concentration was observed in Kaunas-Giraitė (Central Lithuania) population, while the highest was recorded in Plazė population (Western Lithuania). The $DPPH^{\bullet}$ radical scavenging activity of *I. parviflora* populations ranged from 1.43 ± 0.17 to 7.41 ± 0.2 GAE mg/g d. m., min and max values differing significantly ($p < 0.05$)

5.2 times, with an average activity of 3.32 ± 0.74 GAE mg/g d. m. The lowest DPPH$^{\bullet}$ activity was found in Palanga-Botanical Garden (Western Lithuania) population, whereas the highest was recorded in Kaunas-A. Šančiai (Central Lithuania) population. ABTS$^{\bullet+}$ radical scavenging activity was within the range of 57.84 ± 2.28 to 154.07 ± 1.53 µmol/g d. m., min and max values differing significantly ($p < 0.05$) 2.7 times, with a mean value of 107.3 ± 14.51 µmol/g d. m. The lowest ABTS$^{\bullet+}$ activity was found in Vilnius-Šnipiškės population (Eastern Lithuania), while the highest activity was observed in Kaunas-Giraitė population (Central Lithuania). *I. parviflora* populations showed a significant positive correlation between phenols and DPPH$^{\bullet}$ (r = 0.23, $p < 0.05$), between phenols and ABTS$^{\bullet+}$ (r = 0.27, $p < 0.05$), and between DPPH$^{\bullet}$ and ABTS$^{\bullet+}$ (r = 0.61, $p = 0.013$).

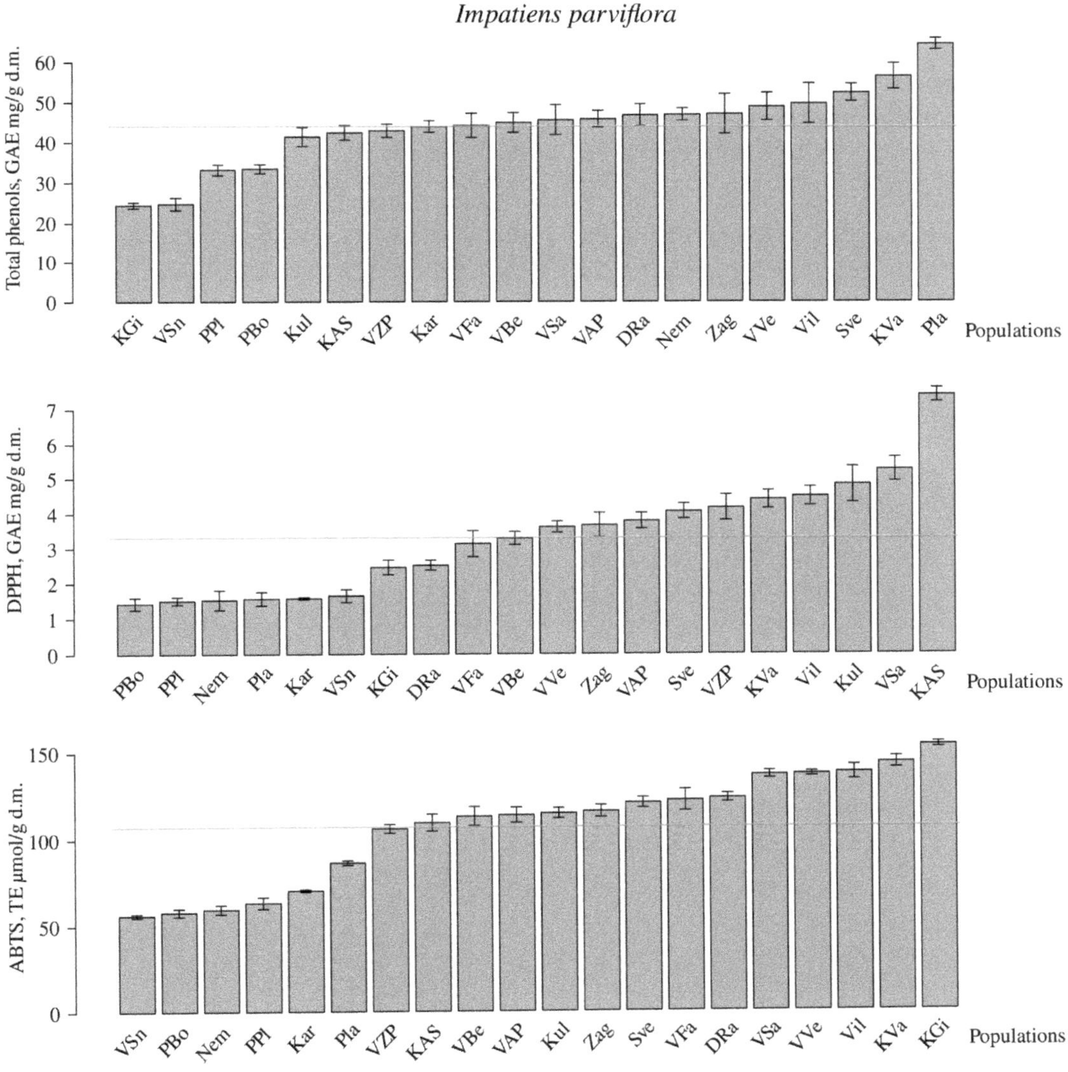

Figure 6. Content of the total phenols (GAE mg/g d. m.), 2,2-Diphenyl-1-picrylhydrazyl (DPPH) radical scavenging activity (GAE mg/g d. m.), and 2,20-azinobis-(3-ethyl-benzothiazoline-6-sulfonic acid) ABTS radical scavenging activity (TE µmol/g d. m.) in the ethanolic extracts of the leaves of *Impatiens parviflora* populations. Columns indicate mean values, and whiskers indicate intervals of confidence.

The mean total phenols content was 2.1 times higher for *I. glandulifera* populations (92.97 ± 8.76) compared to *I. parviflora* populations (44.09 ± 4.49). The mean value of DPPH$^•$ radical scavenging activity was 2.7 times greater in *I. glandulifera* populations (8.82 ± 0. 91) than in *I. parviflora* populations (3.32 ± 0.74). Similarly, ABTS$^{•+}$ radical scavenging activity was 3.3 times higher in *I. glandulifera* populations (351.7 ± 38.04) compared to *I. parviflora* populations (107.3 ± 14.51).

4. Discussion

Previous studies assessing the competitive ability and allelochemicals of *Impatiens* spp. in Europe have in some cases looked at both species [29,56,58,59,62,101,102], with a predominance of studies focusing on *Impatiens glandulifera* [35,54,57,58,61,63,99,103–105] and just a few investigations on *Impatiens parviflora* alone [60,80]. Most allelopathic examinations of *Impatiens* spp. were held in the countries of Central Europe. In addition, the allelopathic and competitive abilities of invasive *Impatiens* spp. were assessed against a small number of neighboring native species [54,57,80], which differed in each study depending on geography/climate: four species in UK [54], three species in another UK assessment [58], four species in Germany [57], five species in Czech [80]. Furthermore, former allelochemical examinations of *Impatiens* spp. were limited in sites. Our study aimed to increase understanding of the competitive ability of invasive balsams by investigating allelopathic potential of 11 co-occurring pristine species of Lithuania.

The allelopathic effects depend on the specific interactions and the concentrations of the allelochemicals involved [106–108]. In various sites, invasive plants and their neighbors differ in abundance, mass of individuals, and many other characteristics. In evaluating allelopathic potential of native and invasive species, two–four different concentrations of the extracts have been examined [28,102]. It was the reason for us to test leaf extracts of *Impatiens* spp. and their neighbors using three different concentrations. Our study demonstrated that aqueous extracts of *Impatiens glandulifera* and *I. parviflora* exhibit concentration-dependent allelopathic effects on the germination, radicle and hypocotyl growth of *Lepidium sativum*. The highest number of statistically significant differences between *I. glandulifera* and/or *I. parviflora* and the co-occurring species was found for *Lepidium sativum* when seeds of this plant were germinated in leaf extracts of 50 mg/mL, followed by the weakest concentration (25 mg/mL). The effect of the extracts of the highest concentration (100 mg/mL) on the biotest plant was not statistically significant between compared *Impatiens* and the other community species. Exceptions were extracts of all concentrations (25, 50, and 100 mg/mL), for which both species of *Impatiens* differed significantly from *Juglans regia* (a species cultivated in Lithuania) in terms of all the biotest parameters (germination, radicle and hypocotyl growth). The highest number of statistically significant cases of difference was found for the effect of leaf extracts on radicle growth of the biotest plant, and the lowest number on hypocotyl growth.

In our study, at all leaf extract concentrations, significantly higher allelopathic potential on radicle growth was characteristic for *I. glandulifera* compared to (a) native species *Alnus glutinosa*, *Callystegia sepium*, and *Urtica dioica*, and (b) alien *Helianthus tuberosus* and *Erigeron annuum*. Based on hypocotyl growth, significantly higher allelopathic potential of *I. glandulifera* was limited to *Urtica dioica* and *Helianthus tuberosus*. Our results are in support of UK investigation findings about the higher competitive ability of *Impatiens glandulifera* compared to co-occurring *Tanacetum vulgare*, *Urtica dioica*, *Chelidonium majus*, and *Arabidopsis thaliana* [54]. Also, our facts correspond to a negative effect of seedlings and juveniles of *I. glandulifera* on the growth of *Geum urbanum*, *Filipendula ulmaria*, *Urtica dioica*, and *Salix fragilis* (investigation performed in Germany) [57]. Response similar to our data direction was observed examining *Impatiens glandulifera* effect on *Trifolium pratense*, *Linum grandiflora*, and *Silene dioica* (UK analyses) [58].

In our study at all leaf extract concentrations, *I. parviflora* showed significantly higher allelopathic potential on radicle growth compared to native species *Urtica dioica* and invasive in Lithuania species *Erigeron annuum*. Facts about allelopathic effects of *Impatiens parviflora* on neighboring species are scarce and, in some cases, opposite to what we got in our experiments. Evaluation of *Impatiens parviflora* potential impacts on native species *Athyrium filix-femina*, *Dryopteris filix-mas*, *Fragaria moschata*, *Luzula luzuloides*, and *Poa nemoralis* (Czech investigations) revealed that the species considered are most likely to prevent *I. parviflora* invasion [80]. In wide scope assessment (18 species investigated, missing *Impatiens* spp.) performed in Germany, it was shown that the native plant community imposed allelopathic effects on the introduced plants. The invasive and non-invasive introduced plants, however, did not differ in their tolerance to such allelopathic effects of the native plant community [18].

Urtica dioica and *Impatiens glandulifera* often co-occur in European riparian herb communities [105]. Evaluation of neighboring species in Lithuanian sites with *Impatiens parviflora* revealed *Urtica dioica* as the most frequently met native species [73,75,77]. In our study, *Urtica dioica* effect on the radicles and hypocotyls of *Lepidium sativum* was significantly lower compared to *Impatiens glandulifera* and *I. parviflora*. Experimental findings of *Impatiens* material collected in other countries (UK, Germany, India)support our data: Among selected species, *Urtica dioica* was the most affected by *Impatiens glandulifera* [104]. Compared to *I. parviflora*, *I. glandulifera* showed stronger inhibitory effects on biotest species, with greater reductions of growth of radicle and hypocotyl, particularly at higher concentrations of the extract. These findings indicate that *I. glandulifera* might have a more pronounced allelopathic potential on co-occurring plants compared to *I. parviflora*.

Juglone, 5-hydroxy-1,4-naphthoquinone, known for its allelopathic properties, based on mitosis inhibition, and creation of reactive oxygen species, is used in allelopathy studies [64,90,102,109,110]. In addition to absolute values of germination and seedling growth of *Lepidium sativum* under the effect of plant leaf extracts, for analyses we have employed juglone index (Table 3) [89] as a cumulative parameter, encompassing numerous data (germination, radicle growth, and hypocotyl growth of biotest species) into one.

In our study, the juglone index provided additional insights into the inhibitory potential of both *Impatiens* species (Table 3). For most species, juglone indexes indicated stronger allelopathic potential of *Impatiens* spp. and their neighboring species compared to juglone (Table 3). Our results agree with the juglone index evaluation for 14 alien woody and 20 alien herbaceous plant species in Hungary (under edaphic and climatic conditions different from the study sites in Lithuania), where the leaf extracts of the same species in both studies showed a higher effect than 1 mM the juglone [102]. In addition to the juglone index, we have introduced analogous indexes—*Impatiens glandulifera* indexes and *I. parviflora* indexes, which are devoted to comparing balsam effects with neighboring species (see Section 2). "*Impatiens glandulifera*-highest index" showed a weaker effect than *I. glandulifera* for all species tested. "*Impatiens parviflora*-highest index" showed that all the species tested had a weaker effect than *I. parviflora*, except *I. glandulifera*. *Impatiens glandulifera* demonstrated the highest juglone index values. These differences underscore the influence of environmental and genetic factors on the biochemical potential of invasive plant species. Our former research on genetic diversity of Lithuanian *Impatiens* spp. populations [74,76] has shown nearly two times the genetic diversity of Lithuanian populations of *Impatiens glandulifera*, compared to *I. parviflora*.

Other researchers have found that naphthoquinones quantity of *Impatiens* species may differ 2.5–37 times, being the highest in *I. glandulifera* [29]. In the competition experiment

(Czech) between three European *Impatiens* species, *Impatiens glandulifera* was the strongest competitor, was the tallest, and produced the greatest biomass. Small size and high plasticity were characteristic of *I. parviflora* [56]. Our results about allelopathic potential and phenolics content, in addition to free radical scavenging ability, confirm the supreme competitive ability of *I. glandulifera* compared to *I. parviflora*.

In our study, differences in phenolics, DPPH•, and ABTS•+ parameters among populations of *I. glandulifera* (Figure 5) or among populations of *I. parviflora* (Figure 6) could be explained by various reasons. Despite the small area of Lithuania, populations of *Impatiens* differ in the extent of genetic diversity, evaluated by a set of molecular markers [72–74,77], and also by various abiotic and biotic factors of the environment [75,76]. Examined in our study, populations of *Impatiens parviflora* near the seacoast had significantly ($p < 0.05$) lower distinct activities of DPPH• and ABTS•+ scavenging compared to populations of Central or Eastern Lithuania (Figure 6). Soil microorganisms might be impaired as important determinants of the allelopathic activity of plants [111]. Other group assessments have shown that disturbances could lead to increased invasibility of alien species [112]. It was shown that the invasion by *I. parviflora* is confounded with the degradation of forest floor communities, associated with nitrification by numerous game animals, and often leads to the expansion of ruderal species, such as *Geranium robertianum*, and especially *Aegopodium podagraria* [80]. Biochemical differences of our populations of *Impatiens* (Figures 5 and 6) could be influenced by distinct pressure of anthropogenic factors registered in sites of the former research (vicinity to the roads, settlements [72,73]).

Finally, our research is significant for several reasons. First, it contributes to the growing body of knowledge regarding the ecological impacts of invasive species, particularly in the context of allelopathy. By elucidating the allelopathic mechanisms employed by *Impatiens parviflora* and *I. glandulifera*, this study may provide insights into how these species interact with native plant communities and the potential consequences for biodiversity. The findings may inform management strategies aimed at controlling the spread of these invasive species, ultimately aiding in the preservation of native ecosystems. Incorporating allelopathic effects of invasive alien plant species into ecological models and management plans can improve the accuracy and effectiveness of restoration efforts, ensuring the long-term sustainability of ecosystems [24,113]. Lastly, this research underscores the importance of understanding the biochemical interactions within plant communities, particularly in the face of increasing global plant invasions driven by human activities.

5. Conclusions

In our study, at all leaf extract concentrations, significantly higher allelopathic potential on radicle growth was characteristic for *I. glandulifera* compared to native species *Alnus glutinosa*, *Callystegia sepium*, and *Urtica dioica*.

At all leaf extract concentrations, *I. parviflora* showed significantly higher allelopathic potential on radicle growth compared to native species *Urtica dioica*.

Impatiens glandulifera had a higher juglone index than *Impatiens parviflora*. *I. glandulifera* also had the highest juglone index compared to all neighboring species studied.

The differences between the populations in the content of phenolic compounds and the activity of DPPH and ABTS radical scavenging were very high (two to three times or more) in case of both *Impatiens* species.

The mean values of total phenolic content, DPPH• and ABTS•+ radical scavenging activity of *I. glandulifera* populations were 2.1, 2.7, and 3.3 times higher, respectively, than those of *I. parviflora* populations.

In general, our results about allelopathic potential and phenolics content, as well as free radical scavenging ability, confirm the supreme competitive ability of *I. glandulifera* compared to *I. parviflora* and native co-occurring species.

Author Contributions: Conceptualization, E.K.; data curation, E.K., L.S. and L.J.; software, T.R., L.S. and L.J.; funding acquisition, E.K.; investigation, L.S., L.J., E.K. and L.S.; methodology, E.K., L.J., V.M., R.B. and V.J.; project administration, E.K.; visualization, L.J., E.K.-K. and T.R.; writing—original draft, E.K., L.J., E.K.-K., R.B., V.J., T.R. and V.M.; writing—review and editing, E.K., L.J., E.K.-K., R.B., V.J. and V.M. All authors have read and agreed to the published version of the manuscript.

Funding: The study was funded by the Research Council of Lithuania (Grant No. SIT-02/2015).

Institutional Review Board Statement: Not applicable.

Data Availability Statement: Data is contained within the article.

Conflicts of Interest: The authors declare no conflicts of interest.

References

1. Richardson, D.M.; Pyšek, P.; Rejmanek, M.; Barbour, M.G.; Panetta, F.D.; West, C.J. Naturalization and invasion of alien plants: Concepts and definitions. *Divers. Distrib.* **2000**, *6*, 93–107. [CrossRef]
2. Weidlich, E.; Florido, F.; Sorrini, T.; Brancalion, P. Controlling invasive plant species in ecological restoration: A global review. *J. Appl. Ecol.* **2020**, *57*, 1806–1817. [CrossRef]
3. Vujanović, D.; Losapio, G.; Milić, S.; Milić, D. The impact of multiple species invasion on soil and plant communities increases with invasive species co-occurrence. *Front. Plant Sci.* **2022**, *13*, 875824. [CrossRef] [PubMed]
4. Baker, H.G. Self-compatibility and establishment after 'long-distance' dispersal. *Evolution* **1955**, *9*, 347–349.
5. Williamson, M.; Brown, K.C. The analysis and modelling of British invasions. *Philos. Trans. R. Soc. Lond. B Biol. Sci.* **1986**, *314*, 505–522.
6. Elton, C.S. *The Ecology of Invasions by Animals and Plants*; Springer: Methuen, MA, USA; London, UK, 1958.
7. Richardson, D.M.; Pyšek, P. Fifty years of invasion ecology—The legacy of Charles Elton. *Divers. Distrib.* **2008**, *14*, 161–168. [CrossRef]
8. Richardson, D.; Pyšek, P. Plant invasions: Merging the concepts of species invasiveness and community invasibility. *Prog. Phys. Geogr. Earth Environ.* **2006**, *30*, 409–431. [CrossRef]
9. Enders, M.; Havemann, F.; Ruland, F.; Bernard-Verdier, M.; Catford, J.A.; Gómez-Aparicio, L.; Haider, S.; Heger, T.; Kueffer, C.; Kühn, I.; et al. A conceptual map of invasion biology: Integrating hypotheses into a consensus network. *Glob. Ecol. Biogeogr.* **2020**, *29*, 978–979. [CrossRef] [PubMed]
10. Ridenour, W.M.; Callaway, R.M. The relative importance of allelopathy in interference: The effects of an invasive weed on a native bunchgrass. *Oecologia* **2001**, *126*, 444–450. [CrossRef] [PubMed]
11. Hierro, J.L.; Callaway, R.M. Allelopathy and exotic plant invasion. *Plant Soil* **2003**, *256*, 29–39. [CrossRef]
12. Bais, H.P.; Vepachedu, R.; Gilroy, S.; Callaway, R.M. Allelopathy and exotic plant invasion: The role of allelochemicals in plant invasions. *Plant Ecol.* **2003**, *167*, 1–12.
13. Callaway, R.M.; Ridenour, W.M. Novel weapons: Invasive success and the evolution of increased competitive ability. *Front. Ecol. Environ.* **2004**, *2*, 436–443. [CrossRef]
14. Rice, E.L. *Allelopathy*; Academic Press: New York, NY, USA, 1984.
15. Inderjit. Soils: Environmental effects on allelochemical activity. *Agron. J.* **2001**, *93*, 79–88. [CrossRef]
16. Inderjit; Seastedt, T.R.; Callaway, R.M.; Pollock, J.L.; Kaur, J. Allelopathy and plant invasions: Traditional, congeneric, and bio-geographical approaches. *Biol. Invasions* **2008**, *10*, 875–890. [CrossRef]
17. Gruntman, M.; Zieger, S.; Tielbörger, K. Invasive success and the evolution of enhanced weaponry. *Oikos* **2015**, *125*, 59–65. [CrossRef]
18. Ning, L.; Yu, F.-H.; van Kleunen, M. Allelopathy of a native grassland community as a potential mechanism of resistance against invasion by introduced plants. *Biol. Invasions* **2016**, *18*, 3481–3493. [CrossRef]
19. Zhu, X.; Li, W.; Shao, H.; Tang, S. Selected aspects of invasive *Solidago canadensis* with an emphasis on its allelopathic abilities: A review. *Chem. Biodivers.* **2022**, *19*, e202200728. [CrossRef] [PubMed]
20. Kalske, A.; Mäkinen, E.; Ramula, S. Allelopathy by the invasive garden lupine inhibits the germination of native herbs. *Botany* **2023**, *101*, 24–29. [CrossRef]

21. Levačić, D.; Perković, L.; Vuković, N.; Jelaska, S.D. Bohemian knotweed *Reynoutria × bohemica* Chrtek et Chrtková seems not to rely heavily on allelopathy for its persistence in invaded sites in the Southwest part of the Zagreb, Croatia. *Plants* **2023**, *12*, 2222. [CrossRef] [PubMed]
22. Kalisz, S.; Kivlin, S.N.; Bialic-Murphy, L. Allelopathy is pervasive in invasive plants. *Biol. Invasions* **2021**, *23*, 367–371. [CrossRef]
23. Ruan, X.; Yang, L.; Yu, M.; Li, Z.; Zhao, Y.; Pan, C.; Jiang, D.; Wang, Q. Temperature dependence of allelopathy duality and its influence on boreal forest succession—A case analysis of *Picea schrenkiana*. *bioRxiv* **2019**. [CrossRef]
24. Zandi, P. A review of significance of allelopathy in anticipating negative climate change effects. *Ann. Univ. Paedagog. Crac. Stud. Naturae* **2023**, *8*, 253–275. [CrossRef]
25. Revillini, D. Allelopathy-selected microbiomes mitigate chemical inhibition of plant performance. *New Phytol.* **2023**, *240*, 2007–2019. [CrossRef]
26. Thorpe, A.; Thelen, G.; Diaconu, A.; Callaway, R. Root exudate is allelopathic in invaded community but not in native community: Field evidence for the novel weapons hypothesis. *J. Ecol.* **2009**, *97*, 641–645. [CrossRef]
27. Kaur, R.; Gonzáles, W.L.; Llambí, L.D.; Soriano, P.J.; Callaway, R.M.; Rout, M.E.; Gallaher, T.J.; Inderjit. Community impacts of *Prosopis juliflora* invasion: Biogeographic and congeneric comparisons. *PLoS ONE* **2012**, *7*, e44966. [CrossRef]
28. Lobon, N.C.; Felix, M.G.; Gallego, J.C.A. Comparison of the allelopathic potential of non-native and native species of mediterranean ecosystems. *Plants* **2023**, *12*, 972. [CrossRef] [PubMed]
29. Lobstein, A.; Brenne, X.; Feist, E.; Metz, N.; Weniger, B.; Anton, R. Quantitative determination of naphthoquinones of *Impatiens* species. *Phytochem. Anal.* **2001**, *12*, 202–205. [CrossRef] [PubMed]
30. Allstadt, A.; Caraco, T.; Molnar, F., Jr.; Korniss, G. Interference competition and invasion: Spatial structure, novel weapons, and resistance zones. *J. Theor. Biol.* **2012**, *306*, 46–60. [CrossRef]
31. Li, Z.H.; Wang, Q.; Ruan, X.; Pan, C.D.; Jiang, D.A. Phenolics and plant allelopathy. *Molecules* **2010**, *15*, 8933–8952. [CrossRef]
32. Palanisamy, C.; Gunasekaran, V.; Dominic, S.; Xuan, T. Phenolic allelochemicals from crops and weed management. In *Plant Phenolics in Sustainable Agriculture*; Lone, R., Shuab, R., Kamili, A., Eds.; Springer: Singapore, 2020; Volume 1, pp. 183–199.
33. Jasim, I.R.; Alwattar, M.T.; Yaqub, H.M. Terpenoids as Natural Allelopathic Compounds in Plants. *Rafidain J. Sci.* **2021**, *30*, 106–116. [CrossRef]
34. Verdeguer, M.; Sánchez-Moreiras, A.M.; Araniti, F. Phytotoxic effects and mechanism of action of essential oils and terpenoids. *Plants* **2020**, *9*, 1571. [CrossRef] [PubMed]
35. Ruckli, R.; Hesse, K.; Glauser, G.; Rusterholz, H.; Baur, B. Inhibitory potential of naphthoquinones leached from leaves and exuded from roots of the invasive plant *Impatiens glandulifera*. *J. Chem. Ecol.* **2014**, *40*, 371–378. [CrossRef]
36. Dong, J.; Cai, L.; Xing, Y.; Yu, J.; Ding, Z. Re-evaluation of $ABTS^{\bullet+}$ assay for total antioxidant capacity of natural products. *Nat. Prod. Commun.* **2015**, *10*, 2169–2172. [CrossRef]
37. Su, B.L.; Zeng, R.; Chen, J.Y.; Chen, C.Y.; Guo, J.H.; Huang, C.G. Antioxidant and antimicrobial properties of various solvent extracts from *Impatiens balsamina* L. stems. *J. Food Sci.* **2012**, *77*, C614–C619. [CrossRef] [PubMed]
38. Zhang, Z.; Liu, Y.; Yuan, L.; Weber, E.; van Kleunen, M. Effect of allelopathy on plant performance: A meta-analysis. *Ecol. Lett.* **2021**, *24*, 348–362. [CrossRef] [PubMed]
39. Ozkan, T.; Uruşak, E.; Appiah, K.; Fujii, Y. First broad screening of allelopathic potential of wild and cultivated plants in Turkey. *Plants* **2019**, *8*, 532. [CrossRef]
40. Beerling, D.J.; Perrins, J.M. Biological flora of British Isles: *Impatiens glandulifera* Royle (*Impatiens roylei* Walp.). *J. Ecol.* **1993**, *81*, 367–382. [CrossRef]
41. Tanner, R.A.; Gange, A.C. Himalayan balsam, *Impatiens glandulifera*: Its ecology, invasion and management. *Weed Res.* **2020**, *60*, 4–7. [CrossRef]
42. Coakley, S.; Petti, C. Impacts of the invasive *Impatiens glandulifera*: Lessons learned from one of Europe's top invasive species. *Biology* **2021**, *10*, 619. [CrossRef] [PubMed]
43. Coombe, D.E. Biological flora of the British Isles, *Impatiens parviflora* DC. *J. Ecol.* **1956**, *44*, 701–713. [CrossRef]
44. Trepl, L. Über *Impatiens parviflora* DC. als Agriophyt in Mitteleuropa. *Diss. Bot.* **1984**, *73*, 1–400.
45. Priede, A. Invasive Neophytes in the Flora of Latvia: Distribution and Dynamics. Ph.D. Dissertation, University of Latvia, Riga, Latvia, 2009; 73p.
46. Chmura, D. *Biology and Ecology of an Invasion of Impatiens parviflora DC in Natural and Semi-Natural Habitats*; Wydawnictwo Naukowe Akademii Techniczno-Humanistycznej w Bielsku-Białej: Bielsko-Biała, Poland, 2014; pp. 1–216.
47. Helsen, K.; Diekmann, M.; Decocq, G.; De Pauw, K.; Govaert, S.; Graae, B.J.; Hagenblad, J.; Liira, J.; Orczewska, A.; Sanczuk, P.; et al. Biological flora of Central Europe: *Impatiens glandulifera* Royle. *Perspect. Plant Ecol. Evol. Syst.* **2021**, *50*, 125609. [CrossRef]
48. Ellenberg, H.; Weber, H.E.; Düll, R.; Wirth, V.; Werner, W.; Paulißen, D. *Zeigerwerte von Pflanzen in Mitteleuropa. 2. Verbesserte und Erweiterte Auflage*; Verlag Erich Goltze KG: Göttingen, Germany, 1992; pp. 1–258.
49. European Union (EU). Commission Implementing Regulation (EU) 2017/1263 of 12 July 2017 Updating the List of Invasive Alien Species of Union Concern Established by Implementing Regulation (EU) 2016/1141 Pursuant to Regulation (EU) No

1143/2014 of the European Parliament and of the Council. *Off. J. Eur. Union* **2017**, *L182*, 37–39. Available online: https://eur-lex.europa.eu/legal-content/EN/TXT/PDF/?uri=uriserv:OJ.L_.2017.182.01.0037.01.ENG (accessed on 6 November 2024).
50. Wagner, V.; Chytrý, M.; Jiménez-Alfaro, B.; Pergl, J.; Hennekens, S.; Biurrun, I.; Pyšek, P. Alien plant invasions in European woodlands. *Divers. Distrib.* **2017**, *23*, 969–981. [CrossRef]
51. Chmura, D.; Sierka, E. Relationships between invasive plant Impatiens parviflora and species richness of forest floor vegetation. *Pol. J. Ecol.* **2006**, *54*, 417–428.
52. Hejda, M.; Sádlo, J.; Kutlvašr, J.; Petřík, P.; Vítková, M.; Vojík, M.; Pyšek, P.; Pergl, J. Impact of invasive and native dominants on species richness and diversity of plant communities. *Preslia* **2021**, *93*, 181–201. [CrossRef]
53. Tokarska-Guzik, B. *The Establishment and Spread of Alien Plant Species (Kenophytes) in the Flora of Poland*; Wydawnictwo Uniwersytetu Śląskiego: Katowice, Poland, 2005.
54. Power, G.; Sánchez Vilas, J. Competition between the invasive *Impatiens glandulifera* and UK native species: The role of soil conditioning and pre-existing resident communities. *Biol. Invasions* **2020**, *22*, 1527–1537. [CrossRef]
55. Helsen, K.; Hagenblad, J.; Acharya, K.P.; Brunet, J.; Cousins, S.A.; Decocq, G.; De Fenne, P.; Kimberley, A.; Kolb, A.; Michaelis, J.; et al. No genetic erosion after five generations for *Impatiens glandulifera* populations across the invaded range in Europe. *BMC Genet.* **2019**, *20*, 20. [CrossRef]
56. Skálová, H.; Jarošik, V.; Dvoračková, S.; Pyšek, P. Effect of intra- and interspecific competition on the performance of native and invasive species of *Impatiens* under varying levels of shade and moisture. *PLoS ONE* **2013**, *8*, e62842. [CrossRef]
57. Bieberich, J.; Lauerer, M.; Drachsler, M.; Heinrichs, J.; Müller, S.; Feldhaar, H. Species- and developmental stage-specific effects of allelopathy and competition of invasive *Impatiens glandulifera* on co-occurring plants. *PLoS ONE* **2018**, *13*, e0205843. [CrossRef]
58. Mynard, P.; Vilas, J. Impacts of the invasive *Impatiens glandulifera* on the growth of three native plants. *Res. Sq.* **2023**. [CrossRef]
59. Vrchotovį, N.; Serį, B.; Krejcovį, J. Allelopathic activity of extracts from *Impatiens* species. *Plant Soil Environ.* **2011**, *57*, 57–60. [CrossRef]
60. Jurová, J.; Matoušková, M.; Wajs-Bonikowska, A.; Kalemba, D.; Renčo, M.; Sedlák, V.; Gorgal'ová, Z.; Poráčová, J.; Šalmún, P.; Grul'ová, D. Potential phytotoxic effect of essential oil of non-native species *Impatiens parviflora* DC. *Plants* **2019**, *8*, 241. [CrossRef] [PubMed]
61. Vieira, M.N.; Winterhalter, P.; Jerz, G. Flavonoids from the flowers of *Impatiens glandulifera* Royle isolated by high performance countercurrent chromatography. *Phytochem. Anal.* **2016**, *27*, 116–125. [CrossRef] [PubMed]
62. Szewczyk, K.; Olech, M. Optimization of extraction method for LC-MS based determination of phenolic acid profiles in different *Impatiens* species. *Phytochem. Lett.* **2017**, *20*, 322–330. [CrossRef]
63. Mikulic-Petkovsek, M.; Ravnjak, E.; Rusjan, D. Identification of phenolic compounds in the invasive plants staghorn sumac and Himalayan balsam: Impact of time and solvent on the extraction of phenolics and extract evaluation on germination inhibition. *Plants* **2024**, *13*, 3339. [CrossRef]
64. Tunaitienė, V.; Patamsytė, J.; Naugžemys, D.; Kleizaitė, V.; Čėsnienė, T.; Rančelis, V.; Žvingila, D. Genetic and allelopathic differences between populations of daisy fleabane *Erigeron annuus* (L.) Pers. (Asteraceae) from disturbed and stable habitats. *Biochem. Syst. Ecol.* **2017**, *70*, 294–303. [CrossRef]
65. Karpavičienė, B.; Danilovienė, J.; Vykertaitė, R. Congeneric comparison of allelopathic and autotoxic effects of four *Solidago* species. *Bot. Serb.* **2019**, *43*, 175–186. [CrossRef]
66. Judžentienė, A.; Būdienė, J.; Labanauskas, L.; Stancelytė, D.; Nedveckytė, I. Allelopathic activity of Canadian goldenrod (*Solidago canadensis* L.) extracts on seed germination and growth of lettuce (*Lactuca sativa* L.) and garden pepper cress (*Lepidium sativum* L.). *Plants* **2023**, *12*, 1421. [CrossRef]
67. Radušienė, J.; Karpavičienė, B.; Vilkickytė, G.; Marksa, M.; Raudonė, L. Comparative analysis of root phenolic profiles and antioxidant activity of five native and invasive *Solidago* L. species. *Plants* **2024**, *13*, 132. [CrossRef] [PubMed]
68. Baležentienė, L. Inhibitory effects of invasive *Heracleum sosnowskyi* on rapeseed and ryegrass germination. *Allelopath. J.* **2012**, *30*, 197–208.
69. Akazawa, M.; Kato-Noguchi, H. Allelopathic characteristic of a noxious weed *Leptochloa chinensis*. *Environ. Control Biol.* **2023**, *61*, 59–62. [CrossRef]
70. Vyšniauskienė, R.; Rančelienė, V.; Naugžemys, D.; Patamsytė, J.; Sinkevičienė, Z.; Butkuvienė, J.; Žvingila, D. Genetic diversity of populations of *Bidens* genera invasive and native species in Lithuania. *Žemdirbystė=Agriculture* **2018**, *105*, 183–190. [CrossRef]
71. Patamsytė, J.; Naugžemys, D.; Čėsnienė, T.; Kleizaitė, V.; Demina, O.N.; Mikhailova, S.I.; Agafonov, V.A.; Žvingila, D. Evaluation and comparison of the genetic structure of *Bunias orientalis* populations in their native range and two non-native ranges. *Plant Ecol.* **2018**, *219*, 101–114. [CrossRef]
72. Zybartaite, L.; Zukauskiene, J.; Jodinskiene, M.; Janssens, S.B.; Paulauskas, A.; Kupcinskiene, E. RAPD analysis of genetic diversity among Lithuanian populations of *Impatiens glandulifera*. *Žemdirbystė=Agriculture* **2011**, *98*, 391–398.
73. Kupcinskiene, E.; Zybartaite, L.; Paulauskas, A. Comparison of genetic diversity of three *Impatiens* species from Central Europe and Baltic region. *Žemdirbystė=Agriculture* **2015**, *102*, 87–94. [CrossRef]

74. Jocienė, L.; Krokaitė, E.; Rekašius, T.; Paulauskas, A.; Kupčinskienė, E. Evaluation of genetic diversity of Himalayan balsam (*Impatiens glandulifera* Royle) populations using microsatellites. *Žemdirbystė=Agriculture* **2022**, *109*, 259–268. [CrossRef]
75. Kupcinskiene, E.; Zybartaite, L.; Janulioniene, R.; Zukauskiene, J.; Paulauskas, A. Molecular diversity of small balsam populations in relation to site characteristics. *Cent. Eur. J. Biol.* **2013**, *8*, 1048–1061. [CrossRef]
76. Jocienė, L.; Stravinskaitė, K.; Krokaitė, E.; Janulionienė, R.; Rekašius, T.; Paulauskas, A.; Marozas, V.; Kupčinskienė, E. AFLP-based genetic structure of Lithuanian populations of small balsam (*Impatiens parviflora* DC.) in relation to habitat characteristics. *Forests* **2022**, *13*, 1228. [CrossRef]
77. Krokaitė, E.; Janulionienė, R.; Jocienė, L.; Rekašius, T.; Rajackaitė, G.; Paulauskas, A.; Marozas, V.; Kupčinskienė, E. Relating invasibility and invasiveness: Case study of *Impatiens parviflora*. *Front. Ecol. Evol.* **2022**, *10*, 845947. [CrossRef]
78. Jarcuška, B.; Slezák, M.; Hrivnák, R.; Senko, D. Invasibility of alien *Impatiens parviflora* in temperate forest understories. *Flora* **2016**, *224*, 14–23. [CrossRef]
79. Bieberich, J.; Müller, S.; Feldhaar, H.; Lauerer, M. Invasive *Impatiens glandulifera*: A driver of changes in native vegetation? *Ecol. Evol.* **2021**, *11*, 1320–1333. [CrossRef]
80. Hejda, M. What is the impact of *Impatiens parviflora* on diversity and composition of herbal layer communities of temperate forests? *PLoS ONE* **2012**, *7*, e39571. [CrossRef] [PubMed]
81. Anderson, N.O.; Jocienė, L.; Krokaitė, E.; Rekašius, T.; Paulauskas, A.; Kupčinskienė, E. Genetic diversity of *Phalaris arundinacea* populations in relation to river regulation in the Merkys basin, Lithuania. *River Res. Appl.* **2018**, *34*, 300–309. [CrossRef]
82. Jocienė, L.; Krokaitė, E.; Rekašius, T.; Juškaitytė, E.; Ielciu, I.; Galanina, O.; Kupčinskienė, E. The molecular evidence for invasive climber *Echinocystis lobata* (Michx.) Torr. & A. Gray in eastern and central Europe. *Diversity* **2023**, *15*, 1084. [CrossRef]
83. Krokaitė, E.; Jocienė, L.; Shakeneva, D.; Rekašius, T.; Valiulis, D.; Kupčinskienė, E. Ionome of Lithuanian populations of purple loosestrife (*Lythrum salicaria*) and its relation to genetic diversity and environmental variables. *Diversity* **2023**, *15*, 418. [CrossRef]
84. Zhao, J.; Yang, Z.; Zou, J.; Li, Q. Allelopathic effects of sesame extracts on seed germination of moso bamboo and identification of potential allelochemicals. *Sci. Rep.* **2022**, *12*, 6661. [CrossRef] [PubMed]
85. Kupčinskienė, E.; Ramoškevičienė, A.; Sliesaravičius, A.; Jankauskaitė, R. Comparison of growth of *Lepidium sativum* L. in the media containing air pollutants of various chemical origin. In *Heavy Metals in the Environment: An Integrated Approach, Proceedings of the 1st International Conference on Metals in the Environment, Vilnius, Lithuania, 15–17 October 1997*; Lovejoy, D.A., Ed.; Institute of Geology: Vilnius, Lithuania, 1999; pp. 179–185.
86. Arminante, F.; De Falco, E.; De Feo, V.; De Martino, L.; Mancini, E.; Quaranta, E. Allelopathic activity of essential oils from Mediterranean Labiatae. *Acta Hortic.* **2006**, *723*, 347–356. [CrossRef]
87. Ravlić, M.; Baličević, R.; Nikolić, M.; Sarajlić, A. Assessment of allelopathic potential of fennel, rue, and sage on weed species hoary cress (*Lepidium draba*). *Not. Bot. Horti Agrobot. Cluj-Napoca* **2016**, *44*, 48–52. [CrossRef]
88. Hovstad, K.A.; Ohlson, M. Physical and chemical effects of litter on plant establishment in semi-natural grasslands. *Plant Ecol.* **2008**, *196*, 251–260. [CrossRef]
89. Szabo, L.G. Juglone index—A possibility for expressing allelopathic potential of plant taxa with various life strategies. *Acta Bot. Hung.* **1999**, *42*, 295–305.
90. Csiszar, A. Allelopathic effects of invasive woody plant species in Hungary. *Acta Silv. Lignaria Hung.* **2009**, *5*, 9–17. [CrossRef]
91. Borowska, E.J.; Czaplicki, S.; Szajdek, A. Effect of bilberry mash treatment on the content of some biologically active compounds and the antioxidant activity of juices. *Acta Aliment.* **2009**, *38*, 281–292.
92. Jamshidi, M.; Shabani, E.; Hashemi, Z.; Ebrahimzadeh, M.A. Evaluation of three methods for the extraction of antioxidants from leaf and aerial parts of *Lythrum salicaria* L. (Lythraceae). *Int. Food Res. J.* **2014**, *21*, 783–788.
93. Szewczyk, K.; Orzelska-Górka, J.; Polakowska, M.; Biała, G. Antinociceptive and antianxiety activity of hydroethanolic extracts of three *Impatiens* species in mice. *Acta Pol. Pharm.* **2018**, *75*, 989–1001. [CrossRef]
94. Folin, O.; Ciocalteu, V. On tyrosine and tryptophane determination in proteins. *J. Biol. Chem.* **1927**, *73*, 627–633. [CrossRef]
95. Wu, C.H.; Murthy, H.N.; Hahn, E.J.; Paek, K.Y. Improved production of caffeic acid derivatives in suspension cultures of *Echinacea purpurea* by medium replenishment strategy. *Arch. Pharm. Res.* **2007**, *30*, 945–949. [CrossRef] [PubMed]
96. Brand-Williams, W.; Cuvelier, M.E.; Berset, C. Use of free radical method to evaluate antioxidant activity. *Lebensm. Wiss. Technol.* **1995**, *28*, 25–30. [CrossRef]
97. Miguel, M.G.; Figueiredo, A.C. Antioxidant activity of medicinal and aromatic plants. *Flavour Fragr. J.* **2010**, *25*, 251–252. [CrossRef]
98. R Core Team. *R: A Language and Environment for Statistical Computing*; R Foundation for Statistical Computing: Vienna, Austria, 2024.
99. Makowski, D.; Ben-Shachar, M.S.; Patil, I.; Lüdecke, D. Methods and algorithms for correlation analysis in R. *J. Open Source Softw.* **2019**, *5*, 2306. [CrossRef]
100. Pohlert, T. PMCMRplus: Calculate Pairwise Multiple Comparisons of Mean Rank Sums Extended; R Package Version 1.9.12. 2024. Available online: https://cran.r-project.org/web/packages/PMCMRplus/PMCMRplus.pdf (accessed on 14 November 2024).

101. Szewczyk, K.; Kalemba, D.; Komsta, Ł.; Nowak, R. Comparison of the essential oil composition of selected *Impatiens* species and its antioxidant activities. *Molecules* **2016**, *21*, 1162. [CrossRef]
102. Csiszár, Á.; Korda, M.; Schmidt, D.; Šporčić, D.; Süle, P.; Teleki, B.; Tiborcz, V.; Zagyvai, G.; Bartha, D. Allelopathic potential of some invasive plant species occurring in Hungary. *Allelopath. J.* **2013**, *31*, 309–318.
103. Szewczyk, K.; Sezai Cicek, S.; Zidorn, C.; Granica, S. Phenolic constituents of theaerial parts of *Impatiens glandulifera* Royle (Balsaminaceae) and their antioxidant activities. *Nat. Prod. Res.* **2019**, *33*, 2851–2855. [CrossRef]
104. Gruntman, M.; Pehl, A.K.; Joshi, S.; Tielbörger, K. Competitive dominance of the invasive plant *Impatiens glandulifera*: Using competitive effect and response with a vigorous neighbour. *Biol. Invasions* **2014**, *16*, 141–151. [CrossRef]
105. Tickner, D.P.; Angold, P.G.; Gurnell, A.M.; Mountford, J.O.; Sparks, T. Hydrology as an influence on invasion: Experimental investigations into competition between the alien *Impatiens glandulifera* and the native *Urtica dioica* in the UK. In *Plant Invasions: Species Ecology and Ecosystem Management*; Backhuys Publishers: Kerkwerve, The Netherlands, 2001; pp. 159–167.
106. Flayyih, T.; Almarie, A. Total phenolic exudation and allelopathic potential of sunflower residues as sustainable weed management. *Iraqi J. Sci.* **2023**, *64*, 2215–2222. [CrossRef]
107. Muhammad, N.; Khan, W.; Majeed, A. Allelopathic potential of *Cassia fistula* against *Lactuca sativa*, *Setaria italica* and *Pennisetum americanum*. *Pure Appl. Biol.* **2018**, *8*, 187–198. [CrossRef]
108. Rys, M.; Saja-Garbarz, D.; Skoczowski, A. Phytotoxic effects of selected herbal extracts on the germination, growth and metabolism of mustard and oilseed rape. *Agronomy* **2022**, *12*, 110. [CrossRef]
109. Hussain, M.I.; Gonzalez-Rodriguez, L.; Reigosa, M.J. Germination and growth response off our plant species to different allelochemical sand herbicides. *Allelopath. J.* **2008**, *22*, 101–110.
110. Babula, P.; Vaverkova, V.; Poborilova, Z.; Ballova, L.; Masarik, M.; Provaznik, I. Phytotoxic action of naphthoquinone juglone demonstrated on lettuce seedling roots. *Plant Physiol. Biochem.* **2014**, *84*, 78–86. [CrossRef] [PubMed]
111. Inderjit. Soil microorganisms: An important determinant of allelopathic activity. *Plant Soil* **2005**, *274*, 227–236. [CrossRef]
112. Hobbs, R.; Humphries, S. An integrated approach to the ecology and management of plant invasions. *Conserv. Biol.* **1995**, *9*, 761–770. [CrossRef]
113. Hierro, J.L.; Callaway, R.M. The ecological importance of allelopathy. *Ann. Rev. Ecol. Evol. Syst.* **2021**, *52*, 25–45. [CrossRef]

diversity

MDPI

Article

Forest Attribute Dynamics in Secondary Forests: Insights for Advancing Ecological Restoration and Transformative Territorial Management in the Amazon

Carlos H. Rodríguez-León [1], Armando Sterling [1,2,*], Amelia Trujillo-Briñez [1,3], Yerson D. Suárez-Córdoba [1,3] and Lilia L. Roa-Fuentes [4,*]

[1] Programa Modelos de Funcionamiento y Sostenibilidad, Instituto Amazónico de Investigaciones Científicas SINCHI, Florencia 180001, Colombia; crodriguez@sinchi.org.co (C.H.R.-L.); ameliatrujillo2196@gmail.com (A.T.-B.); yersuarez1@gmail.com (Y.D.S.-C.)
[2] Programa de Biología, Facultad de Ciencias Básicas, Universidad de la Amazonía, Florencia 180001, Colombia
[3] Programa de Ingeniería Agroecológica, Facultad de Ingeniería, Universidad de la Amazonía, Florencia 180001, Colombia
[4] Departamento de Ecología y Territorio, Facultad de Estudios Ambientales y Rurales, Pontifica Universidad Javeriana, Bogotá 110231, Colombia
* Correspondence: asterling@sinchi.org.co (A.S.); lilia.roa@javeriana.edu.co (L.L.R.-F.)

Abstract: The Amazon ecosystem plays a vital role in global climate regulation and biodiversity conservation but faces escalating threats from deforestation and degradation. The resulting secondary forests (SFs) provide a promising opportunity for Transformative Territorial Management, fostering restoration and enhancing conservation values. This study evaluated aboveground biomass (AGB), species diversity, forest structure, and soil properties in SFs of the Colombian Amazon along a chronosequence, from early to mature successional stages, in landscapes of mountains and of hills to identify key indicators for effective restoration management. The results show a consistent increase in AGB, species diversity, forest structure, and soil quality with forest age, though recovery patterns varied between both landscapes evaluated. Topographic differences influenced successional dynamics, with mountainous landscapes showing faster early recovery compared to the steadier, linear growth observed in hill areas. In hills, AGB at 10 years reached 12.65% of the biomass expected in a mature forest, increasing to nearly 42% by 40 years of abandonment, at a rate of 0.708 Mg C ha^{-1} $year^{-1}$. In contrast, in the mountain landscape, AGB at 10 years reached approximately 8.35% of the carbon in a mature forest and increased to nearly 63.55% at 40 years. Forest age and soil properties emerged as primary drivers of AGB recovery, while diversity and forest structure played indirect but significant roles. In hill areas, soil conservation practices are critical for maintaining steady growth, whereas mountain regions benefit from assisted natural regeneration (ANR) to accelerate recovery. These findings highlight the importance of prioritizing the management of SFs as a central strategy for achieving restoration goals. Such practices are essential to enhance the ecological resilience of SFs and ensure their long-term sustainability, fostering their role as key contributors to restoration efforts and the provision of ecosystem services.

Keywords: secondary succession; natural restoration; chronosequence; aboveground biomass; carbon sequestration; landscape configuration; tropical forest

1. Introduction

The Amazon is one of the richest and most biodiverse biomes on Earth, with vast forests that remain largely undisturbed by direct human use [1]. It plays a crucial role in

https://doi.org/10.3390/d17010039

climate regulation as a major carbon reservoir [2] and serves as a vital refuge for biodiversity, housing nearly half of the world's tropical forest species [3]. Despite its immense ecological significance, the Amazon biome is severely threatened by human activities, which compromise its integrity and resilience. These pressures have led to habitat loss, landscape fragmentation, and the disruption of fundamental ecological processes [4]. In the Colombian Amazon, deforestation is intensified by colonization processes that convert forests into pastures, trigger forest fires, and promote the establishment of illicit crops, further aggravated by illegal logging and mining activities [5]. These disturbances reduce biodiversity and undermine ecosystems' capacity to recover from natural disruptions, adapt to climate change, and continue providing essential services, such as carbon storage in biomass [6].

Widespread deforestation has transformed Amazonian ecosystems into a fragmented mosaic of secondary forests (SFs), which currently cover about 4.1% of the Amazon biome. Evidence suggests that these SFs are caught in a destructive "Degradative Cycle" of regrowth followed by clearing, often occurring within five years or less. This cycle is particularly concerning in regions with high deforestation rates, where forest recovery is significantly hindered [7]. In the Colombian Amazon, 50% of SFs are found in areas heavily impacted by human activity [8]. Data show that 70% of these regenerating SFs are subsequently converted back into pastures, perpetuating the "Degradative Cycle" and severely compromising their ability to regenerate [8].

The management of SFs is essential to harness their functional potential while preventing them from falling into the previously mentioned "Degradative Cycle". Ecological restoration, through assisted natural regeneration and enrichment planting, can serve as a key management strategy for enhancing the ecological and functional recovery of SFs. In addition, the ecological restoration of SFs can mitigate the adverse effects of deforestation and aligns with global initiatives such as the Bonn Challenge, which aims to restore 350 million hectares (Mha) by 2030, and the Kunming-Montreal Global Biodiversity Framework, targeting the rehabilitation of 30% of degraded areas by the same year. Achieving these ambitious goals presents a significant challenge for Colombia. The National Restoration Strategy, launched in 2024, sets an objective to restore 735,000 hectares by 2026, employing a comprehensive approach that adapts restoration actions to various contexts, including full, partial, and productive recovery [9]. Given the extensive distribution of SFs, effectively managing these areas is critical to achieving national and global restoration targets [10,11]. However, large-scale tree planting is prohibitively expensive for developing countries and often fails to restore native biodiversity effectively [12]. Therefore, it is vital to identify scenarios where the restoration of SFs can be optimized [12,13].

Restoring SFs requires integrating the existing knowledge of ecological processes such as natural regeneration and succession. The rate and success of recovering the composition, structure, and function of these forests depend on factors like forest type, climatic conditions, land-use history, and landscape configuration [14,15]. Consequently, restoration strategies must be tailored to these conditions and may include actions like Assisted Natural Regeneration (ANR), which promotes forest recovery by removing barriers to natural regeneration and enhancing biodiversity and resilience [16–18]. Other strategies involve enriching SFs with native species to accelerate succession and improve ecosystem functionality [19,20], implementing fire and livestock protection measures, and ensuring continuous monitoring.

This research aims to enhance the existing body of evidence on the ecological attributes that contribute to forest integrity in Amazonian ecosystems in Colombia [21], providing a foundation for strategies focused on Transformative Territorial Management centered around the ecological restoration of SFs. Previous studies have demonstrated that plant species richness increases as the duration since the cessation of productive activities lengthens, with a threshold influenced by the vulnerability of the plant community [22]. Additionally, SFs have been observed to exhibit trends toward biotic homogenization within plant communities [10]. Despite these findings, there remains a limited understanding of how aboveground plant biomass changes over time and how various environmental factors influence these dynamics. Evidence indicates that the speed and time required to recover aboveground biomass (AGB) are affected by factors such as forest type, climatic conditions, land-use history, seed dispersal, and the establishment of new trees [23–26]. The dominance or relative abundance of specific tree species groups, as opposed to shrub groups, also plays a significant role in shaping diversity, biomass stocks, and other functional attributes of SFs during succession [27]. Moreover, functional plant groups, such as nitrogen fixers, have been shown to enhance soil nutrient availability [28], positively impacting long-term biomass growth and optimizing soil carbon sequestration [29]. Furthermore, understanding how factors such as previous management intensity and landscape configuration influence soil and vegetation recovery is crucial for informing restoration efforts. Such insights are essential for developing effective restoration strategies and ensuring their successful implementation [15]. Therefore, this study aimed to assess the successional trends in the recovery of forest aboveground biomass carbon, tree diversity, and forest structure in a chronosequence of tropical SFs following the abandonment of cattle ranching according to two types of landscape (hill and mountain). We also evaluated the recovery of biomass in response to changes in forest age, species diversity, forest structure, and soil properties, which are key aspects for designing future strategies aimed at transformative ecological restoration in Colombian deforested areas.

2. Materials and Methods

2.1. Study Area and Sampling Design

The study area is located in the department of Caquetá, in the northwestern Colombian Amazon, covering the municipalities of Florencia (1°36′50″ N; 75°36′46″ W), Morelia (1°29′09″ N; 75°43′28″ W), Belén de los Andaquíes (1°24′59.1″ N; 75°52′21.2″ W), San José del Fragua (1°19′52″ N; 75°58′28″ W), and Albania (1°19′44″ N; 75°52′42″ W) (Figure 1). The region's climate is classified as humid tropical, with an average annual precipitation of 3235 mm, concentrated during a rainy season from March to June, and a mean temperature of 25° [30]. The predominant soils are Inceptisols and Oxisols, according to USDA classification, characterized by fine texture, drainage limitations, high acidity (pH ranging from 4.5 to 5.8), high aluminum saturation, and low base saturation. These soils have limited contents of carbon, phosphorus, potassium, and magnesium, affecting their fertility and aeration capacity [31].

This region comprises two main landscapes: hill and mountain. Hills, with altitudes below 300 m and slopes ranging from 7 to 12%, represent a typical undulating landscape of the Amazonian plains, primarily used for extensive cattle grazing and agriculture, with a high level of degradation and remnants of natural forests. In contrast, mountains, with altitudes above 300 m and slopes between 12.1 and 75%, form part of the Andean–Amazon transition and are dominated by natural forests, pastures, crops, and SFs [31,32].

Sampling sites were selected based on the age of land abandonment (chronosequence) and grouped into five successional categories in each landscape type: (i) sites <10 years, representing early successional stages with shrub and pioneer vegetation; (ii) 11–20 years, corresponding to young SFs; (iii) 21–30 years, representing intermediate SFs; (iv) 31–40 years, reflecting advanced SF stages; and (v) old-growth forest (OF), characterizing mature forests similar to those reported by Rodríguez-León et al. [32].

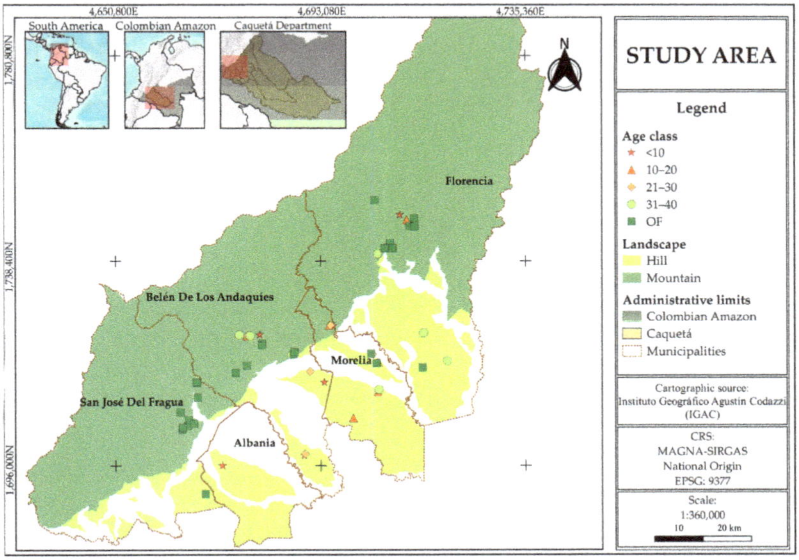

Figure 1. Study area and distribution of forest sampling plots in Caquetá, Colombian Amazon.

2.2. Vegetation Sampling

The data were collected from 54 plots of 50 × 50 m (0.25 ha) distributed across hill and mountain landscapes, selected based on the age of land abandonment (chronosequence). The identification of the plots was carried out through interviews with farm owners, who provided information about land-use history [32,33], and complemented with observations of specific vegetation attributes, such as floristic composition, species density, trunk diameter, and basal area [34]. Of the 54 sampled plots, 29 correspond to SFs aged between 5 and 40 years, while 25 represent OFs, distributed across various municipalities in the region (Table 1).

Table 1. Number of forest plots sampled in secondary and old-growth forests in the Colombian Amazon.

Landscape	Site	Number of Plots in Secondary Forests	Stand Age (Years)			Number of Plots in Old-Growth Forests
			Mean	Min	Max	
Hill	Albania	1	8	8	8	0
	Belén De Los Andaquíes	4	19	7	30	0
	Florencia	3	35	35	35	0
	Morelia	5	22	12	35	2
	San José Del Fragua	0	--	--	--	1
Mountain	Belén De Los Andaquíes	6	26	7	40	6
	Florencia	6	20	8	40	7
	Morelia	3	15	5	25	0
	San José Del Fragua	1	35	35	35	9
	Total	29				25

Within each plot, all trees with a diameter at breast height (DBH) ≥ 1 cm were marked and identified to the species and family levels. The collected samples were processed, taxonomically determined, and stored at the "Colombian Amazon Herbarium-COAH" of the Sinchi Institute in Bogotá, Colombia, ensuring the traceability of taxonomic information [10,22].

2.3. Aboveground Biomass

To estimate the AGB of trees, the allometric equation developed by Chave et al. [35] was used. This equation is widely applied in tropical forest studies due to its ability to capture local variations in forest conditions [29,36]. The equation applied is as follows (Equation (1)):

$$\text{AGB} = 0.0673 \times \left(\text{WD} \times \text{DBH}^2 \times \text{HT}\right)^{0.976} \quad (1)$$

where AGB represents the aboveground biomass in kilograms, WD is the wood density (g cm^{-3}), DBH is the diameter at breast height (cm), and HT is the total tree height (m). This model has proven suitable for the study area as it incorporates local species data, ensuring accurate forest biomass estimates [29,35].

To calculate the carbon stored in AGB, wood carbon fractions (Wood CFs) were determined using the approach proposed by Doraisami et al. [37]. This method establishes conversion factors based on the wood density (WD) using the following equation (Equation (2)):

$$\text{Wood CF} = 49.3 + (-3.5 \times \text{WD}) \quad (2)$$

In this equation, WD is the wood density in g cm^{-3}, while coefficients −3.5 and 49.3 represent the relationship between wood density and carbon fraction, respectively.

The calculated Wood CF values were multiplied by the estimated AGB for each tree to obtain the carbon stored in AGB. Finally, AGB values for all trees within each plot were summed, and the results are reported in Mg C ha^{-1}.

2.4. Forest Diversity and Structure Indices

This study considered two key dimensions of biodiversity: taxonomic diversity and forest structure. To evaluate taxonomic diversity, three Hill numbers were used, essential tools for measuring effective diversity by considering species relative abundances [38,39]. Hill numbers are defined by three orders of "q" reflecting different degrees of sensitivity to relative abundance: species richness (q = 0), which includes all species equally; exponential of Shannon entropy (q = 1), which assigns less weight to rare species, considering both richness and evenness within a community (Equation (3)); and inverse Simpson diversity (q = 2), which assigns more weight to dominant species, assigning less importance to rare ones (Equation (4)). These indices were calculated using the "iNEXT" package in R [38].

$$\text{Exponential of Shannon} - \text{Weaver's diversity index (ExpH)} = \exp\left(-\sum_{i=0}^{s} P_i \ln P_i\right) \quad (3)$$

$$\text{Inverse of Simpson's dominance index } (1/D) = \frac{1}{\sum_{i=0}^{s} P_i^2} \quad (4)$$

In these equations, P_i represents the proportion of individuals of species i relative to the total number of individuals in the community, calculated as $P_i = \frac{n_i}{N}$, where n_i is the number of individuals of species i and N is the total number of individuals of all species. S is the total number of species in the respective forest plot. Higher values in these indices indicate a more equitable species distribution within the community [40,41].

To quantify β-diversity and assess differences in species composition between plots, the Chao–Jaccard dissimilarity index was used. This index accounts for both observed and unobserved shared species among communities, reducing sampling bias (Equation (5)).

$$\text{Chao} - \text{Jaccard index (CJI)} = \text{U} \times \text{V} \div (U + V - (U \times V)) \quad (5)$$

In this equation, U and V represent the relative abundances of shared species in the compared communities. Values close to 1 indicate high similarity, while values near 0 reflect marked differences [27]. These calculations were performed using the CommEcol package in R [42].

The forest structure was evaluated through various structural indices, including the mean diameter at breast height (DBH, cm), the mean total tree height (Height, m), the average wood density (WD, g cm^{-3}), and the total basal area (BA, m^2 ha^{-1}) of all woody species per plot. These indicators provide a detailed perspective on the size and vertical distribution of trees in the forest community [43].

Additionally, to analyze the horizontal structure of the communities, the Importance Value Index (IVI) was used. This index combines relative abundance, relative dominance, and relative frequency of each species in the community (Equation (6)).

$$\mathrm{IVI}(\%) = \frac{\mathrm{Abun}(\%) + \mathrm{Dom}(\%) + \mathrm{Frec}(\%)}{3} \tag{6}$$

In this equation, Abun (%) is relative abundance, Dom (%) is relative dominance, and Frec (%) is relative frequency. IVI helps identify the species with the greatest ecological importance within the community [44]. Calculations were carried out using the BiodiversityR package in R [45].

2.5. Soil Properties and Landscape-Related Parameters

In each plot, three soil samples were randomly collected at a depth of 0 to 20 cm of the mineral soil layer on the same day as vegetation sampling. Following the methods described in Garate-Quispe et al. [36] and Coelho et al. [15], fifteen soil properties were analyzed using standardized laboratory techniques detailed in previous studies [10,32]. These included the organic matter content (OM, %) determined by the Walkley–Black method using potassium dichromate, while the total nitrogen concentration (N, %) was determined using the Kjeldahl method. Other analyses included pH, measured in a saturation paste using the conductimetric method of the USDA, and soil moisture saturation (SMS, %) determined gravimetrically in a saturation paste following the USDA method. The soil texture (percentages of clay, sand, and silt) was determined using the Bouyoucos method. Additionally, the effective cation exchange capacity (CEC, meq 100 g^{-1}), exchangeable potassium (E_K, mg kg^{-1}), and exchangeable calcium (E_Ca, mg kg^{-1}) were determined by ammonium acetate extraction and quantified by ICP-OES spectroscopy. The iron concentration (Fe, mg kg-1) was measured using Mehlich I acid extraction, while the available phosphorus (P, mg kg^{-1}) was determined by Bray II extraction. Magnesium (Mg_S, %) and aluminum (Al_S, %) saturation levels were calculated as the fraction represented by each relative to the effective cation exchange capacity (CEC, meq 100 g^{-1}), expressed as a percentage. Finally, the soil organic carbon (SOC, Mg ha^{-1}) was calculated using the method of Maia et al. [46], estimating the carbon concentration (g g^{-1}) multiplied by the bulk density (kg cm^{-3}) and soil layer thickness (m).

In addition, two landscape-related parameters were evaluated in each plot as follows: slope (%) and elevation (m) according to previous studies [10,22]. The slope was recorded with a laser hypsometer (Forestry Pro II), while the elevation was measured with a digital altimeter (GPSMAP 64CSX) similar to that reported by Rodríguez et al. [22].

2.6. Data Analysis

Generalized linear models (GLMs) were applied to evaluate the effect of forest age, landscape, and their interaction on AGB variable, as well as those related to species diversity, forest structure, and the physical–chemical properties of the soil using the lme function from the nlme package [47] in R language v. 4.3.3 [48] and the interface in InfoStat v.2020 [49]. A normal distribution was employed for all variables, except for tree species richness, where a Poisson distribution was used. The species diversity and forest structure variables were fitted for normal distribution with logarithmic transformation. The assumptions of normality and homogeneity of variances were validated through an exploratory analysis of the model's residuals ($n = 54$), with quantile–quantile plots (QQ-plot) for normality and residuals vs. predicted plots (fitted-plot) for homoscedasticity. Corrections for heteroscedasticity for species diversity and forest structure variables were made by age class using the varIdent function [47] from the nlme package in R. When significant differences were found, a Fisher's LSD test at 95% confidence was used to separate the means of the studied effects.

To model the recovery of biomass, biodiversity, and forest structure over the chronosequence, simple and quadratic linear regression functions were applied using the lm function, along with nonlinear regression models, such as logistic, monomolecular, and Gompertz, using the nls function from the stats package [50] in R, and Chapman–Richards using the sar_chapman function from the sars package [51] in R. The evaluation of the models included coefficients of determination (R^2) and residual analysis to verify the quality of the fit, similar to that has been reported in previous studies [29,36].

A non-metric multidimensional scaling (nMDS) analysis (50 iterations) was conducted based on the Chao–Jaccard similarity index to visualize species composition across different age classes and landscapes using the metaMDS function from the vegan package in R [52]. The accepted stress level was ≤ 0.2, following the criteria of Oberleitner et al. [53] and Chao et al. [54]. A Permutational Multivariate Analysis of Variance Using Distance Matrices (PERMANOVA) based on Chao–Jaccard distance and 9999 permutations was performed using the adonis2 function from the vegan package in R [52]. The pairwise.adonis function with 9999 permutations from the pairwiseAdonis package in R [55] was used to conduct pairwise multi-level comparisons of the studied effects.

Finally, a Pearson correlation analysis and simple linear regression using the pairs.panels function from the psych package in R [56] were employed to study the bivariate relationships between the studied variables, and structural equation models (SEMs) were used to analyze direct and indirect causal relationships among forest age, soil properties, species diversity, floristic structure, and carbon in aboveground biomass. The SEM were conducted using the sem and semPaths functions from the lavaan package in R [57], utilizing latent variables that grouped the most relevant components and applying logarithmic transformations to improve model fit, similar to what has been reported in previous studies [15,22,29,36]. The multivariate analyses in R were performed using the interface in RStudio v. 2024.04.1 [58].

3. Results

3.1. Effects of Age and Landscape on the Recovery of AGB, Tree Diversity, and Forest Structure

The results of this study show a clear recovery pattern in forest structural and diversity attributes throughout succession (Figure 2 and Supplementary Figure S1). The GLMs revealed a significant effect of abandonment age on all forest attributes. Diameter at breast height, tree height, wood density, Shannon exponential diversity, and Simpson's inverse diversity varied significantly across abandonment ages but not between landscape units ($p > 0.05$). Basal area, AGB, and tree species richness showed significant variations both across age classes and at the landscape level (both, $p < 0.05$). Additionally, the interaction between abandonment age and landscape was only significant for tree species richness ($F = 12.20$, $p < 0.0001$) (Supplementary Figure S1). This finding indicates that the recovery of tree species richness varied depending on both the successional stage and landscape type, being more pronounced in mountainous landscapes during advanced stages of succession.

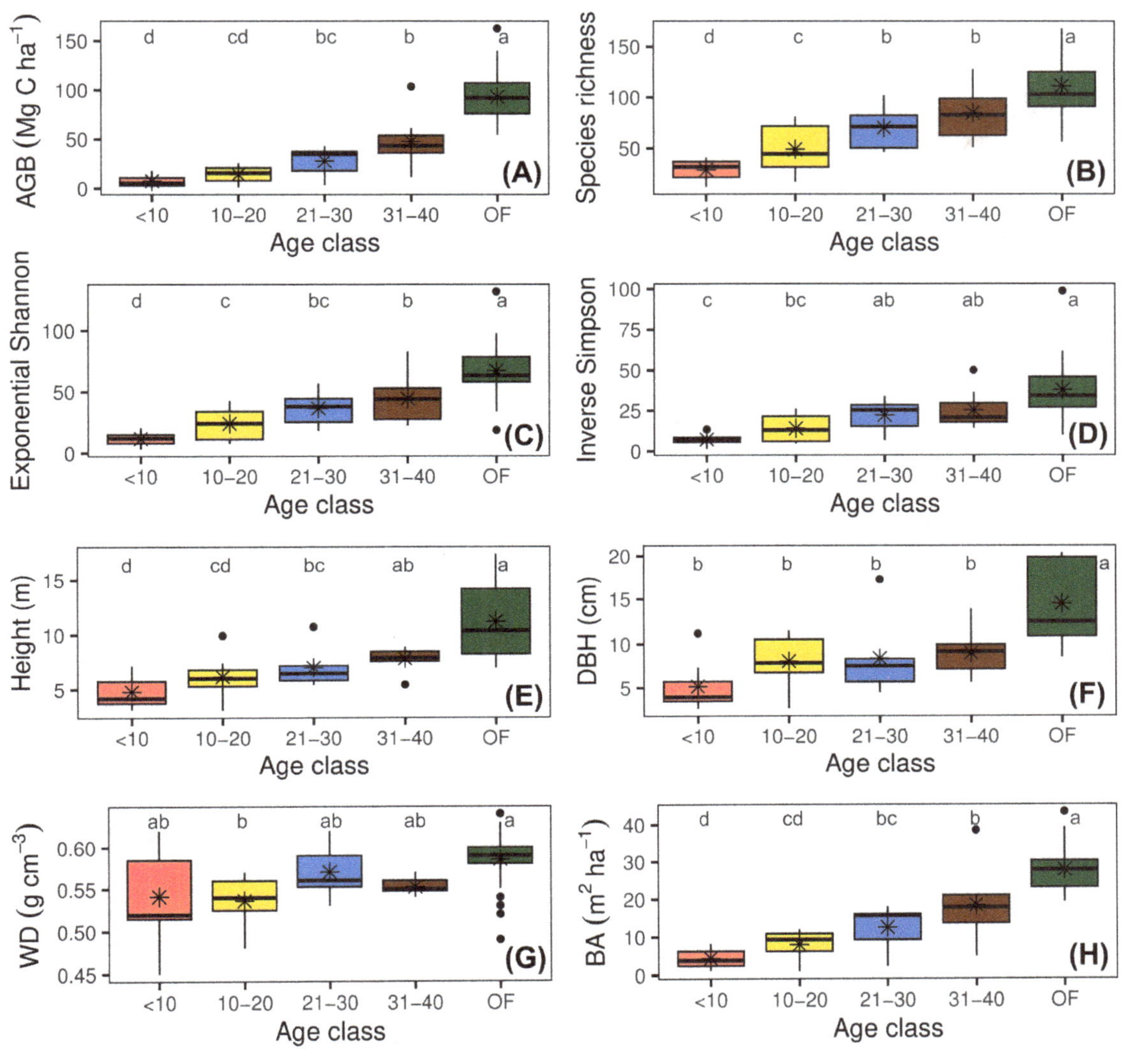

Figure 2. Changes in forest structural and diversity attributes across four age classes of secondary forests and in reference forest following the abandonment of cattle ranching. (**A**) Aboveground biomass stock (AGB); (**B**) tree species richness; (**C**) exponential Shannon diversity (ExpH); (**D**) inverse Simpson diversity (1/D); (**E**) mean tree height (Height); (**F**) diameter at breast height (DBH); (**G**) wood density (WD); and (**H**) basal area (BA). Different letters indicate significant differences among age classes (Fisher's LSD test; $p < 0.05$).

The AGB increased from 7.60 Mg C ha^{-1} in young vegetation plots (<10 years) to 84.08 Mg C ha^{-1} in OF plots (Figure 2A). This pattern was also evident in BA, which varied from 4.32 m^2 ha^{-1} in young forests to 26.32 m^2 ha^{-1} in mature forests (Figure 2H). Similarly, DBH increased from 5.30 cm in young vegetation plots to 12.81 cm in reference forests (Figure 2F), while height showed an upward trend with age, rising from 4.80 m in young vegetation plots to 9.62 m in reference forests (Figure 2E).

The WD showed a slight increase in more mature vegetation plots, reaching a maximum of 0.59 g cm^{-3} in reference vegetation plots (OF) (Figure 2G).

The mean tree species richness increased from 28.46 species in young vegetation plots to 115.72 in reference plots (Figure 2B). This trend was similar to what was observed in ExpH and 1/D, which reflect greater evenness in species distribution during advanced stages of succession (Figure 2C,D).

At the landscape level, higher averages of BA (15.57 m^2 ha^{-1}), AGB (42.57 Mg C ha^{-1}), and tree species richness (66.90) were observed in mountains compared to those observed in hills.

The rate of AGB (Mg C ha^{-1}) accumulation throughout succession followed distinct patterns depending on the landscape type (Figure 3). In the hill landscape, AGB increased linearly (R^2 = 0.85), reaching 12.65% of the biomass expected in a mature forest after 10 years and nearly 42% at 40 years of abandonment. This increase, at a rate of 0.708 Mg C ha^{-1} $year^{-1}$, showed consistent growth over time. In contrast, in the mountainous landscape, the change in AGB followed a quadratic linear trend (R^2 = 0.75), with a faster increase in the initial stages. In the mountainous landscape, AGB at 10 years reached approximately 8.35% of the carbon in a mature forest and increased to nearly 63.55% at 40 years. These findings suggest that the mountainous landscape provides a favorable environment for rapid recovery in the early stages, while the hill landscape shows a uniform, linear growth pattern.

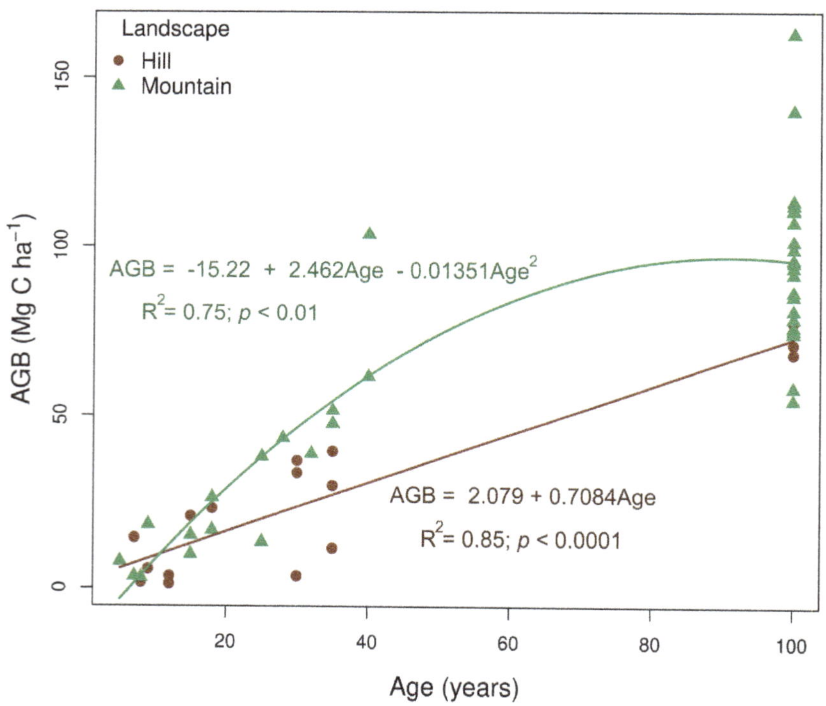

Figure 3. Relationship between aboveground biomass carbon (AGB, in Mg C ha^{-1}) and forest age in two landscape types (hill and mountain) across a chronosequence of natural regeneration following the abandonment of cattle ranching in the Colombian, deforested Amazon. The solid lines represent the fitted linear regression models for each landscape.

On the other hand, the recovery trajectories of various forest structural and diversity attributes, modeled using the Chapman–Richards functions and quadratic regression (Supplementary Figure S2), revealed how AGB, tree species richness, ExpH, 1/D, and BA followed a sigmoidal growth pattern, with AGB and BA showing the highest explanatory power (R^2 = 0.77 and 0.73, respectively). Meanwhile, mean tree height, DBH, and WD exhibited quadratic trends with R^2 values of 0.5, 0.47, and 0.21, respectively. Overall, a stabilization of growth curves for floristic attributes in terms of diversity and structure was observed after 80 years of abandonment of degraded pastures.

3.2. Soil Property Dynamics

The comparative analysis between hills and mountains revealed significant differences in several soil properties (Supplementary Table S1). Soils in hills showed a slightly higher SOC content, with 77.01 Mg ha^{-1} compared to 75.46 Mg ha^{-1} in mountains ($p > 0.05$). However, the pH was significantly higher in hill soils (4.46) than in mountain soils (4.15), indicating that hill soils are less acidic. This lower acidity is associated with a reduced aluminum saturation (Al_S), which was 75.06% in hill soils compared to 59.97% in mountain soils. Hill soils also exhibited a higher cation exchange capacity (CEC) (7.58 meq 100 g^{-1}) compared to mountain soils (5.08 meq 100 g^{-1}), reinforcing their ability to retain essential nutrients. On the other hand, mountain soils have a higher sand content (64.25%) and a lower clay percentage (24.68%), which could limit their capacity to retain water and nutrients.

The age of the plots' vegetation influenced SOC accumulation. This parameter reached its maximum value in the plots' aged vegetation between 10 and 20 years (81.90 Mg ha^{-1}) and remained stable in older plots' vegetation (OF, 80.83 Mg ha^{-1}). Similarly, soil pH decreased with age, from 4.47 in young plots' vegetation (<10 years) to 4.12 in older plots' vegetation (>40 years), where a higher aluminum saturation (Al_S) (66.35%) was evident.

The CEC increased with plot vegetation age, reaching a maximum value of 7.86 meq 100 g^{-1} in the 31–40-year age class, indicating a greater nutrient retention capacity in older vegetation soils. Similarly, organic matter (OM) and total nitrogen (N) increased during the early stages of succession and stabilized in vegetation plots >40 years old. Regarding soil texture, the clay content increased with vegetation age, reaching 39.75% in soils aged 31–40 years, while the proportion of sand decreased throughout the chronosequence.

3.3. Changes in Dominance and Composition of Tree Species Throughout Succession

The five most abundant species showed differential dominance patterns throughout successional stages in the hill and mountain landscapes (Supplementary Table S2). In early successional stages (<10 years), pioneer species dominated in both landscapes, with *Piptocoma discolor* (Kunth) Pruski (IVI = 33.6%) and *Trattinnickia boliviana* (Swart) Daly (26.3%) in hills, while in mountains, *Henriettea fascicularis* (Sw.) M.Gómez (47.4%) and *Bellucia grossularioides* (L.) Triana (21.2%) were predominant. In intermediate successional stages (11 to <30 years), species such as *H. fascicularis* and *Jacaranda copaia* (Aubl.) D. Don dominated in hills, whereas in mountains, *Cyathea lasiosora* (Kuhn) Domin and *Graffenrieda conostegioides* Triana were prominent. In both stages, pioneer species began to lose importance to secondary species. In advanced succession (31 to <40 years), dominant species in hills included *Tapirira guianensis* Aubl. (IVI = 20.1%) and *Inga thibaudiana* DC. (15.2%), while in mountains, *Casearia arborea* (Rich.) Urb. (7.5%) and *T. guianensis* (20.1%) were significant, indicating a transition towards a more mature community. In the OF plots, the composition was dominated by species typical of mature forests, such as *Pseudosenefeldera inclinata* (Müll.Arg.) Esser (IVI = 18.4% in hills and 8.8% in mountains) and *Virola elongata* (Benth.) Warb.

The results of the PERMANOVA and nMDS ordinations show significant variation in species composition both among age classes ($F = 3.52$; $p < 0.001$) and landscape types

($F = 3.20$; $p < 0.001$) (Figure 4 and Supplementary Table S3). Although the interaction age_class*landscape was not significant ($p > 0.05$), a greater divergence in the species composition was observed between SF and the reference forest in the hill landscape compared to the mountain (Supplementary Figure S3 and Supplementary Table S3). Consequently, post hoc pairwise comparisons revealed that tree species composition varied significantly between early regeneration stages (<10 years) and the oldest vegetation plots (OFs). In this regard, the community composition of the four SF categories differed significantly from the reference forest, and the SFs aged 31–40 years was significantly different from the SFs aged <10 and 10–20 years ($p < 0.05$) (Supplementary Table S4).

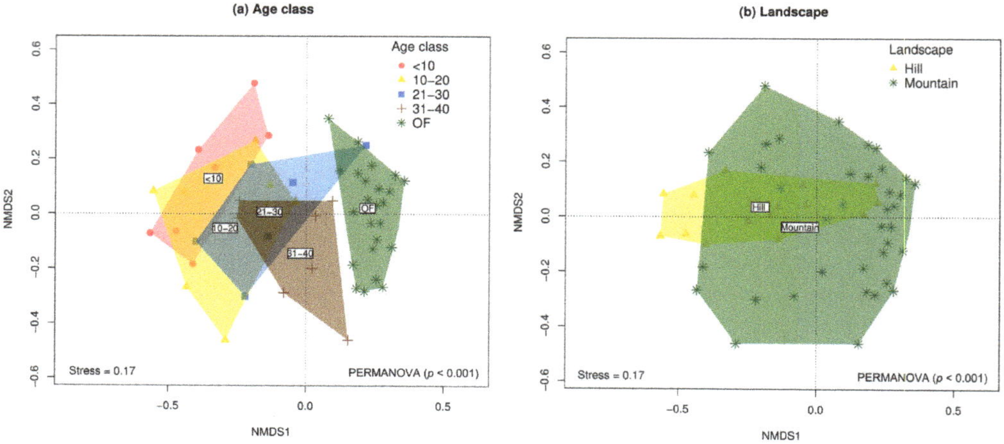

Figure 4. Non-metric multidimensional scaling (nMDS) ordinations of tree community composition (beta diversity) based on Chao–Jaccard similarity between natural regeneration plots following abandonment of cattle ranching and reference forests in the Colombian, deforested Amazon. (**a**) Age classes. (**b**) Landscapes.

3.4. Bivariate Relationships Between Biomass, Diversity, Structure, and Soil Properties

Figure 5 reveals a strong positive correlation between forest age (Age) and various structural and diversity metrics, such as aboveground biomass carbon ($r = 0.87$, $p < 0.0001$), tree species richness ($r = 0.77$, $p < 0.0001$), and ExpH ($r = 0.71$, $p < 0.0001$), indicating that forest aging promotes both carbon accumulation and diversity. Similarly, BA showed a significant correlation with Age ($r = 0.85$, $p < 0.0001$) and AGB ($r = 0.97$, $p < 0.0001$). Community-weighted mean diameter (CWM.DBH) and height (CWM.Height) were also positively correlated with forest age ($r = 0.61$ and $r = 0.62$, respectively; both, $p < 0.01$). On the other hand, although community-weighted mean wood density (CWM.WD) showed moderate positive correlations with Age ($r = 0.52$, $p = 0.0025$) and AGB ($r = 0.49$, $p = 0.0049$), its relationship with other structural metrics, such as CWM.DBH and CWM.Height, was less evident (both, $r < 0.4$; $p < 0.05$).

On the other hand, significant correlations between soil properties and AGB were observed (Supplementary Figure S4). Thus, a significant negative correlation between AGB and soil pH was evidenced, and a significant positive correlation with exchangeable magnesium (Mg_S). SOC showed no significant correlation with AGB, but strong positive correlations with SMS, OM, and N were observed. Soil moisture saturation presented positive correlations with Clay and Silt and negative correlations with Sand and Mg_S, highlighting the ability of soils with a higher proportion of clay and silt to retain water. Cation exchange capacity correlated positively with SOC, Clay, and Al_S and negatively with Sand and Mg_S. Aluminum saturation (Al_S) showed a negative correlation with

Mg_S, evidencing a possible competitive interaction between these elements in soils with high CEC.

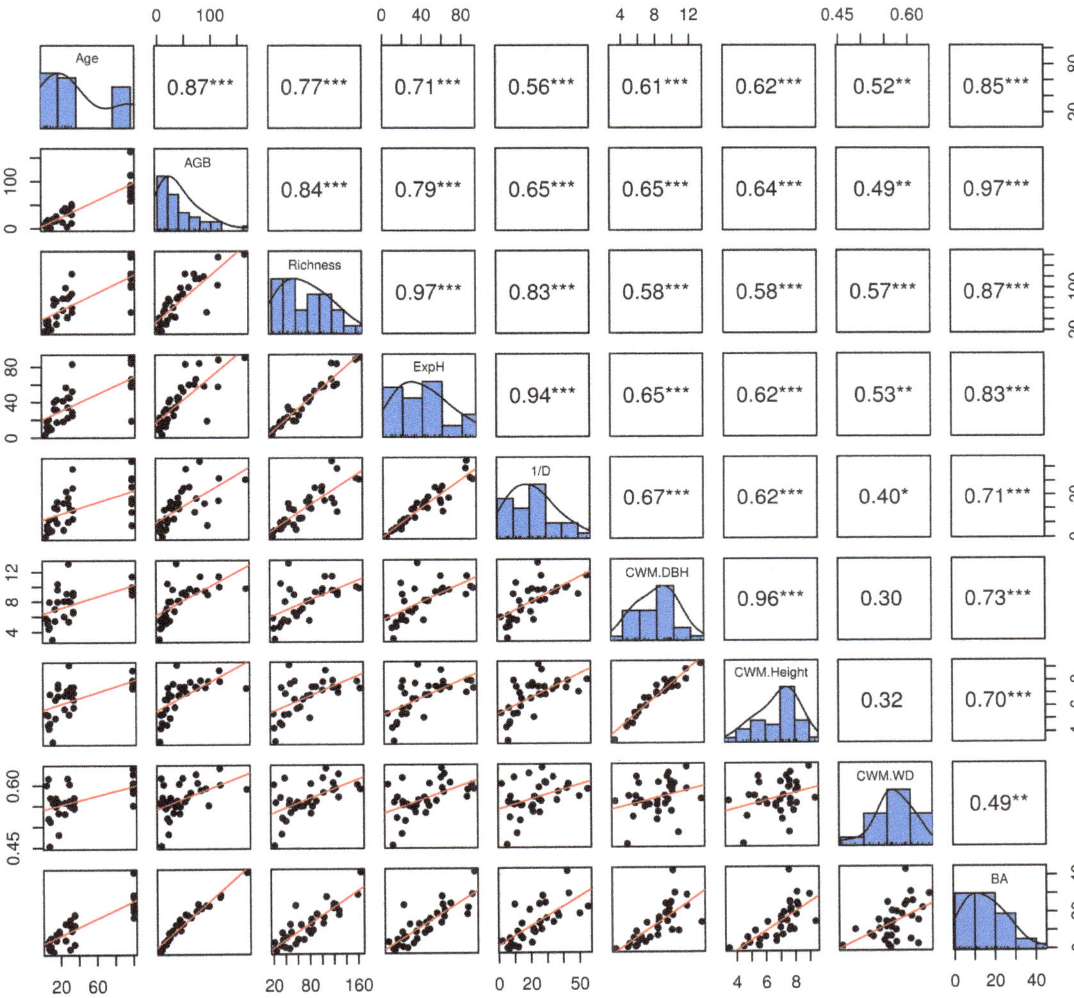

Figure 5. Pairwise correlation matrix among the key variables forest age (Age), aboveground biomass carbon (AGB), tree species richness (Richness), exponential Shannon diversity index (ExpH), Simpson's diversity index (1/D), community-weighted mean diameter at breast height (CWM.DBH), community-weighted mean tree height (CWM.Height), community-weighted mean wood density (CWM.WD), and basal area (BA).

3.5. Influence of Age, Diversity, Structure, Soil Properties and Landscape-Related Parameters on Biomass

The SEM showed a good fit compared to the baseline model and converged (CFI = 0.78, lower AIC and BIC values, and Test statistic = 142.74, $p < 0.001$) (Figure 6 and Supplementary Table S5). The SEM revealed a significant positive effect of forest age (AGE) ($E = 0.74$, $p < 0.001$), but not of tree diversity (DIV) or forest structure (STR) on AGB. However, a slightly negative effect of soil properties (SOI) on AGB was observed ($E = -0.23$, $p = 0.024$). In contrast, SOI showed no significant relationship with STR (covariance = 0.192, $p = 0.312$) or DIV (covariance = 0.218, $p = 0.253$) (Figure 6). The age-mediated effect on STR was significant and positive (covariance = 0.69, $p = 0.001$), indicating that STR improves with AGE. Similarly, the relationship between DIV and STR was significant (covariance = 0.72, $p = 0.001$), highlighting the importance of diversity in shaping STR. The covariance between AGE and DIV was positive and highly significant (covariance = 0.78, $p < 0.001$), suggesting an increase in diversity as ecological succession progresses. These results demonstrate that SFs' age and diversity play a predominant positive role in forest structure over time. No significant effect of the landscape-related parameters (elevation or slope) on AGB was observed. On the contrary, the inclusion of these variables increased the SEM parameterization and reduced its fit; therefore, they were excluded from the final model.

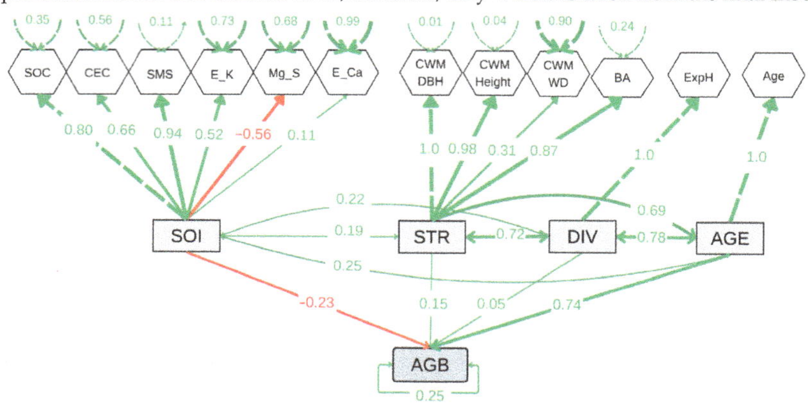

Figure 6. Structural equation model showing the relationships among aboveground biomass carbon (AGB), forest age (AGE), tree diversity (DIV), soil properties (SOI), and forest structure (STR). DIV, STR, and SOI are latent variables constructed from observed indicators such as the log-transformed exponential Shannon index (ExpH), log-transformed community-weighted mean diameter at breast height (CWM.DBH), log-transformed community-weighted mean tree height (CWM.Height), log-transformed community-weighted mean wood density (CWM.WD), log-transformed mean basal area (BA), log-transformed soil organic carbon (SOC), cation exchange capacity (CEC), soil moisture saturation (SMS), log-transformed exchangeable potassium (E_K), exchangeable magnesium (Mg_S), log-transformed exchangeable calcium (E_Ca), and log-transformed forest age (Age).

4. Discussion

4.1. Aboveground Biomass and Forest Structure Recovery

The recovery of carbon in AGB observed in this study follows a similar pattern to that reported in other studies of tropical SFs. The increase from 7.60 Mg C ha^{-1} in young vegetation plots to 84.08 Mg C ha^{-1} in mature vegetation plots aligns with research highlighting these ecosystems' capacity to rapidly accumulate carbon after the abandonment of agricultural or mining activities [14,29,36,53,59]. Tropical forests are crucial for climate change mitigation, storing more than 60% of terrestrial carbon in their AGB [60]. The differential recovery patterns between hills and mountains reinforce that topographic conditions influence forest recovery dynamics. In mountainous landscapes, where AGB growth follows a quadratic pattern, accumulation is more rapid in the early successional stages. Such a pattern may be due to higher moisture retention and better microclimatic conditions [61]. In contrast, AGB growth in hills is linear, reflecting a steadier but less efficient accumulation [33]. However, it is possible that the relatively small sample size (i.e., hills with n = 16) may not accurately represent the population, which could indicate that the rate of change in AGB is less pronounced than that shown in the simple linear model presented in Figure 3. This is particularly due to the limited data available on AGB in older vegetation plots (OFs) within the hilly landscape.

The improvement in structural parameters, such as DBH and BA, as the forest ages advances supports the idea that more mature forest communities develop greater structural complexity, facilitating biomass accumulation [62,63]. Additionally, the increase in WD suggests a transition toward slower-growing species with denser wood, which has also been observed in other studies [14,64]. The vegetation age of the plots has a significant impact on the accumulation of SOC and CEC. This indicates that the early stages of succession are marked by the rapid incorporation of organic matter and cations from plant residues. At the same time, soil pH decreases with age, reflecting an acidification process driven by the accumulation of organic acids and the leaching of basic cations. This process is further accompanied by an increase in aluminum saturation in older vegetation plots, which may restrict nutrient availability, especially in hill landscapes [15,29,32]

Soils with higher clay content and SMS were positively correlated with these structural parameters (DBH and Height), indicating that more fertile soils with better water retention facilitate the establishment of larger, taller trees [35]. Despite soil variation, the relationship between soil properties and biomass was limited, suggesting that stand structure and species composition play a more significant role in biomass recovery [65]. These findings are consistent with previous studies indicating that, while soil properties are important, their impact on biomass is indirect, mediated by structure and diversity [66].

4.2. Species Composition and Diversity

The increase in tree diversity with forest age observed in this study reflects the dynamics of ecological succession. Tree species richness rises from 28.46 in young vegetation plots to 115.72 in mature vegetation plots, consistent with the literature reporting a rise in diversity throughout secondary succession [67]. This change is evident in diversity indices, such as the ExpH and 1/D, which indicate a more equitable distribution of species in older forests [68].

Floristic composition varies between landscapes, with higher initial diversity in mountainous areas than hills. However, it is important to note that the smaller sample size considered for the hill region (n = 16) could increase the risk of overlooking trivial differences in species composition compared to the mountainous region (n = 38), which could explain the high overlap observed in Figure 4b. The last results could be influenced by the greater availability of propagules and favorable conditions in mountainous landscapes [69] and the topographic conditions of these ecosystems [67]. The dominance of pioneer species, such as *P. discolor* and *H. fascicularis*, in early successional stages is a typical pattern in tropical SFs [59,70]. The transition to shade-tolerant species in later successional stages suggests a process of ecological stabilization, supporting the initial floristic composition hypothesis [53,71].

Notably, the interaction between forest age and landscape significantly impacted species richness, indicating that landscape configuration can modulate the recovery of diversity [72]. These shifts in composition align with theories of ecological assembly and environmental filtering, where landscape heterogeneity and local conditions play a crucial role in species selection [73].

4.3. Limitations and Implications of Research in Ecological Restoration

This study provides valuable insights for developing restoration strategies in the Amazon. The rapid recovery of biomass and diversity observed in mountain landscapes highlights their high potential for restoration projects, given their ability to accumulate carbon and biodiversity more rapidly [59]. Additionally, improvements in soil properties with forest age emphasize the importance of implementing management practices that promote organic matter accumulation and nutrient stability [33].

This study underscores the importance that SF restoration consider diversity and forest structure, to promote efficient recovery of carbon and biodiversity [74]. In less favorable landscapes, restoration strategies such as planting key species may be critical to accelerating the recovery of forest functionality [67,75]. Species composition varies significantly between hill and mountains landscapes, necessitating tailored strategies. Assisted Natural Regeneration and SF enrichment must adapt to the ecological characteristics of each landscape. Assisted Natural Regeneration practices in hill areas should focus on maintaining and promoting existing diversity, while enrichment may be needed in mountainous regions to accelerate the recovery of more diverse compositions. Local factors such as topography, soil, and microclimate should be integral to planning ANR interventions. Structural variables such as DBH, BA, and AGB show significant differences based on SF age, emphasizing the need for age-specific interventions. Early-stage (<10 years) interventions should promote rapid growth and stabilization of pioneer species, such as protecting against disturbances or planting fast-growing species. In intermediate stages (11–30 years), facilitating transitions toward greater structural complexity through secondary species enrichment and ecological interaction promotion is critical. Advanced stages (31–40 years) should focus on conserving forest diversity and stability, aiming to restore structures resembling mature forests through density management and the protection of key species. The linear growth of AGB in hills and the faster quadratic pattern in mountains indicate that interventions must be tailored to these dynamics. In hills, practices should aim to maintain long-term uniform growth through soil conservation and the establishment of ecological corridors. In mountains, leveraging favorable conditions for rapid growth requires early-stage ANR and the protection of key species that promote accelerated recovery. Differences in WD and forest structure underscore the need for interventions prioritizing ecological functions, such as carbon storage and biodiversity protection, to accelerate ecosystem recovery effectively.

In early successional SFs, pioneer species are essential for initiating regeneration but gradually lose importance as the SFs matures, giving way to secondary species. Management should encourage this natural transition to facilitate ecological succession. It is essential to recognize that dominant pioneer species play a crucial role in stabilizing ecosystems during early stages, facilitating the transition to more diverse and functionally complex communities that are critical for maximizing forest resilience to future disturbances [64,67]. In contrast, the inclusion of shade-tolerant species during advanced successional stages can accelerate the recovery of forest structure and function, enhancing ecosystem services such as carbon sequestration and water regulation [71,76]. In intermediate and advanced stages, introducing or protecting secondary species can enhance forest structure and functionality, contributing to more stable and resilient ecosystems. Significant differences in β-diversity between plot ages and landscapes highlight the importance of maintaining species composition heterogeneity for ecosystem resilience. Assisted Natural Regeneration strategies should focus on conserving this diversity, potentially through planting or protecting species that maximize β-diversity, particularly in stages where diversity tends to decline. Our results allow implementers and decision-makers to design a Transformative Ecological Restoration Program involving SFs.

Our findings emphasize the necessity of managing SFs through ecological restoration as a central component of Transformative Territorial Management (TTM). The use and management of SFs in the Amazon Basin, particularly in the Colombian Amazon, requires comprehensive planning that involves: (1) Generating and utilizing knowledge about SF integrity within transformed landscapes; (2) Prioritizing SFs as strategic areas for ecological restoration, accounting for socio-economic and environmental contexts [18]; and (3) Implementing evidence-based restoration practices to enhance conservation value and biodiversity [13,77]. These efforts aim to ensure SF functionality and the provision of essential ecosystem services [14], while strengthening ecological connectivity between the Andes and the Amazon, as well as across regions within the Amazon [69]. In addition to achieving the goals set for restoration, both from international and national demands, and beyond these demands, understanding the need to ensure the viability of life on the planet.

5. Conclusions

The natural recovery of AGB in secondary Amazonian forests is influenced mainly by forest age and soil properties. Among these factors, forest age stands out as the key driver, mediating significant increases in structural metrics such as BA, DBH, and WD. These changes reflect the transition toward more mature forest communities with greater carbon storage capacity. Similarly, tree diversity increases with succession, indicating a more equitable species distribution in mature forests.

Comparatively, mountainous landscapes demonstrate faster AGB recovery during early successional stages, while hilly areas exhibit steadier, linear growth. These differences highlight the critical role of topographic conditions in shaping successional dynamics. Soil properties also improve with SF age, promoting the accumulation of organic matter and nutrient retention as is expected in the natural regeneration process. However, the observed increase in aluminum saturation and declining pH levels indicate an ongoing acidification process that could limit the availability of certain nutrients, highlighting the soil limitation conditions of the Amazon region.

These findings provide key insights to guide adaptive restoration strategies that consider and integrate landscape heterogeneity. Such adapted strategies are essential to maximize carbon sequestration, enhance ecological resilience, and ensure the long-term sustainability of SFs in the Amazon. The results highlight the complexity of natural regeneration and underscore the need of achieve Transformative Territorial Management of SFs through future-oriented restoration strategies. Assisted natural regeneration, aimed at accelerating the recovery of carbon (C) stocks, must be tailored to the specific conditions of each landscape and successional stage, prioritizing biodiversity, connectivity, and continuous monitoring. By adopting these tailored approaches, the effective and sustainable recovery of SFs in the Amazon can be achieved, contributing to global restoration goals and the ecological health of the region.

Supplementary Materials: The following supporting information can be downloaded at: https://www.mdpi.com/article/10.3390/d17010039/s1, Figure S1: Interaction between forest age and landscape on structural and diversity attributes. (A) Aboveground biomass carbon (AGB), (B) tree species richness, (C) exponential Shannon diversity (ExpH), (D) inverse Simpson diversity (1/D), (E) mean tree height (Height), (F) diameter at breast height (DBH), (G) wood density (WD), and (H) basal area (BA). Different letters indicate significant differences between age classes and landscapes ($p < 0.05$); Figure S2: Modeled trends of forest attributes recovery. (A) Aboveground biomass carbon (AGB), (B) tree Species richness, (C) exponential Shannon diversity (ExpH), (D) inverse Simpson diversity (1/D), (E) mean tree height (Height), (F) diameter at breast height (DBH), (G) wood density (WD), and (H) basal area (BA); Figure S3: Non-metric multidimensional scaling (nMDS) ordinations of tree community composition (beta diversity) based on Chao–Jaccard similarity between natural regeneration plots following abandonment of cattle ranching and reference forests in two landscape units in the Colombian deforested Amazon. (A) Hills. (B) Mountains. In both plots, hulls are overlaid for each variable; Figure S4: Pairwise correlation matrix among key soil and biomass variables, including aboveground biomass carbon (AGB), soil organic carbon (SOC), pH, cation exchange capacity (CEC), mean moisture saturation (SMS), organic matter (MO), nitrogen (N), clay content (Clay), sand content (Sand), silt content (Silt), exchangeable potassium (E_K), exchangeable calcium (E_Ca), available iron (Fe), available phosphorus (P), exchangeable magnesium (Mg_S), and aluminum saturation (Al_S); Table S1: Mean values of the evaluated soil variables in five forest age class and two landscape types (hill and mountain). Soil organic carbon (SOC, Mg ha^{-1}), pH in water (pH), cation exchange capacity (CEC, meq 100 g^{-1}), soil moisture saturation (SMS, %), organic matter (OM, %), total nitrogen (N, %), clay content (Clay, %), sand content (Sand, %), silt content (Silt, %), exchangeable potassium (E_K, mg kg^{-1}), exchangeable calcium (E_Ca, mg kg^{-1}), available iron (Fe, mg kg^{-1}), available phosphorus (P, mg kg^{-1}), exchangeable magnesium (Mg_S, %), and exchangeable aluminum (Al_S, %); Table S2: Comparison of the five species with the highest importance value index (IVI) of the tree community by forest stand age and landscape type (hill and mountain) across a chronosequence in degraded lands of the Amazon; Table S3: Results of analysis of variance (PERMANOVA) tests on natural regeneration of tree communities across a chronosequence in degraded lands of the Amazon. Landscape was treated as a fixed effect and stand age (years) was used as a covariate; Table S4: Post hoc analysis results of floristic similarity between age classes using pairwise Adonis with Chao index; Table S5: Details of structural equation model.

Author Contributions: Conceptualization, C.H.R.-L., A.S., and L.L.R.-F.; methodology, C.H.R.-L., A.S., and L.L.R.-F.; software, A.S.; validation, A.S., C.H.R.-L., and L.L.R.-F.; formal analysis, A.S.; investigation, C.H.R.-L., L.L.R.-F., and A.S.; resources, C.H.R.-L.; data curation, A.T.-B., Y.D.S.-C., and A.S.; writing—original draft preparation, C.H.R.-L., A.S., A.T.-B., Y.D.S.-C., and L.L.R.-F.; writing—review and editing, C.H.R.-L., L.L.R.-F., A.S., and Y.D.S.-C.; visualization, A.S. and Y.D.S.-C.; supervision, C.H.R.-L., A.S., and L.L.R.-F.; project administration, C.H.R.-L.; funding acquisition, C.H.R.-L. All authors have read and agreed to the published version of the manuscript.

Funding: This research was part of the project: "Restauración de áreas disturbadas por implementación de sistemas productivos agropecuarios en zonas de alta intervención en el Caquetá", funded by the Fondo de Ciencia, Tecnología e Innovación FCTeI—SGR, the Amazonian Scientific Research Institute Sinchi, the Government of Caquetá, the Universidad de la Amazonía, the Asociación de Reforestadores y Cultivadores de Caucho del Caquetá ASOHECA, and the Federación Departamental de Ganaderos del Caquetá FEDEGANGA. Contract 60/2013; and by the Government of Colombia through the project BPIN 202300000000285 "Investigación científica transformativa para potenciar el bienestar, la conservación y la gobernanza ambiental en la Amazonia colombiana Amazonas, Caquetá, Guainía, Guaviare, Meta, Putumayo, Vaupés".

Institutional Review Board Statement: Not applicable.

Data Availability Statement: Data are available from the authors upon request.

Acknowledgments: The authors thank all the farmers in the study area for their help and support during the fieldwork.

Conflicts of Interest: The authors declare no conflicts of interest. The funders had no role in the design of the study; in the collection, analyses, or interpretation of the data; in the writing of the manuscript, or in the decision to publish the results.

References

1. Potapov, P.; Hansen, M.C.; Laestadius, L.; Turubanova, S.; Yaroshenko, A.; Thies, C.; Smith, W.; Zhuravleva, I.; Komarova, A.; Minnemeyer, S.; et al. The Last Frontiers of Wilderness: Tracking Loss of Intact Forest Landscapes from 2000 to 2013. *Sci. Adv.* **2017**, *3*, e1600821. [CrossRef]
2. Mitchard, E.T.A. The Tropical Forest Carbon Cycle and Climate Change. *Nature* **2018**, *559*, 527–534. [CrossRef]
3. Cooper, D.L.M.; Lewis, S.L.; Sullivan, M.J.P.; Prado, P.I.; ter Steege, H.; Barbier, N.; Slik, F.; Sonké, B.; Ewango, C.E.N.; Adu-Bredu, S.; et al. Consistent Patterns of Common Species across Tropical Tree Communities. *Nature* **2024**, *625*, 728–734. [CrossRef]
4. Tito, R.; Salinas, N.; Cosio, E.G.; Boza Espinoza, T.E.; Muñiz, J.G.; Aragón, S.; Nina, A.; Roman-Cuesta, R.M. Secondary Forests in Peru: Differential Provision of Ecosystem Services Compared to Other Post-Deforestation Forest Transitions. *Ecol. Soc.* **2022**, *27*, art12. [CrossRef]
5. Armenteras, D.; Murcia, U.; González, T.M.; Barón, O.J.; Arias, J.E. Scenarios of Land Use and Land Cover Change for NW Amazonia: Impact on Forest Intactness. *Glob. Ecol. Conserv.* **2019**, *17*, e00567. [CrossRef]
6. Csillik, O.; Asner, G.P. Aboveground Carbon Emissions from Gold Mining in the Peruvian Amazon. *Environ. Res. Lett.* **2020**, *15*, 014006. [CrossRef]

7. Smith, C.C.; Healey, J.R.; Berenguer, E.; Young, P.J.; Taylor, B.; Elias, F.; Espírito-Santo, F.; Barlow, J. Old-Growth Forest Loss and Secondary Forest Recovery across Amazonian Countries. *Environ. Res. Lett.* **2021**, *16*, 085009. [CrossRef]
8. SIAT-AC Sistema de Información Ambiental Territorial de La Amazonia Colombiana—SIAT-AC. Available online: https://siatac.co/ (accessed on 25 October 2024).
9. Gann, G.D.; McDonald, T.; Walder, B.; Aronson, J.; Nelson, C.R.; Jonson, J.; Hallett, J.G.; Eisenberg, C.; Guariguata, M.R.; Liu, J.; et al. International Principles and Standards for the Practice of Ecological Restoration. Second Edition. *Restor. Ecol.* **2019**, *27*, S3–S46. [CrossRef]
10. Rodríguez-León, C.H.; Roa-Fuentes, L.L.; Sterling, A.; Suárez, J.C. Plant Biodiversity Homogenization across the Chronosequence in Highly Fragmented Landscapes in the Colombian Andean–Amazonian Transition. *Forests* **2022**, *13*, 1422. [CrossRef]
11. Pain, A.; Marquardt, K.; Khatri, D. Secondary Forests and Agrarian Transitions: Insights from Nepal and Peru. *Hum. Ecol.* **2021**, *49*, 249–258. [CrossRef]
12. Williams, B.A.; Beyer, H.L.; Fagan, M.E.; Chazdon, R.L.; Schmoeller, M.; Sprenkle-Hyppolite, S.; Griscom, B.W.; Watson, J.E.M.; Tedesco, A.M.; Gonzalez-Roglich, M.; et al. Global Potential for Natural Regeneration in Deforested Tropical Regions. *Nature* **2024**, *636*, 131–137. [CrossRef] [PubMed]
13. Crouzeilles, R.; Ferreira, M.S.; Chazdon, R.L.; Lindenmayer, D.B.; Sansevero, J.B.B.; Monteiro, L.; Iribarrem, A.; Latawiec, A.E.; Strassburg, B.B.N. Ecological Restoration Success Is Higher for Natural Regeneration than for Active Restoration in Tropical Forests. *Sci. Adv.* **2017**, *3*, e1701345. [CrossRef]
14. Poorter, L.; Bongers, F.; Aide, T.M.; Almeyda Zambrano, A.M.; Balvanera, P.; Becknell, J.M.; Boukili, V.; Brancalion, P.H.S.; Broadbent, E.N.; Chazdon, R.L.; et al. Biomass Resilience of Neotropical Secondary Forests. *Nature* **2016**, *530*, 211–214. [CrossRef]
15. Coelho, A.J.P.; Mancini Teixeira, H.; Verweij, P.; Matos, F.A.R.; Villa, P.M.; Meira-Neto, J.A.A.; Teixeira, H.M.; Verweij, P.; Matos, F.A.R.; Villa, P.M.; et al. Functional Richness Mediates Landscape and Management Effects on Tree Biomass and Soil Fertility during Secondary Forest Succession. *Ecol. Indic.* **2024**, *162*, 112029. [CrossRef]
16. Lamb, D.; Erskine, P.D.; Parrotta, J.A. Restoration of Degraded Tropical Forest Landscapes. *Science* **2005**, *310*, 1628–1632. [CrossRef]
17. Shono, K.; Cadaweng, E.A.; Durst, P.B. Application of Assisted Natural Regeneration to Restore Degraded Tropical Forestlands. *Restor. Ecol.* **2007**, *15*, 620–626. [CrossRef]
18. Chazdon, R.L.; Guariguata, M.R. Natural Regeneration as a Tool for Large-scale Forest Restoration in the Tropics: Prospects and Challenges. *Biotropica* **2016**, *48*, 716–730. [CrossRef]
19. Parrotta, J.A.; Knowles, O.H. Restoration of Tropical Moist Forests on Bauxite-Mined Lands in the Brazilian Amazon. *Restor. Ecol.* **1999**, *7*, 103–116. [CrossRef]
20. Holl, K.D.; Aide, T.M. When and Where to Actively Restore Ecosystems? *For. Ecol. Manag.* **2011**, *261*, 1558–1563. [CrossRef]
21. Rosenfield, M.F.; Jakovac, C.C.; Vieira, D.L.M.; Poorter, L.; Brancalion, P.H.S.; Vieira, I.C.G.; de Almeida, D.R.A.; Massoca, P.; Schietti, J.; Albernaz, A.L.M.; et al. Ecological Integrity of Tropical Secondary Forests: Concepts and Indicators. *Biol. Rev.* **2023**, *98*, 662–676. [CrossRef] [PubMed]
22. Rodríguez-León, C.H.; Roa-Fuentes, L.L.; Sterling, A.; Suárez, J.C. Plant-Community Vulnerability in Highly Fragmented Landscapes Is Higher in Secondary Forests Than in Old Growth Forests in the Andean–Amazonian Transition. *Plants* **2022**, *11*, 3284. [CrossRef]
23. Aryal, D.R.; De Jong, B.H.J.; Ochoa-Gaona, S.; Esparza-Olguin, L.; Mendoza-Vega, J. Carbon Stocks and Changes in Tropical Secondary Forests of Southern Mexico. *Agric. Ecosyst. Environ.* **2014**, *195*, 220–230. [CrossRef]
24. Bauters, M.; Vercleyen, O.; Vanlauwe, B.; Six, J.; Bonyoma, B.; Badjoko, H.; Hubau, W.; Hoyt, A.; Boudin, M.; Verbeeck, H.; et al. Long-term Recovery of the Functional Community Assembly and Carbon Pools in an African Tropical Forest Succession. *Biotropica* **2019**, *51*, 319–329. [CrossRef]
25. Palma, A.C.; Goosem, M.; Fensham, R.J.; Goosem, S.; Preece, N.D.; Stevenson, P.R.; Laurance, S.G.W. Dispersal and Recruitment Limitations in Secondary Forests. *J. Veg. Sci.* **2021**, *32*, e12975. [CrossRef]
26. Cecilio Rebola, L.; Pandolfo Paz, C.; Valenzuela Gamarra, L.; Burslem, D.F.R.P. Land Use Intensity Determines Soil Properties and Biomass Recovery after Abandonment of Agricultural Land in an Amazonian Biodiversity Hotspot. *Sci. Total Environ.* **2021**, *801*, 149487. [CrossRef] [PubMed]
27. Rozendaal, D.M.A.; Bongers, F.; Aide, T.M.; Alvarez-Dávila, E.; Ascarrunz, N.; Balvanera, P.; Becknell, J.M.; Bentos, T.V.; Brancalion, P.H.S.; Cabral, G.A.L.; et al. Biodiversity Recovery of Neotropical Secondary Forests. *Sci. Adv.* **2019**, *5*, eaau3114. [CrossRef]
28. Lai, H.R.; Hall, J.S.; Batterman, S.A.; Turner, B.L.; van Breugel, M. Nitrogen Fixer Abundance Has No Effect on Biomass Recovery during Tropical Secondary Forest Succession. *J. Ecol.* **2018**, *106*, 1415–1427. [CrossRef]
29. Aryal, D.R.; De Jong, B.H.J.; Sánchez-Silva, S.; Haas-Ek, A.; Esparza-Olguin, L.; Ochoa-Gaona, S.; Ghimire, R.; Morales-Ruiz, D.E. Biomass Recovery along a Tropical Forest Succession: Trends on Tree Diversity, Wood Traits and Stand Structure. *For. Ecol. Manag.* **2024**, *555*, 121709. [CrossRef]

30. Murad, C.A.; Pearse, J. Landsat Study of Deforestation in the Amazon Region of Colombia: Departments of Caquetá and Putumayo. *Remote Sens. Appl. Soc. Environ.* **2018**, *11*, 161–171. [CrossRef]
31. IGAC. *Estudio General de Suelos y Zonificación de Tierras Departamento de Caquetá. Escala 1:100.000*; Imprenta Nacional de Colombia: Bogotá, Colombia, 2014; ISBN 978-958-8323-73-2.
32. Rodríguez-León, C.H.; Peña-Venegas, C.P.; Sterling, A.; Castro, D.; Mahecha-Virguez, L.K.; Virguez-Díaz, Y.R.; Silva-Olaya, A.M. Soil Quality Restoration during the Natural Succession of Abandoned Cattle Pastures in Deforested Landscapes in the Colombian Amazon. *Agronomy* **2021**, *11*, 2484. [CrossRef]
33. Teixeira, H.M.; Cardoso, I.M.; Bianchi, F.J.J.A.; da Cruz Silva, A.; Jamme, D.; Peña-Claros, M. Linking Vegetation and Soil Functions during Secondary Forest Succession in the Atlantic Forest. *For. Ecol. Manag.* **2020**, *457*, 117696. [CrossRef]
34. Norden, N.; Angarita, H.A.; Bongers, F.; Martínez-Ramos, M.; Granzow-de la Cerda, I.; van Breugel, M.; Lebrija-Trejos, E.; Meave, J.A.; Vandermeer, J.; Williamson, G.B.; et al. Successional Dynamics in Neotropical Forests Are as Uncertain as They Are Predictable. *Proc. Natl. Acad. Sci. USA* **2015**, *112*, 8013–8018. [CrossRef] [PubMed]
35. Chave, J.; Réjou-Méchain, M.; Búrquez, A.; Chidumayo, E.; Colgan, M.S.; Delitti, W.B.C.; Duque, A.; Eid, T.; Fearnside, P.M.; Goodman, R.C.; et al. Improved Allometric Models to Estimate the Aboveground Biomass of Tropical Trees. *Glob. Chang. Biol.* **2014**, *20*, 3177–3190. [CrossRef]
36. Garate-Quispe, J.; Herrera-Machaca, M.; Pareja Auquipata, V.; Alarcón Aguirre, G.; Baez Quispe, S.; Carpio-Vargas, E.E. Resilience of Aboveground Biomass of Secondary Forests Following the Abandonment of Gold Mining Activity in the Southeastern Peruvian Amazon. *Diversity* **2024**, *16*, 233. [CrossRef]
37. Doraisami, M.; Domke, G.M.; Martin, A.R. Improving Wood Carbon Fractions for Multiscale Forest Carbon Estimation. *Carbon Balance Manag.* **2024**, *19*, 25. [CrossRef]
38. Hsieh, T.C.; Ma, K.H.; Chao, A. INEXT: An R Package for Rarefaction and Extrapolation of Species Diversity (Hill Numbers). *Methods Ecol. Evol.* **2016**, *7*, 1451–1456. [CrossRef]
39. Rother, D.C.; Liboni, A.P.; Magnago, L.F.S.; Chao, A.; Chazdon, R.L.; Rodrigues, R.R. Ecological Restoration Increases Conservation of Taxonomic and Functional Beta Diversity of Woody Plants in a Tropical Fragmented Landscape. *For. Ecol. Manag.* **2019**, *451*, 117538. [CrossRef]
40. Gotelli, N.J.; Chao, A. Measuring and Estimating Species Richness, Species Diversity, and Biotic Similarity from Sampling Data. In *Encyclopedia of Biodiversity*; Elsevier: Amsterdam, The Netherlands, 2013; pp. 195–211.
41. Chao, A.; Kubota, Y.; Zelený, D.; Chiu, C.; Li, C.; Kusumoto, B.; Yasuhara, M.; Thorn, S.; Wei, C.; Costello, M.J.; et al. Quantifying Sample Completeness and Comparing Diversities among Assemblages. *Ecol. Res.* **2020**, *35*, 292–314. [CrossRef]
42. Melo, A.S. CommEcol: Community Ecology Analyses. CRAN Contributed Packages. 2016. Available online: https://cran.r-project.org/web/packages/CommEcol/CommEcol.pdf (accessed on 30 October 2024).
43. Su, L.; Heydari, M.; Omidipour, R.; Soheili, F.; Cheraghi, J.; Manuel Villa, P.; Prévosto, B. Stand Structural Diversity and Elevation Rather than Functional Diversity Drive Aboveground Biomass in Historically Disturbed Semiarid Oak Forests. *For. Ecol. Manag.* **2023**, *543*, 121139. [CrossRef]
44. Curtis, J.T.; McIntosh, R.P. An Upland Forest Continuum in the Prairie-Forest Border Region of Wisconsin. *Ecology* **1951**, *32*, 476–496. [CrossRef]
45. Kindt, R. BiodiversityR: Package for Community Ecology and Suitability Analysis. CRAN Contributed Packages. 2024. Available online: https://cran.r-project.org/web/packages/BiodiversityR/index.html (accessed on 30 October 2024).
46. Maia, S.M.F.; Ogle, S.M.; Cerri, C.C.; Cerri, C.E.P. Changes in Soil Organic Carbon Storage under Different Agricultural Management Systems in the Southwest Amazon Region of Brazil. *Soil Tillage Res.* **2010**, *106*, 177–184. [CrossRef]
47. Pinheiro, J.; Bates, D.; DebRoy, S.; Sarkar, D. Nlme: Linear and Nonlinear Mixed Effects Models. R Package Version 3.1-131.1. 2018. Available online: https://cran.r-project.org/web/packages/nlme/index.html (accessed on 30 October 2024).
48. R Core Team. *R: A Language and Environment for Statistical Computing*, v.4.3.3; The Comprehensive R Archive Network: Viena, Austria, 2024.
49. Di Rienzo, J.A.; Casanoves, F.; Balzarini, M.G.; Gonzalez, L.; Tablada, M.; Robledo, C.W. *InfoStat*, Universidad Nacional de Córdoba: Córdoba, Argentina, 2020.
50. Worldwide, R.C.T. Contributors Package The R Stats Package Version 4.3.3. 2024. Available online: https://CRAN.R-project.org/package=STAT (accessed on 30 October 2024).
51. Matthews, T.J.; Guilhaumon, F.; Cazelles, K. Package Sars: Fit and Compare Species-Area Relationship Models Using Multimodel Inference—Version 1.3.7. 2024. Available online: https://cran.r-project.org/web/packages/sars/sars.pdf (accessed on 30 October 2024).
52. Oksanen, J.; Blanchet, G.; Friendly, M.; Kindt, R.; Legendre, P.; McGlinn, D.; Minchin, P.; O'Hara, R.; Simpson, G.; Solymos, P.; et al. *Package 'Vegan'*, Community Ecology Package Version 2.5-7; The Comprehensive R Archive Network: Viena, Austria, 2018.

53. Oberleitner, F.; Egger, C.; Oberdorfer, S.; Dullinger, S.; Wanek, W.; Hietz, P. Recovery of Aboveground Biomass, Species Richness and Composition in Tropical Secondary Forests in SW Costa Rica. *For. Ecol. Manag.* **2021**, *479*, 118580. [CrossRef]
54. Chao, A.; Chazdon, R.L.; Colwell, R.K.; Shen, T. A New Statistical Approach for Assessing Similarity of Species Composition with Incidence and Abundance Data. *Ecol. Lett.* **2005**, *8*, 148–159. [CrossRef]
55. Martinez-Arbizu, P. *Package: "PairwiseAdonis"*, Pairwise Multilevel Comparison Using Adonis Package Version: 0.0.1; The Comprehensive R Archive Network: Viena, Austria, 2017.
56. Revelle, W. *Package: Psych*, Procedures for Psychological, Psychometric, and Personality Research—Version 2.4.6.26; The Comprehensive R Archive Network: Viena, Austri, 2024.
57. Rosseel, Y. Lavaan: An R Package for Structural Equation Modeling. *J. Stat. Softw.* **2012**, *48*, 1–36. [CrossRef]
58. RStudio Team. *RStudio: Integrated Development Environment for R*; RStudio Team: Boston, MA, USA, 2024.
59. Kalamandeen, M.; Gloor, E.; Johnson, I.; Agard, S.; Katow, M.; Vanbrooke, A.; Ashley, D.; Batterman, S.A.; Ziv, G.; Holder-Collins, K.; et al. Limited Biomass Recovery from Gold Mining in Amazonian Forests. *J. Appl. Ecol.* **2020**, *57*, 1730–1740. [CrossRef]
60. Poorter, L.; Craven, D.; Jakovac, C.C.; van der Sande, M.T.; Amissah, L.; Bongers, F.; Chazdon, R.L.; Farrior, C.E.; Kambach, S.; Meave, J.A.; et al. Multidimensional Tropical Forest Recovery. *Science* **2021**, *374*, 1370–1376. [CrossRef] [PubMed]
61. Kim, S.; Zang, H.; Mortimer, P.; Shi, L.; Li, Y.; Xu, J.; Ostermann, A. Tree Species and Recovery Time Drives Soil Restoration after Mining: A Chronosequence Study. *Land Degrad. Dev.* **2018**, *29*, 1738–1747. [CrossRef]
62. Mensah, S.; Noulèkoun, F.; Salako, V.K.; Lokossou, C.S.M.J.; Akouété, P.; Seifert, T.; Glèlè Kakaï, R. Structural and Taxonomic Diversity Predict Above-ground Biomass Better than Functional Measures of Maximum Height in Mixed-species Forests. *Appl. Veg. Sci.* **2023**, *26*, e12732. [CrossRef]
63. Gris, D.; Casagrande, J.C.; Marques, M.R.; Oldeland, J.; Damasceno-Júnior, G.A. Periodic Flooding and Edaphic Factors Shape Erythrina Fusca Dominance in Riparian Forests in the Pantanal Wetland. *Trop. Ecol.* **2024**, *65*, 224–238. [CrossRef]
64. Bello, C.; Galetti, M.; Pizo, M.A.; Magnago, L.F.S.; Rocha, M.F.; Lima, R.A.F.; Peres, C.A.; Ovaskainen, O.; Jordano, P. Defaunation Affects Carbon Storage in Tropical Forests. *Sci. Adv.* **2015**, *1*, 6310. [CrossRef]
65. Ali, A.; Lin, S.-L.; He, J.-K.; Kong, F.-M.; Yu, J.-H.; Jiang, H.-S. Climate and Soils Determine Aboveground Biomass Indirectly via Species Diversity and Stand Structural Complexity in Tropical Forests. *For. Ecol. Manag.* **2019**, *432*, 823–831. [CrossRef]
66. Gavito, M.E.; Paz, H.; Barragán, F.; Siddique, I.; Arreola-Villa, F.; Pineda-García, F.; Balvanera, P. Indicators of Integrative Recovery of Vegetation, Soil and Microclimate in Successional Fields of a Tropical Dry Forest. *For. Ecol. Manag.* **2021**, *479*, 118526. [CrossRef]
67. Rodrigues, A.C.; Villa, P.M.; Silla, F.; Gomes, L.P.; Meira-Neto, J.A.A.; Torres, C.M.M.E.; Neri, A.V. Functional Composition Enhances Aboveground Carbon Stock during Tropical Late-Secondary Forest Succession. *Plant Biosyst.—An Int. J. Deal. with all Asp. Plant Biol.* **2023**, *157*, 1–11. [CrossRef]
68. Coelho, A.J.P.; Ribeiro Matos, F.A.; Villa, P.M.; Heringer, G.; Pontara, V.; de Paula Almado, R.; Alves Meira-Neto, J.A. Multiple Drivers Influence Tree Species Diversity and Above-Ground Carbon Stock in Second-Growth Atlantic Forests: Implications for Passive Restoration. *J. Environ. Manag.* **2022**, *318*, 115588. [CrossRef] [PubMed]
69. Arroyo-Rodríguez, V.; Rito, K.F.; Farfán, M.; Navia, I.C.; Mora, F.; Arreola-Villa, F.; Balvanera, P.; Bongers, F.; Castellanos-Castro, C.; Catharino, E.L.M.; et al. Landscape-Scale Forest Cover Drives the Predictability of Forest Regeneration across the Neotropics. *Proc. R. Soc. B Biol. Sci.* **2023**, *290*, 20222203. [CrossRef]
70. Coelho, A.J.P.; Villa, P.M.; Matos, F.A.R.; Heringer, G.; Bueno, M.L.; de Paula Almado, R.; Meira-Neto, J.A.A. Atlantic Forest Recovery after Long-Term Eucalyptus Plantations: The Role of Zoochoric and Shade-Tolerant Tree Species on Carbon Stock. *For. Ecol. Manag.* **2022**, *503*, 119789. [CrossRef]
71. Poorter, L.; Amissah, L.; Bongers, F.; Hordijk, I.; Kok, J.; Laurance, S.G.W.; Lohbeck, M.; Martínez-Ramos, M.; Matsuo, T.; Meave, J.A.; et al. Successional Theories. *Biol. Rev.* **2023**, *98*, 2049–2077. [CrossRef]
72. Matos, F.A.R.; Magnago, L.F.S.; Aquila Chan Miranda, C.; de Menezes, L.F.T.; Gastauer, M.; Safar, N.V.H.; Schaefer, C.E.G.R.; da Silva, M.P.; Simonelli, M.; Edwards, F.A.; et al. Secondary Forest Fragments Offer Important Carbon and Biodiversity Cobenefits. *Glob. Chang. Biol.* **2020**, *26*, 509–522. [CrossRef]
73. Laliberté, E.; Legendre, P. A Distance-based Framework for Measuring Functional Diversity from Multiple Traits. *Ecology* **2010**, *91*, 299–305. [CrossRef] [PubMed]
74. Mori, A.S.; Dee, L.E.; Gonzalez, A.; Ohashi, H.; Cowles, J.; Wright, A.J.; Loreau, M.; Hautier, Y.; Newbold, T.; Reich, P.B.; et al. Biodiversity–Productivity Relationships Are Key to Nature-Based Climate Solutions. *Nat. Clim. Chang.* **2021**, *11*, 543–550. [CrossRef]
75. Aerts, R.; Honnay, O. Forest Restoration, Biodiversity and Ecosystem Functioning. *BMC Ecol.* **2011**, *11*, 29. [CrossRef] [PubMed]

76. Balvanera, P.; Pfisterer, A.B.; Buchmann, N.; He, J.; Nakashizuka, T.; Raffaelli, D.; Schmid, B. Quantifying the Evidence for Biodiversity Effects on Ecosystem Functioning and Services. *Ecol. Lett.* **2006**, *9*, 1146–1156. [CrossRef]
77. Crouzeilles, R.; Beyer, H.L.; Monteiro, L.M.; Feltran-Barbieri, R.; Pessôa, A.C.M.; Barros, F.S.M.; Lindenmayer, D.B.; Lino, E.D.S.M.; Grelle, C.E.V.; Chazdon, R.L.; et al. Achieving Cost-effective Landscape-scale Forest Restoration through Targeted Natural Regeneration. *Conserv. Lett.* **2020**, *13*, e12709. [CrossRef]

diversity

MDPI

Article

Diversity in Burned Pinyon–Juniper Woodlands Across Fire and Soil Parent Material Gradients

Scott R. Abella [1,*], Lindsay P. Chiquoine [1], Elizabeth C. Bailey [1], Shelley L. Porter [1], Cassandra D. Morrison [1], Calvin A. Farris [2] and Jennifer E. Fox [3]

1 School of Life Sciences, University of Nevada Las Vegas, Las Vegas, NV 89154-4004, USA; lindsay.chiquoine@unlv.edu (L.P.C.)
2 National Park Service, Pacific West Region Fire Ecology Program, Klamath Falls, OR 97601, USA
3 National Park Service, Grand Canyon-Parashant National Monument, St. George, UT 84790, USA; jennifer_fox@nps.gov
* Correspondence: scott.abella@unlv.edu

Abstract: Co-varying disturbance and environmental gradients can shape vegetation dynamics and increase the diversity of plant communities and their features. Pinyon–juniper woodlands are widespread in semi-arid climates of western North America, encompassing extensive environmental gradients, and a knowledge gap is how the diversity in features of these communities changes across co-varying gradients in fire history and soil. In pinyon–juniper communities spanning soil parent materials (basalt, limestone) and recent fire histories (0–4 prescribed fires or managed wildfires and 5–43 years since fire) in Grand Canyon-Parashant National Monument (Arizona, USA), we examined variation at 25 sites in three categories of plant community features including fuels, tree structure, and understory vegetation. Based on ordinations, canonical correlation analysis, and permutation tests, plant community features varied primarily with the number of fires, soil coarseness and chemistry, and additionally with tree structure for understory vegetation. Fire and soil variables accounted for 33% of the variance in fuels and tree structure, and together with tree structure, 56% of the variance in understories. The cover of the non-native annual *Bromus tectorum* was higher where fires had occurred more recently. In turn, *B. tectorum* was positively associated with the percentage of dead trees and negatively associated with native forb species richness. Based on a dendroecological analysis of 127 *Pinus monophylla* and *Juniperus osteosperma* trees, only 18% of trees presently around our study sites originated before the 1870s (Euro-American settlement) and <2% originated before the 1820s. Increasing contemporary fire activity facilitated by the National Park Service since the 1980s corresponded with increasing tree mortality and open-structured stands, apparently more closely resembling pre-settlement conditions. Using physical geography, such as soil parent material, as a landscape template shows promise for (i) incorporating diversity in long-term community change serving as a baseline for vegetation management, (ii) customizing applying treatments to unique conditions on different soil types, and (iii) benchmarking monitoring metrics of vegetation management effectiveness to levels scaled to biophysical variation across the landscape.

Keywords: disturbance; environmental gradient; fuels; managed wildfire; national park; prescribed fire; resource objective wildfire; spatio-temporal gradient; vegetation change

https://doi.org/10.3390/d17020088

1. Introduction

Co-varying environmental and disturbance gradients can be associated with a diversity of species and plant community features across landscapes [1–3]. For example, soils derived from different parent materials can vary in texture and chemistry, leading to differences in community structure and species assemblages associated with particular parent materials [4–6]. Disturbances occurring across environmental gradients can further increase diversity, such as through reducing competitive dominants to enable species coexistence [7]. Although disturbances can accentuate vegetational differences across environmental gradients, they could instead homogenize vegetative conditions [8]. Depending on pre-disturbance conditions, for example, fires either differentially or similarly affecting woodland tree structure across soil types could in turn enhance or reduce heterogeneity in understory plant communities across landscapes [9,10].

We examined the diversity of plant community features across gradients of soil parent materials and the history of fire activity within the last 43 years in a landscape of pinyon–juniper woodlands containing the small trees *Pinus monophylla* (hereafter *Pinus*) and *Juniperus osteosperma* (hereafter *Juniperus*). Pinyon–juniper communities containing various tree species within these genera occupy 20 million hectares across a range of soils in semi-arid climates in western North America [11]. One of the largest challenges for ecological research, ecological restoration, and fire and conservation management in pinyon–juniper communities is understanding and incorporating long-term community change and response to disturbances, including fires [12]. A commonly used structural classification includes three broad types of pinyon–juniper communities [13]. First, persistent pinyon–juniper woodlands are thought to occur on sites with infrequent stand-replacing fires (e.g., return intervals of 300–500+ years), where sparse fine fuel limited fire spread and where contemporary communities have many old trees or are in stages of re-establishment since the last severe disturbance. Second, wooded shrublands are thought to contain moderate densities of trees and shrubs and potentially mixed-severity fire regimes including surface fires or periodic crown fires. Third, pinyon–juniper savannas historically had low tree density and more frequent surface fires partly fueled by herbaceous plants. It is also thought that around many pinyon–juniper communities, portions of landscapes historically contained few or no trees, such as grasslands or sagebrush (*Artemisia* spp.)-dominated shrublands because of climatic, edaphic, or fire constraints on tree recruitment [14].

Since the mid-1800s to early 1900s, land use change associated with settlement occurred in pinyon–juniper communities, such as the introduction of non-native plants (e.g., the annual grass *Bromus tectorum*), woodcutting (for fuelwood or uses such as fence posts), livestock grazing that reduced fine fuels and potentially fire activity, and the intentional suppression of fires in some areas [15–17]. Decreased fire activity, livestock grazing potentially reducing herbaceous competition with tree seedlings, and possibly changing climatic conditions (warming after the mid-1800s) are thought to be associated with two general types of tree structural change in some formerly more open pinyon–juniper communities: infilling and expansion [18]. Infilling involves increased tree density in historically more open pinyon–juniper communities. Expansion involves the establishment of trees in areas that historically had few or no trees.

Although some wildlife benefit from dense pinyon–juniper stands, increasing tree density has been associated with degraded habitat quality for open habitat-affiliated wildlife and pollinators, lowered understory productivity, soil erosion, and increased risk of severe wildfires often followed by accelerated non-native plant invasion [19]. To address these negative effects, mechanical tree cutting, prescribed fires, or allowing wildfires to burn if the effects are considered desirable (managed wildfires) are being adopted as restoration or management strategies in pinyon–juniper communities thought to have experienced increased tree density [12].

Many authors have noted that these pinyon–juniper communities, as among the most widespread vegetation types in western North America, occupy a range of landforms and soils and may exhibit high variability in changes in tree structure in the last 100+ years and responses to contemporary fires, e.g., [20–22]. In fact, a recent review of pinyon–juniper research highlighted understanding diversity in these communities, including after fires across environmental gradients, as a major knowledge gap and research need [23]. Across the dual gradients of soil parent material and fire activity, we examined variation in pinyon–juniper communities in three categories of characteristics, which were fuels, tree structure, and understory vegetation. We asked the following question: which combinations of soil and fire history variables are most strongly associated with variation in post-fire pinyon–juniper community characteristics? Our results are intended to advance the understanding of diversity in post-disturbance vegetation across environmental gradients and to facilitate incorporating variation in physical geography when applying vegetation management treatments.

2. Materials and Methods

2.1. Study Area

We conducted this study in the 84,357 ha portion administered by the National Park Service of the 411,652 ha Grand Canyon-Parashant National Monument, Arizona, USA (Figure 1). Within the Colorado Plateau ecoregion, our 25 study plots (detailed below) were in pinyon–juniper communities at elevations of 1800–2000 m. The Köppen climate classification for the study area is cold semi-arid. In a pinyon–juniper community at an elevation of 1878 m in the study area, precipitation averaged 398 mm/year (165–664 mm/year range; Yellow John Mountain, Arizona; 1988 through 2023 records; Western Regional Climate Center, Reno, NV). Daily temperature ranges averaged −9–10 °C in January and 10–32 °C in July. In summarizing land use history, Ireland et al. [24] noted that the Southern Paiutes inhabited the area before Euro-American land use change primarily beginning in the 1870s with livestock grazing. Livestock likely reduced fine fuels and potentially fire activity. At least two sawmills operated in the area by the early and mid-1900s, but they are thought to have focused on *Pinus ponderosa* sawtimber rather than on the smaller *Pinus* and *Juniperus* trees at lower elevations. In *P. ponderosa* forests at higher elevations or moister sites around pinyon–juniper communities near our study area, surface fires were frequent (fire return intervals averaging 3–5 years for all fire-scarred trees and 7–23 years for fires scarring $\geq$ 25% of trees) in the 1600s until the 1880s [24]. After the 1880s, fire became uncommon through the 1900s, which Ireland et al. [24] hypothesized was primarily related to a combination of livestock grazing that reduced fine fuels and a management policy of fire exclusion. Although fire history in *P. ponderosa* forests suggests that spreading fires were frequent before Euro-American settlement and virtually ceased thereafter, pre-settlement fire regimes are less clear in the pinyon–juniper communities. However, pinyon–juniper communities similarly experienced little widespread fire since the late 1800s [25] until land managers began incorporating prescribed and managed wildfires in the decades preceding our study.

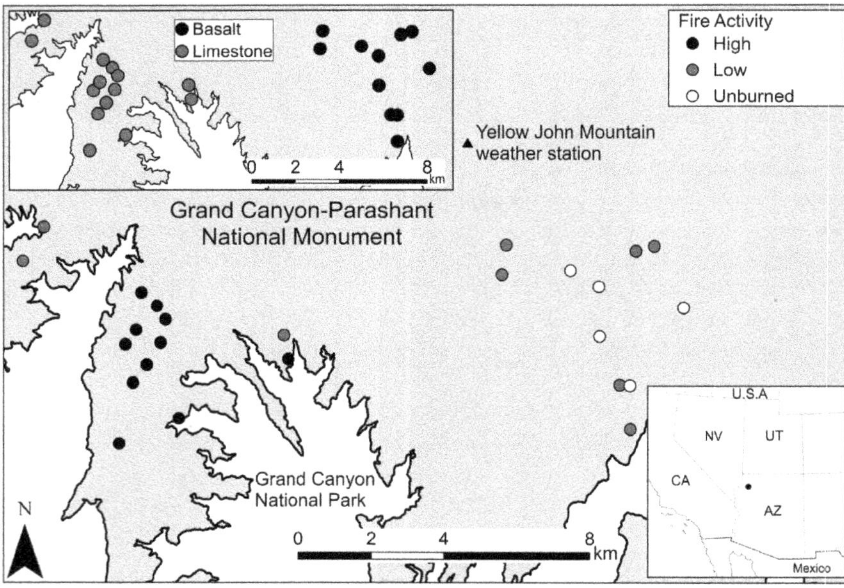

Figure 1. Location of 25 study plots (classified according to fire activity in the main image and according to soil parent material in the inset) in pinyon–juniper communities in Grand Canyon-Parashant National Monument, Arizona, USA. The fire activity classification integrates the number of fires and time since fire and is detailed in Figure S1. Plots burning the most times (up to four times) and with more recent fires were in the high fire activity group. Plots experiencing fewer fires and with a longer time since fire (up to 43 years) were in the low fire activity group. Plots with no history of fire since at least the 1970s were classified as unburned (Table S1).

2.2. Data Collection and Sample Processing

Using paper and digital records provided by the National Park Service (Grand Canyon-Parashant National Monument, St. George, UT, USA) of prescribed fires and wildfires occurring from 1980 (when records were available) to 2018 (the most recent fire) in pinyon–juniper communities in the study area, we created polygons representing different combinations of time since the most recent fire (years) and number of fires incurred. Polygons of fires ranged from ~0.1–1600 ha. The size range resulted from factors such as physical barriers to fire spread (e.g., changes in topography or fuels [26]), changes in weather during fires, and anthropogenic fire management activities (ignitions and suppression depending on fire locations and behavior). We did not separate prescribed fires and wildfires for two reasons, namely that most wildfires became managed fires under a policy of allowing wildfires to continue burning if meeting fire management goals, termed resource objective wildfires [27], and it was not possible to distinguish between their influences in our study. For example, some plots had been burned by one or more prescribed fires as well as wildfires, or one or the other but with varying years since fire. Although this did not enable distinguishing between fire types, the variable fire histories among plots enabled evaluating communities across a diverse gradient of fire history (Table S1, Figure 2).

Figure 2. Example photographs from 6 plots (of 25 total) sampled to study variation across pinyon–juniper communities with different fire histories and soil parent materials in Grand Canyon-Parashant National Monument, Arizona, USA. For the example plots on basalt soil, from top to bottom photographs, fire history included 1 fire and time since fire (TSF) = 19 years, 1 fire and TSF = 28 years, and unburned. For the example plots on limestone soil, from top to bottom photographs, fire history included 4 fires and TSF = 5 years, 2 fires and TSF = 13 years, and 1 fire and TSF = 43 years. The photographs were taken in summer 2023 by University of Nevada Las Vegas research staff.

Within fire history polygons, we randomly selected polygons and one point (using Universal Transverse Mercator coordinates, zone 12) within each selected polygon at which to establish a plot. We sampled a total of 25 plots (5 in 2022 and 20 in 2023 in mid--to-late summer from the end of June to early October). During years of data collection in our study, precipitation was 116% of the average in 2022 (462 mm) and 2023 (464 mm). Sampling included 5 unburned plots (assigned a 50-year time since fire to reflect no fire since at least the 1970s) and 20 plots burned 1–4 times between 1980 and 2022–2023 (Figure 1, Table S1).

To identify soil parent material for each plot, we used a soil survey in which soil taxonomic units (typically series or two closely associated series with similar parent materials) were mapped at a 1:24,000 scale [26]. We classified parent material as either basalt or limestone. Soils derived from basalt were mapped as Argiustolls and Haplusterts at the great group level in U.S. soil taxonomy, while limestone soils were Paleustolls [26].

To sample fuels, trees, and understory vegetation on each of the 25 plots, we followed the National Park Service fire monitoring handbook [28]. In summary, in each 20 m × 30 m (600 m^2) plot, we measured the number of intercepts of fine (herbaceous) and woody fuels and depth of the Oi (undecomposed litter) and Oea (duff) organic soil horizons along four 15 m long transects. From the intercepts and diameters measured for sound and rotten pieces of woody fuel, we calculated average woody fuel weight as kg/m^2 for each plot. Along two 30 m point–line intercept transects running across the bottom and top of each plot, we counted the number of points (out of 100 evenly spaced points) that intersected live and dead fine and woody fuels (overlapping fuels, if present at a point, were tabulated separately). From these transects, we calculated the frequency of fine and woody fuels.

In the entire 600 m^2 of each plot for each live tree or snag (≥1 cm in diameter), we recorded the species and trunk diameter at 1.4 m, or at the root collar where the trunk arose from the root system near ground level, for trees that branched below a height of 1.4 m (mostly *Juniperus*). We also counted the number of seedling-size live individuals (<1 cm in diameter at 1.4 m). We randomly selected trees of each conifer taxa (*Pinus* and *Juniperus*) by size class (≥ 1 < 10 cm, ≥ 10 < 20 cm, and ≥20 cm in diameter at a coring height of 0.4 m) from which to collect an increment core using 5 mm diameter increment borers. We sought to collect at least two cores per size class for each species on each plot. If there were insufficient trees within a plot, we collected cores from trees within 20 m of plots or accepted fewer trees from a plot. Cores were mounted, sanded increasingly fine using up to 2000-grit sandpaper, and years to pith was estimated using a pith locator if cores missed the pith. We used the imaging software CooRecorder 9.4 (Cybis Elektronik & Data AB, Saltsjöbaden, Sweden) to scan cores at a resolution ≥94 dots/mm to estimate the age of each tree. Cores are assumed to represent the minimum ages of trees, not including the time required to reach coring height, which averaged 9 years in one study [20].

To sample understory communities, we counted the number of shrubs by species in a 5 m × 30 m (150 m^2) transect across the center of each plot [28]. We used cover classes (0.1, 0.5, 1% intervals to 5% cover, thereafter 5% intervals to 100%) to categorize the aerial cover of each vascular plant species rooted in 1 m × 1 m quadrats (10 per plot, with 5 evenly spaced along the 30 m centerline and 5 along the northern plot line). We sampled in mid-to-late summer to correspond with peak plant cover, and to avoid excluding earlier flowering species (especially annuals), we included senesced annual–biennial plants in cover categorizations. Nomenclature and the classification of species by growth form (e.g., annual forb) and native or non-native status to the U.S. follow the Plants Database [29].

We sampled site factors on each plot including topography and soils. From the center of each plot, we used a clinometer to measure slope gradient and aspect. As the maximum slope gradient was only 25% and most (84%) plots had slope gradients ≤ 10% (hence minimal or no slope aspect), we did not include these variables further in analyses. On the western and eastern edges of each plot (1 m outside of plots), we collected three subsamples (each 1 m apart) of the 0–5 cm mineral soil using a 4 cm diameter metal corer. Each subsample was 63 cm^3, totaling 378 cm^3 for the 6 subsamples/plot which were composited for each plot. We sieved air-dried samples through a 2 mm sieve and measured the weight of coarse fragments (gravel > 2 mm in diameter) and bulk density including volume occupied by gravel. On the < 2 mm fraction, we analyzed samples for texture (hydrometer method) and pH and electrical conductivity (both using 1:1 soil/H_2O extracts [30]). Compared with soil derived from basalt, limestone soils were coarser (sandier with more gravel), had lower bulk density, and had higher electrical conductivity and pH (Table S2).

2.3. Statistical Analysis

We prepared a database including eight habitat variables (time since fire and number of fires and soil variables) and 18 univariate response variables in three categories including fuels, tree structure, and understory vegetation (Table S1). We derived further habitat variables using cluster analysis and principal component analysis (PCA). In addition to the binary classification of soil parent material (basalt, limestone), we derived a three-category fire activity variable (high, low, unburned) by applying cluster analysis (Euclidean distance and Ward's linkage method, 7.8% chaining, and 6006 sum of squares) to the fire variables of time since fire and number of fires a plot had received. We then applied PCA (cross-products matrix computed using correlation) separately to the fire and soil variables to generate PCs for use in the analyses described below (Figures S1 and S2).

We conducted complementary multivariate analyses on the fuel, tree, and understory response variables. For each of the categories of response variables, we ordinated them using PCA and included habitat variables (including the PCs described above) as potential correlates with ordination axes. Next, we performed canonical correlation to examine associations between each group of response variables with habitat variables. To both reduce multicollinearity and exceed sample sizes (N > 3× total variables) relative to the number of variables recommended by McGarigal et al. [31], we performed analyses on subsets of the variables by removing strongly correlated (Pearson $r > |0.70|$) variables within the respective response and habitat variable sets, retaining the habitat and response variables most strongly correlated with each other (Table S3). We analyzed understory species composition data (relative cover as cover of $\text{species}_i / \Sigma$ cover of all species on each plot) using non-metric multidimensional scaling ordination (Sørensen distance, random starting coordinates, and 250 runs with real and randomized data). We then used multi-response permutation procedures (Sørensen distance, $n_i / \Sigma\, n$ group weighting) to compare species composition across fire activity and soil parent material categories. To identify individual species associated with these categories, we used indicator species analysis [32]. We performed ordinations, cluster analysis, multi-response permutation procedures, and indicator species analysis in PC-ORD 7.10 [33]. We performed canonical correlation in SAS 9.4 [34].

3. Results

3.1. Tree Age Distribution

At the coring height of 0.4 m, 82% of trees sampled at study sites were younger than 150 years old and had reached coring height after the 1870s, approximately the time of settlement (Figure 3). Over half the sampled trees reached coring height after the 1920s. Only seven (5%) of the 127 sampled trees were in the oldest age class of ≥200 years, corresponding to reaching coring height before 1822.

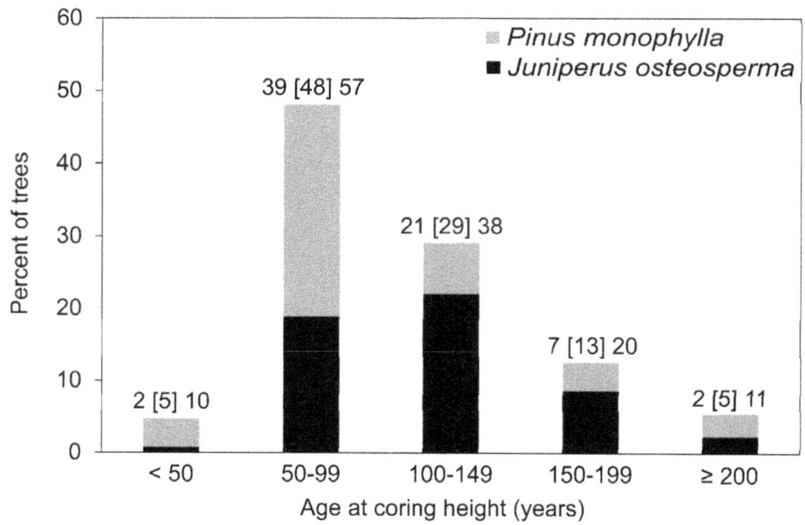

Figure 3. Age structure for trees sampled at 25 sites in pinyon–juniper communities in Grand Canyon-Parashant National Monument, Arizona, USA. Representing the portion in that age class of total trees for all ages combined for both species, numbers at the top of bars are the lower 95% confidence limit [observed %] and upper 95% confidence limit. Observed % represents the % of the total trees (n = 127) present in each age class. The confidence limits are asymmetrical for percentages. For *J. osteosperma*, 67 trees were sampled including 1, 24, 28, 11, and 3 trees from the youngest to the oldest age class. For *P. monophylla*, 60 trees were sampled including 5, 37, 9, 5, and 4 trees from the youngest to the oldest age class. Coring height was 0.4 m.

3.2. *Ordinations*

3.2.1. Fuels

In the PCA ordination of fuels, variation occurred along a primary gradient of fuel depth (Oi and Oea organic horizon thickness with loadings of 0.57 and 0.62, respectively) and woody fuel weight (0.51 loading) on PC1 that accounted for 38% of the total variance in fuels (Figure 4). The secondary gradient was litter frequency (0.77 loading for woody and 0.56 for fine litter frequency) on PC2 representing 24% of variance. Increasing fuel amounts and frequencies were correlated with coarser soils, lower soil bulk density, and higher soil electrical conductivity, all generally associated with limestone soil. Fire variables were less closely associated with variation in fuels, with an exception being that the number of fires and woody litter frequency were positively associated, as also reflected in bivariate correlation (Pearson r = 0.47; Table S3).

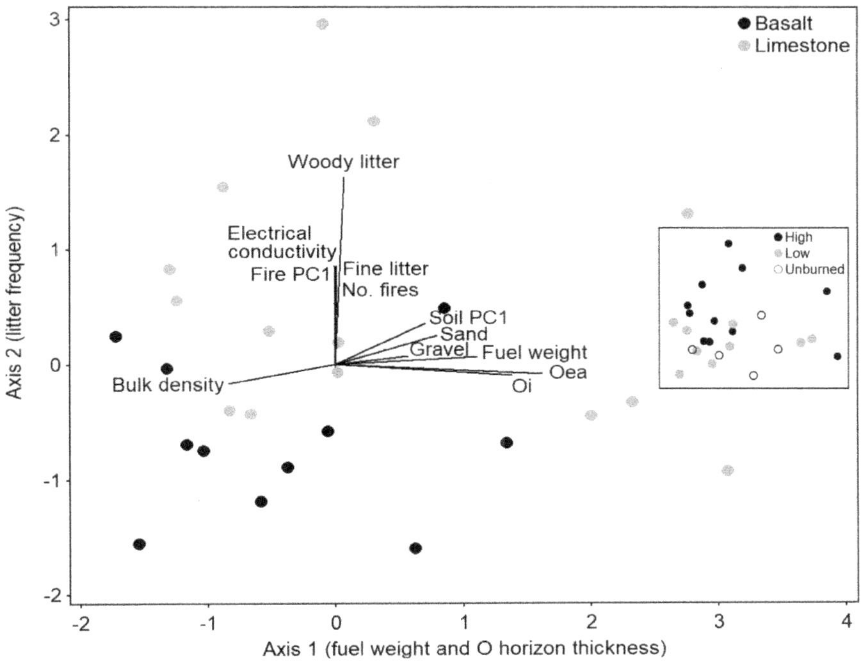

Figure 4. Principal component analysis of fuel variables in pinyon–juniper communities in Grand Canyon-Parashant National Monument, Arizona, USA. Points in the ordination are plots, symbolized according to soil parent material in the main image (and by fire activity in the inset), arranged based on their similarity to each other among fuel variables. Vectors ($r^2 \geq 0.20$) display fuel variables (woody fuel weight and O horizon thickness for Axis 1 and fine and woody litter frequency for Axis 2) most strongly correlated with their own ordination axes and the habitat variables most strongly correlated with the fuel variables. For habitat variables, soil properties (e.g., bulk density) are for the 0–5 cm mineral soil. The soil and fire principal components (PCs) are composite variables from Figures S1 and S2. Soil properties on basalt compared with limestone soil are shown in Table S2.

3.2.2. Tree Structure

A primary gradient of variation in tree structure in the PCA ordination included the percent of dead *Juniperus* trees (0.59 loading) and live *Juniperus* trees/ha (−0.49) along PC1 (35% of total variance; Figure 5). A secondary gradient portrayed variation in the percent of dead *Pinus* trees (−0.56 loading) and densities of *Pinus* (−0.49) and *Juniperus* (−0.49) seedlings along PC2 (27% of variance). The percentage of dead trees of both tree genera increased with increasing fire activity, which was also associated with limestone soil. Higher live tree densities were associated with less or no fire activity. Tree seedling density was negatively associated with higher soil pH, also indicated in bivariate correlations (Pearson $r = -0.44$ for *Juniperus* and -0.71 for *Pinus*; Table S3).

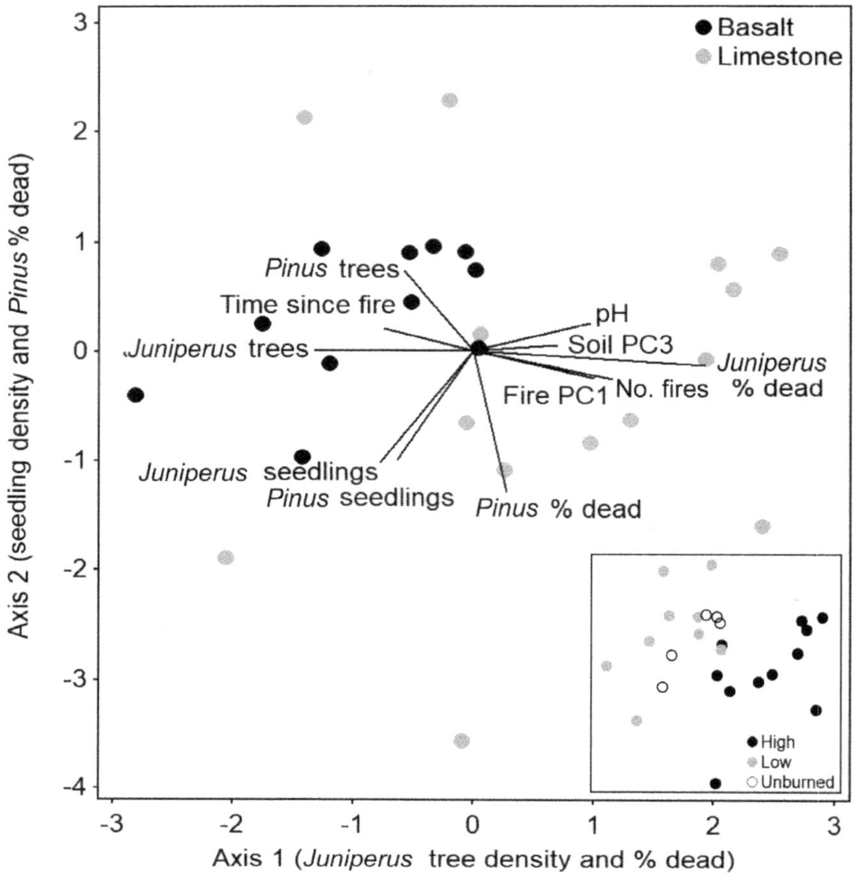

Figure 5. Principal component analysis of tree variables in pinyon–juniper communities in Grand Canyon-Parashant National Monument, Arizona, USA. Points in the ordination are plots, symbolized according to soil parent material in the main image (and by fire activity in the inset), arranged based on their similarity to each other among tree variables. Vectors ($r^2 \geq 0.20$) display tree variables most strongly correlated with their own ordination axes and the habitat variables most strongly correlated with tree variables. For tree variables, seedlings and trees are individuals/ha and % dead is for trees. For habitat variables, soil properties (e.g., pH) are for the 0–5 cm mineral soil. The soil and fire principal components (PCs) are composite variables from Figures S1 and S2.

3.2.3. Understory Plant Communities

For univariate understory variables in the PCA ordination, PC1 extracted 36% of the total variance and portrayed a gradient of native forb species richness (−0.54 loading), shrub density (0.52) and cover (0.49), and native perennial grass cover (−0.28; Figure 6). Habitat variables most strongly associated with PC1 included time since fire, the number of fires, soil electrical conductivity, and *Juniperus* live tree density and % dead. Supporting the PC1 gradient, bivariate correlations showed that native forb species richness declined as the number of fires increased (and with recency of fire) and as *Juniperus* % dead trees increased (Table S3). Conversely, shrub density increased as the number of fires increased and as the *Juniperus* % dead increased (Table S3). Native forb species richness and shrub density were themselves negatively correlated (Pearson $r = -0.55$, Table S3).

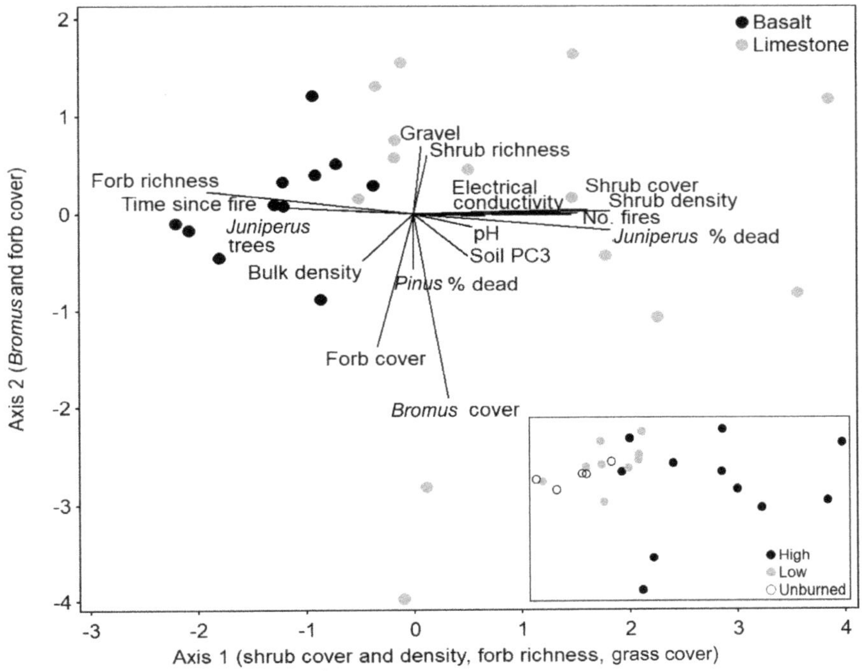

Figure 6. Principal component analysis of understory variables in pinyon–juniper communities in Grand Canyon-Parashant National Monument, Arizona, USA. Points in the ordination are plots, symbolized according to soil parent material in the main image (and by fire activity in the inset), arranged based on their similarity to each other among understory variables (grass cover is for native perennials; *Bromus* is the non-native annual *B. tectorum*). Vectors ($r^2 \geq 0.20$) display understory variables most strongly correlated with their own ordination axes and the habitat variables most strongly correlated with understory variables. For habitat variables, soil properties (e.g., pH) are for the 0–5 cm mineral soil (PC = principal component from Figure S2). Tree variables are for individuals/ha or the % of trees that are dead.

The second gradient (PC2: 23% of total variance) in the understory PCA ordination reflected variation in non-native *Bromus tectorum* cover (−0.68 loading) and native forb cover (−0.57). In bivariate correlation analysis, *B. tectorum* cover increased with more recent fire activity, a higher percentage of dead trees, and higher pH generally associated with subsets of limestone soil (Table S3).

Non-metric multidimensional scaling ordination revealed the separation of species composition of understory communities according to fire activity and soil parent material (Figure 7). The two-axis ordination had a stress of 23.7 and represented 55% of the variance in the original data (30% for Axis 1 and 25% for Axis 2). A variety of understory species (e.g., *Purshia mexicana*, *Bouteloua gracilis*) and univariate understory metrics (e.g., forb species richness) were correlated with understory species compositional variation. For example, communities with higher relative cover of the major native perennial grasses *B. gracilis* and *Koeleria macrantha* were associated with higher native forb species richness. Numerous fuel, tree, soil, and fire variables were also associated with community compositional variation. For example, sandy, gravelly soil, an increasing number of fires, and higher percentage of dead *Juniperus* trees were associated with shrub-dominated communities (e.g., *Artemisia tridentata*).

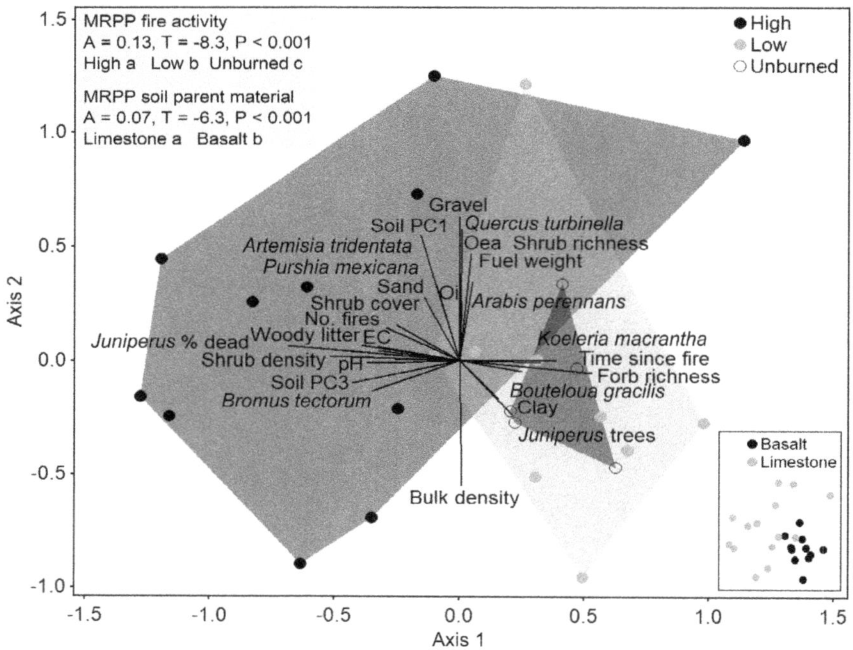

Figure 7. Non-metric multidimensional scaling ordination of understory species composition (relative cover) in pinyon–juniper communities in Grand Canyon-Parashant National Monument, Arizona, USA. Points in the ordination are plots, symbolized according to fire activity in the main image (and by soil parent material in the inset), arranged based on their similarity to each other in understory species composition. Vectors ($r^2 \geq 0.20$) display understory species, fuel and understory metrics (e.g., shrub density), and the habitat variables most strongly correlated with species compositional variation. For habitat variables, soil properties (e.g., EC, which stands for electrical conductivity) are for the 0–5 cm mineral soil (PC = principal component from Figure S2). Tree variables are for individuals/ha or the % of trees that are dead. The inset in the top left shows multi-response permutation procedure (MRPP) tests of the null hypothesis of no difference in species composition (relative cover) across fire activity and soil parent material categories (categories not followed by shared letters differ at $p < 0.017$ in Bonferroni-corrected pairwise comparisons).

The multi-response permutation procedure tests supported the separation of community compositional groupings portrayed in the ordination according to gradients of fire activity and soil parent material. Each fire activity category (high, low, unburned) and soil parent material (basalt, limestone) displayed unique community composition (Figure 7 inset).

3.3. Indicator Species Analysis of Understory Communities

Several understory species were significantly associated with fire activity categories and soil parent material types (Table 1). For shrub density, *Gutierrezia sarothrae* and *Purshia mexicana* were associated with high fire activity, with *P. mexicana* also associated with limestone soil. No shrubs were associated with low or no fire activity. For cover, no native herbaceous species were associated with any fires but rather were associated with unburned plots (e.g., the perennial grass *Bouteloua gracilis* and perennial forb *Menodora scabra*). In contrast, the non-native annual grass *Bromus tectorum* was associated with high fire activity. Results for fire activity were qualitatively similar when only two groups at a time were compared. For example, when only high fire activity and unburned plots were compared with each other, the only significant indicator species among native herbaceous plants were for unburned plots.

Table 1. Indicator species analysis assessing species associated with fire activity groups and types of soil parent material in Grand Canyon-Parashant National Monument, Arizona, USA. Separately for the density of shrubs and cover of all vascular plant growth forms, the group that a species was significantly (indicator value ≥ 50 with $p < 0.05$) associated with is shown. Indicator values range from 0 (no association of a species with any group) to 100 (maximum association). All indicator species were native, except for the non-native *Bromus tectorum* (*).

Species	Group	Indicator Value (*p*)
Shrub density		
Fire activity		
Gutierrezia sarothrae	High	54 (0.041)
Purshia mexicana	High	78 (<0.001)
None	Low	
None	Unburned	
Soil parent material		
Artemisia tridentata	Limestone	64 (0.039)
Purshia mexicana	Limestone	89 (<0.001)
None	Basalt	
Cover		
Fire activity		
Artemisia tridentata (S) [1]	High	61 (0.016)
Bromus tectorum * (AG)	High	85 (0.001)
Juniperus osteosperma (T)	Low	69 (0.008)
Bouteloua gracilis (PG)	Unburned	57 (0.017)
Erigeron divergens (BF)	Unburned	55 (0.025)
Hymenoxys cooperi (PF)	Unburned	73 (0.002)
Menodora scabra (PF)	Unburned	68 (0.002)
Phlox spp.	Unburned	63 (0.007)
Soil parent material		
Artemisia tridentata (S)	Limestone	57 (0.008)
Bromus tectorum * (AG)	Limestone	80 (0.002)
Allium acuminatum (PF)	Basalt	55 (0.003)
Bouteloua gracilis (PG)	Basalt	66 (0.003)
Hymenoxys cooperi (PF)	Basalt	100 (<0.001)
Hymenopappus filifolius (PF)	Basalt	52 (0.029)
Koeleria macrantha (PG)	Basalt	62 (0.010)
Lupinus brevicaulis (AF)	Basalt	70 (0.003)
Menodora scabra (PF)	Basalt	64 (<0.001)
Penstemon linarioides (PF)	Basalt	65 (0.010)
Phlox spp. (PF)	Basalt	65 (0.014)

[1] AF, annual forb; AG, annual grass; BF, biennial forb; PF, perennial forb; PG, perennial grass; S, shrub; and T, tree.

3.4. Canonical Correlation Between Sets of Variables

Canonical correlation extracted dominant gradients within sets of habitat and response variables and revealed that combinations of habitat variables accounted for a third to over half of the variance in sets of response variables (Table 2). Within the set of soil–fire habitat variables, joint gradients in soil chemistry (electrical conductivity and pH) or coarseness (sand and gravel content) and the number of fires represented about half of the variance in habitat variables within the first two variates.

Table 2. Canonical correlation analysis of habitat variables with sets (fuel, trees, understory) of response variables in pinyon–juniper communities, Grand Canyon-Parashant National Monument, Arizona, USA.

Variate	Habitat [1]	Response [1]	Canonical Correlation
	—— Fire-soil ——	———— Fuel ————	
1	EC 0.90 No. fires 0.63 26% [3]	Fine litter 0.75 Woody litter 0.73 23%	0.82 ± 0.07 [2] 15%
2	Gravel 0.85 Sand 0.74 22%	Oea 0.79 Fuel weight 0.59 29%	0.79 ± 0.08 18%
	—— Fire-soil ——	———— Trees ————	
1	pH 0.83 No. fires 0.74 33%	*Juniperus* % dead 0.87 *Pinus* seedlings/ha −0.57 28%	0.93 ± 0.03 24%
2	Sand 0.60 No. fires 0.56 19%	*Pinus* seedlings/ha 0.71 *Juniperus* % dead 0.42 17%	0.71 ± 0.10 9%
	—— Fire-soil-trees ——	——— Understory ———	
1	No. fires 0.86 EC 0.62 38%	SR native forb −0.83 Shrubs/ha 0.63 38%	0.92 ± 0.03 33%
2	Gravel −0.64 *Juniperus* trees/ha −0.49 20%	*Bromus tectorum* cover 0.81 SR shrub −0.66 35%	0.71 ± 0.10 23%

[1] The habitat and response columns show their variables followed by their correlations with their own canonical variates. The two variables exhibiting the largest $|r|$ are shown for each variate in the respective habitat or response variates. Oea is duff thickness, EC is electrical conductivity, and SR is species richness. [2] Mean ± standard error of the canonical correlation. [3] Rows with percents display the standardized percent variance portrayed by their own canonical variates for their own variables (habitat and response variables separately) and the standardized percent variance in the response canonical variates accounted for by the habitat canonical variates (canonical correlation column). Totaling up the % variance for variates 1 and 2 represents the cumulative variance portrayed for the respective habitat or response variates and for the response variates by the habitat variates in the canonical correlation column.

Within fuel response variables, the first variate represented a gradient in litter frequency and the second represented fuel depth and weight, cumulatively portraying 52% of the variance in fuels. Fire and soil variables accounted for 33% of the variance in fuels.

For the tree structure response variables, their variates portrayed combinations of gradients in the percent of dead *Juniperus* and the density of live *Juniperus* trees and *Pinus* seedlings. Soil and fire variables also accounted for 33% of the variance in tree structure.

Within the set of understory variables, variate 1 portrayed a gradient of forb species richness and shrub density and variate 2 non-native *Bromus tectorum* cover and shrub species richness. Native forb richness and *B. tectorum* cover showed opposing relationships with the variates, consistent with a negative bivariate correlation (Pearson $r = -0.55$) between *B. tectorum* cover and native forb richness (Table S3). The understory variates cumulatively represented 73% of the variance in the understory variables. In turn, a set of habitat variables including the number of fires, soil electrical conductivity and coarseness, and *Juniperus* tree density portrayed 56% of the variance in understory variables.

4. Discussion

4.1. Study Limitations

Four of the main limitations of our study included its retrospective nature, the comparison of only two soil parent material types, challenges with cross-dating trees of the study species, and the one-time sampling limiting inference of potential inter-annual fluctuations in community characteristics. We sought to accommodate these limitations in several ways. First, the retrospective nature of our study and limited information on the behaviors of many fires (e.g., wildfires burning before observations) precluded us from differentiating among different types of fires. We sought to address this by grouping plots into broader fire activity categories, focusing on the degree of fire activity a site experienced within the last 43 years rather than on responses to specific types of fires. Future research could further differentiate the effects of particular types of fires, such as fire seasonality [35], including experimentally applying fire treatments across soil parent materials or environmental gradients to assess potential interactions [36,37].

The second limitation was that we only compared two soil parent material types, and there are many more in American Southwest conifer forests [38]. However, the two parent materials (basalt and limestone) we compared predominated in our study area [39] and are broadly distributed in southwestern conifer forests, e.g., [40–42]. Moreover, by collecting and analyzing continuous soil variables, we were able to represent variability in soil properties within and across the broad parent material types that could be further explored. For example, increasing amounts of chert within a matrix of limestone can lower pH and modify other properties of soils derived primarily from limestone [26].

The third limitation was challenges also extensively noted in the literature with cross-dating *Juniperus* trees, resulting in many studies including ours to focus on ring counts or minimum ages for likely partially dated trees [20,43,44]. There can also be variation among sites in the time trees require to grow to coring height [20]. This can be estimated by sampling growth rate to coring height in contemporary seedlings, but this may not match the time required in previous centuries for the oldest trees. Retrospectively determining this typically necessitates cutting cross-sections at the root collar (and comparing with the age at coring height) via the destructive sampling of old trees, which was not appropriate in our study in a national park setting. We sought to minimize limitations with cross-dating by grouping trees into 50-year age classes to enable approximate comparisons between broad time periods, such as circa pre- or post-1870s representing general land use change associated with settlement in the study region [24].

The fourth limitation of our study was that it was a "snapshot" of plant community features at one time and was not intended to portray potential inter-annual fluctuations in plant community conditions. Annual plants may be among the community components we examined most sensitive to inter-annual fluctuations, and we sought to address this by at least including senesced annuals in our community assessments. Beatley [45] noted that senesced annuals in a semi-arid climate can remain standing for 1–2 years. Precipitation in our 2022–2023 sampling was near average, within 16% of the 36-year average.

4.2. Long-Term Changes in Tree Structure

The analysis of tree age structure, combined with prior research [13,14,19], suggests that the study area was more open with fewer trees in the 1800s and experienced a marked increase in tree density in the 1900s. Compared with *Pinus ponderosa* trees readily recording surface fire scars, the difficulty of reconstructing fire history in pinyon–juniper communities complicates interpreting whether the tree density increase may represent ongoing recovery from a large historical wildfire or climatic or anthropogenic influences associated with settlement [46]. While Ireland et al. [24] found that surface fires were frequent before the 1870s in *P. ponderosa* forests adjacent to our study area, these fires did not necessarily burn through pinyon–juniper sites and further work would be required to attempt fire history reconstruction there. Recovery of a mature pinyon–juniper community from stand-replacing fire can be a multi-century process because of lags (often 30+ years) in tree recruitment due to factors such as a lack of seed sources, few nurse plants to facilitate seedlings, and the semi-arid climate rarely conducive to tree establishment [44,47,48].

Another scenario of change is that at least moderately frequent historical fires (e.g., multiple fires/century) limited tree establishment to only portions of landscapes and that mosaics of pinyon–juniper communities were interspersed with shrublands or herbaceous-dominated patches [18]. After settlement with both livestock grazing and fire suppression potentially reducing fire activity, perhaps coupled with reduced herbaceous competition with tree seedlings or climatic changes, tree establishment increased [12]. The processes of infilling and expansion of trees in historically more open habitats have been observed across millions of hectares of western North America [49]. Further research to reconstruct historical reference conditions and ranges of variability across environmental gradients [4] would be required to understand long-term change in pinyon–juniper communities in our study area. Our results do, however, support the notion that a major increase in tree density occurred over the last 150 years, forming the basis for the National Park Service beginning to use prescribed fires and managed wildfires to reduce tree density under a management goal of reestablishing more open habitats (Figure 2).

4.3. Variation in Community Characteristics Along the Fire Activity Gradient

The major variation in plant community features along the fire activity gradient was that as fire activity increased, the percentage of dead trees, total shrub density and cover, and non-native *Bromus tectorum* cover all increased, while the density of live trees and native forb species richness decreased. Although modeling individual tree survival was beyond the scope of our study, we found that the percentage of dead trees was higher on plots burned the most times. Effects of repeated fires on tree mortality in western conifer forests have varied substantially with factors such as the amount of fuel accumulation between fires affecting fire severity, tree population density-dependent effects hinging on how much initial fires reduce tree density, and additional disturbances such as insects or droughts, e.g., [50,51]. Modeling fire-related individual tree mortality is a research need in pinyon–juniper communities and could help refine predicting the effects of single or multiple fires varying in characteristics [52].

Shrub density and cover increased as fire activity and the percentage of dead trees increased. A common, though not universal, pattern reported in the literature for pinyon–juniper communities after stand-replacing fires is that shrub abundance can initially decline within the first post-fire decade, then increase to or exceed pre-fire levels at 50+ years after a fire before declining again as trees predominate [48,53,54]. After multiple fires, shrub response can vary with factors such as fire severity, tree canopy cover or the duration that trees have been at increased density on a site if long-term tree dominance has increased, and traits of the constituent shrub species [21]. In our study, indicator species analysis revealed that no shrub species were associated with low or no fire activity. In contrast, *Gutierrezia sarothrae* was associated with high fire activity, consistent with post-fire increases in this species previously observed in pinyon–juniper communities by Huffman et al. [15,44]. *Gutierrezia sarothrae* is often a relatively short-lived (4–7 years) prolific seed producer that can form dense stands [55,56]. Another shrub species we found associated with high fire activity, the evergreen *Purshia mexicana*, was previously described as having markedly lower density and frequency 13–15 years post-fire in burned areas relative to unburned pinyon–juniper communities in Arizona [57]. When using cover, we found that *Artemisia tridentata* was also significantly associated with high fire activity. One of the most extensively studied shrub species in pinyon–juniper communities, *A. tridentata*, has generally declined in the first 1–12 years after fires, then often increased to or exceeded pre-fire abundance, at least if not competitively reduced by trees, e.g., [58–60]. *Artemisia tridentata* regenerates only via seed and can produce copious but short-lived seed crops, making the species vulnerable to declines if fires are frequent or sufficiently large to limit seed dispersal [61]. These observations exemplify the utility of further autecological investigations of species' natural histories for planning and potentially better predicting the influences of burning in different seasons, frequencies, and severities.

The negative relationship of native forb species richness with fire activity and no forb species associated with fire in indicator species analysis was surprising to us, given that forbs were fire-stimulated in several earlier studies in pinyon–juniper communities [53–55,62] and in other western North American forests, e.g., [63–65]. There are several possibilities as to why forbs were limited at sites with a history of fires in our study. It is possible that the minimum time since fire of five years in our study did not detect an earlier post-fire increase if one had occurred but dissipated. The post-fire increase in forbs can be quite transient in just the early years post-fire in pinyon–juniper communities [53–55]. However, Bates et al. [17] found that forb abundance remained elevated for 4–8 years following prescribed fire. Similarly, Everett and Ward [62] found that post-fire forb increases were sustained for at least five years. It is possible that weather during our 2022–2023 study period was not favorable for forbs, although precipitation was slightly above average at 116% both years. Another possibility is that forbs were competitively suppressed by shrubs and non-native annual grasses, which were both associated with high fire activity. Prior correlational studies have found negative associations between the non-native annual *Bromus tectorum* and forbs in burned pinyon–juniper communities, e.g., [66,67]. A competition experiment revealed that *B. tectorum* competitively suppresses many native forbs [68].

We found that increasing cover of *Bromus tectorum* was associated with high fire activity, recent fire occurrences, and higher percentages of dead trees. Post-fire chronosequence studies in pinyon–juniper communities have reported that *B. tectorum* cover was highest in younger burns and declined by six [53] or eight years post-fire [54]. Experimentally imposing prescribed fire also revealed that *B. tectorum* abundance increased within five years relative to pre-fire or unburned areas, sometimes following an initial lag [15,16,49]. Additionally, after wildfires in the Great Basin Desert in Utah, Ott et al. [66] found that native herbaceous species (especially forbs) increased the first post-fire year but sharply declined in years 2–3 concomitant with rising *B. tectorum* cover. Collectively, these studies and our results suggest that further assessing the response of non-native annual grasses to variations in fire management strategies, as well as how control treatments (e.g., herbicides) may modify the responses, is warranted to determine if reducing non-native annual grasses can enable native herbaceous species to respond more positively to fires.

Other than litter frequency, variation in ground fuels (O horizon depth and woody fuel weight) was more closely associated with soils than with fire variables. A minimal relationship between fuel depth and fire activity could relate to the balance of fires consuming the O horizon while also transferring material to the ground in our study's 5+-year post-fire period, such as needle drops from damaged or dead trees [69]. This may also account for the observed greater litter frequency in our study where fire activity had been greatest. Total fuel loads are primarily driven by heavy pieces of coarse woody debris [70]. The creation of coarse woody debris is influenced by the degree of tree mortality and how long snags remain standing [44]. In California, Reed et al. [69] found that 33% of *Pinus monophylla* snags fell within five years of tree death, though remaining snags could persist much longer. Owing to slow wood decomposition in the semi-arid climate, woody ground fuel loads can primarily vary with the degree of consumption in future fires [70].

4.4. Variation Along the Soil Parent Material Gradient

On coarse soil with higher pH or electrical conductivity (generally limestone soil), the percentage of dead *Juniperus* trees, non-native grass cover, and shrub density increased. In contrast, the density of *Pinus* and *Juniperus* seedlings and species richness of native forbs decreased. Finer-textured soils supported more native herbaceous plants, which generally occurred on basalt soil with less fire activity.

Based on our soil analysis coupled with prior research, primary differences between soils derived from the two broad parent material types include moisture availability, mechanical properties such as shrink–swell activity, and nutrient availability [36,41,71]. With an average of 398 mm/year of precipitation, our study area was just above a commonly cited threshold of 370 mm below which the inverse soil texture effect can occur [72]. In arid regions receiving less than 370 mm of annual precipitation, sandy soil can promote infiltration into deeper layers to reduce evaporative loss from surface soil, resulting in greater water availability and plant growth relative to finer-textured soil [72]. With increasing precipitation and humidity, these benefits on sandy soil can be outweighed by the lower water-holding capacity of sandy soil and loss of water to deep drainage, resulting in coarser soil generally being driest in moister climates [72].

Soil parent material and fire could have interacted in several ways in the co-varying soil–fire gradient context of our study. Fire activity was more prevalent on limestone soil, a function of greater fuel loads likely resulting in greater wildfire spread and prescribed fires under a National Park Service vegetation management goal emphasizing re-establishing open stands. As an example of soil parent material potentially modifying fire outcomes for plant growth, limestone soil can be less susceptible (compared with basalt soil with more clay) to the frost heaving of tree seedlings [42,73]. However, the greater mortality of fire-susceptible tree seedlings [50] from more prevalent fire activity on limestone soil could account for the lower density of tree seedlings inhabiting limestone soil.

In our study, we were able to examine variation across co-varying soil–fire gradients. Partitioning the relative causal influences of soil and fire would require experimental approaches we encourage in future research, given the diversity of soil types experiencing fire in pinyon–juniper communities. In prior experimental research in higher elevation *Pinus ponderosa* forests in Arizona, we found that effects of treatments uniformly applied across soil parent materials varied among soils derived from limestone, basalt, and benmorite [36]. Similar experimentation could be a next research step to explore the predictability of fire responses across diverse soil types in pinyon–juniper communities.

4.5. Considerations for General Patterns of Vegetation Dynamics

Our finding of associations of heterogeneity in community features with joint soil–fire gradients adds to the global literature focused on variability across dual environmental and disturbance gradients. As one of the major types of environmental gradients, soil parent materials can lead to the genesis of a diversity of soils [74]. Understanding how soil parent materials can interact with disturbances and shape vegetation responses to them continues to be an active area of research globally, including as climates and disturbance regimes change [1,3,6]. For example, in Belgium, Verstraeten et al. [8] found that species composition and the diversity of plant communities recovering from forest cutting varied across soil parent materials differing in pH. In Japan, rates of forest recovery following landslide disturbances were higher on soil derived from colluvium [75]. Considering fire disturbances specifically, Staver et al. [3] found that fire activity was greater on moister finer-textured soil, where herbaceous biomass was also higher than on drier sandier soil in South African savannas. In Canadian boreal forests, pathways of post-wildfire community change varied across gradients in soil drainage and organic matter content [9,76,77]. This sampling of studies highlights the variety of ways different soil gradients can influence vegetation dynamics and post-disturbance community conditions among global habitats. These observations support linking environment and disturbance as jointly or independently varying gradients for studying the diversity of plant community dynamics, e.g., [2,5,7].

4.6. Applications for Vegetation Management

Variation in plant community features across gradients in soil and fire could be applied several ways to vegetation management. For example, knowledge of this variation could be used to prioritize management actions on sites responding most positively to management activities or that are most deviated from reference or desired ecological conditions [6,36]. In our study, for example, the high cover of the non-native annual *Bromus tectorum* was associated with limestone soil and high fire activity. This annual is one of the highest-priority non-native plants for management because it can alter fuels and fire behavior, compete with native plants, and reduce overall carrying capacity for wildlife [19]. Control treatments for *B. tectorum* in our study could be prioritized on limestone soil and where fires have recently occurred or are planned. On the other hand, early detection monitoring and the treatment of small incipient populations could be prioritized on basalt soil and where fire activity was less prevalent. Variation in the longevity of fire effects on fuels and community features across soils is another key consideration. The persistence of fire effects can vary with factors such as fire severity and rates of tree recruitment [78]. For example, we found that tree seedling density was negatively correlated with coarse-textured soil and higher pH, suggesting that soil-driven seedling density could affect how long post-fire stands remain open.

More generally, our results illustrate the potential for incorporating heterogeneity in plant communities across the physical geography template of landscapes to improve vegetation conservation and management. We provide three examples. First, reference conditions, or baseline conditions against which to set or evaluate management effectiveness, can vary across soil types and geomorphic features [4,5,20,43]. A next step in this research in our study could be reconstructing pre-settlement forest structure and multi-century woodland change across soil types. Second, management treatments could be customized to the unique ecological communities and anticipated responses spanning spatial biophysical gradients across the landscape [2,6,7,74]. Different soil types could have variable management needs and may respond heterogeneously to treatments [1,22,36]. Third, given differences in baseline conditions and the anticipated magnitude of change, metrics for monitoring vegetation management effectiveness could be tailored to quantitative levels scaled to communities differing in productivity, structure, and diversity across soil types [8,9,75]. Each of these three examples highlight how further research augmenting the existing literature could advance incorporating variation in physical geography across the landscape as a framework for understanding and managing vegetation diversity.

Supplementary Materials: The following supporting information can be downloaded at: https://www.mdpi.com/article/10.3390/d17020088/s1, Figure S1: Principal component analysis of fire variables; Figure S2: Principal component analysis of 0–5 cm mineral soil variables; Table S1: Dataset; Table S2: Comparison of 0–5 cm mineral soil properties between soil parent materials; Table S3: Pairwise Pearson correlations between habitat and plant community variables.

Author Contributions: Conceptualization, S.R.A., L.P.C., E.C.B., C.A.F., and J.E.F.; field data collection, L.P.C., E.C.B., S.L.P., and C.D.M.; formal analysis, S.R.A.; writing—original draft preparation, S.R.A.;

writing—review and editing, all authors; funding acquisition, C.A.F., J.E.F., and S.R.A. All authors have read and agreed to the published version of the manuscript.

Funding: This research was funded through a cooperative agreement (P21AC12070-00) between the National Park Service and the University of Nevada Las Vegas.

Institutional Review Board Statement: Not applicable.

Data Availability Statement: Data associated with this study are in Table S1.

Acknowledgments: We thank John Foley and Jeremie Gamiao (National Park Service) for providing perspectives on land use and fire history and ecological interpretations; Grace Friedmann, Sarah Galera, Katie Kline, and Jacob Luna (University of Nevada Las Vegas [UNLV]) for help collecting and processing field data; Cosset Hernandez Peña (UNLV) for performing soil assays; Sarah Galera and Jacob Luna (UNLV) for help with dendroecology; and two anonymous reviewers for comments on the manuscript.

Conflicts of Interest: The authors declare no known conflicts of interest.

References

1. Eviner, V.T.; Hawkes, C.V. Embracing variability in the application of plant-soil interactions to the restoration of communities and ecosystems. *Restor. Ecol.* **2008**, *16*, 713–729. [CrossRef]
2. Boyle, F.B.; Abella, S.R.; Shelburne, V.B. An ecosystem classification approach to assessing forest change in the southern Appalachian Mountains. *For. Ecol. Manag.* **2014**, *323*, 85–97. [CrossRef]
3. Staver, A.C.; Botha, J.; Hedin, L. Soils and fire jointly determine vegetation structure in an African savanna. *New Phytol.* **2017**, *216*, 1151–1160. [CrossRef] [PubMed]
4. Abella, S.R.; Denton, C.W. Spatial variation in reference conditions: Historical tree density and pattern on a *Pinus ponderosa* landscape. *Can. J. For. Res.* **2009**, *39*, 2391–2403. [CrossRef]
5. Laliberté, E.; Grace, J.B.; Huston, M.A.; Lambers, H.; Teste, F.P.; Turner, B.L.; Wardle, D.A. How does pedogenesis drive plant diversity? *Trends Ecol. Evol.* **2013**, *28*, 331–340. [CrossRef]
6. Craigg, T.L.; Adams, P.W.; Bennett, K.A. Soil matters: Improving forest landscape planning and management for diverse objectives with soils information and expertise. *J. For.* **2015**, *113*, 343–353. [CrossRef]
7. Kirkman, L.K.; Goebel, P.C.; Palik, B.J.; West, L.T. Predicting plant species diversity in a longleaf pine landscape. *Écoscience* **2004**, *11*, 80–93. [CrossRef]
8. Verstraeten, G.; Baeten, L.; Van den Broeck, T.; De Frenne, P.; Demey, A.; Tack, W.; Muys, B.; Verheyen, K. Temporal changes in forest plant communities at different site types. *Appl. Veg. Sci.* **2013**, *16*, 237–247. [CrossRef]
9. Taylor, A.R.; Chen, H.Y.H. Multiple successional pathways of boreal forest stands in central Canada. *Ecography* **2011**, *34*, 208–219. [CrossRef]
10. Leonard, J.M.; Medina, A.L.; Neary, D.G.; Tecle, A. The influence of parent material on vegetation response 15 years after the Dude Fire, Arizona. *Forests* **2015**, *6*, 613–635. [CrossRef]
11. Evans, R.A. *Management of Pinyon-Juniper Woodlands*; General Technical Report INT, 249; U.S. Forest Service, Intermountain Research Station: Ogden, UT, USA, 1988.
12. Floyd, M.L.; Romme, W.H. Ecological restoration priorities and opportunities in piñon-juniper woodlands. *Ecol. Restor.* **2012**, *30*, 37–49. [CrossRef]
13. Romme, W.H.; Allen, C.D.; Bailey, J.D.; Baker, W.L.; Bestelmeyer, B.T.; Brown, P.M.; Eisenhart, K.S.; Floyd, M.L.; Huffman, D.W.; Jacobs, B.F.; et al. Historical and modern disturbance regimes, stand structures, and landscape dynamics in piñon-juniper vegetation of the western United States. *Rangel. Ecol. Manag.* **2009**, *62*, 203–222. [CrossRef]
14. Chambers, J.C.; Strand, E.K.; Ellsworth, L.M.; Tortorelli, C.M.; Urza, A.K.; Crist, M.R.; Miller, R.F.; Reeves, M.C.; Short, K.C.; Williams, C.L. Review of fuel treatment effects on fuels, fire behavior and ecological resilience in sagebrush (*Artemisia* spp.) ecosystems in the western U.S. *Fire Ecol.* **2024**, *20*, 32. [CrossRef]
15. Huffman, D.W.; Stoddard, M.T.; Springer, J.D.; Crouse, J.E.; Chancellor, W.W. Understory plant community responses to hazardous fuels reduction treatments in pinyon-juniper woodlands of Arizona, USA. *For. Ecol. Manag.* **2013**, *289*, 473–488. [CrossRef]
16. Roundy, B.A.; Miller, R.F.; Tausch, R.J.; Young, K.; Hulet, A.; Rau, B.; Jessop, B.; Chambers, J.C.; Eggett, D. Understory cover responses to piñon-juniper treatments across tree dominance gradients in the Great Basin. *Rangel. Ecol. Manag.* **2014**, *67*, 482–494. [CrossRef]
17. Bates, J.D.; Davies, K.W.; Hulet, A.; Miller, R.F.; Roundy, B. Sage grouse groceries: Forb response to piñon-juniper treatments. *Rangel. Ecol. Manag.* **2017**, *70*, 106–115. [CrossRef]

18. Miller, R.F.; Tausch, R.J. *The role of fire in juniper and pinyon woodlands: A descriptive analysis*; Miscellaneous Publication No. 11; Galley, K.E.M., Wilson, T.P., Eds.; Tall Timbers Research Station: Tallahassee, FL, USA, 2001.
19. Miller, R.F.; Chambers, J.C.; Evers, L.; Williams, C.J.; Snyder, K.A.; Roundy, B.A.; Pierson, F.B. *The Ecology, History, Ecohydrology, and Management of Pinyon and Juniper Woodlands in the Great Basin Desert and Northern Colorado Plateau of the Western United States*; General Technical Report RMRS-GTR-403; U.S. Forest Service, Rocky Mountain Research Station: Fort Collins, CO, USA, 2019.
20. Miller, R.F.; Heyerdahl, E.K. Fine-scale variation of historical fire regimes in sagebrush-steppe and juniper woodland: An example from California, USA. *Int. J. Wildland Fire* **2008**, *17*, 245–254. [CrossRef]
21. Miller, R.F.; Chambers, J.C.; Pyke, D.A.; Pierson, F.B.; Williams, C.J. *A Review of Fire Effects on Vegetation and Soils in the Great Basin Region: Response and Ecological Site Characteristics*; General Technical Report RMRS-GTR-308; U.S. Forest Service, Rocky Mountain Research Station: Fort Collins, CO, USA, 2013.
22. Rau, B.M.; Chambers, J.C.; Pyke, D.A.; Roundy, B.A.; Schupp, E.W.; Doescher, P.; Caldwell, T.G. Soil resources influence vegetation and response to fire and fire-surrogate treatments in sagebrush-steppe ecosystems. *Rangel. Ecol. Manag.* **2014**, *67*, 506–521. [CrossRef]
23. Hartsell, J.A.; Copeland, S.M.; Munson, S.M.; Butterfield, B.J.; Bradford, J.B. Gaps and hotspots in the state of knowledge of pinyon-juniper communities. *For. Ecol. Manag.* **2020**, *455*, 117628. [CrossRef]
24. Ireland, K.B.; Stan, A.B.; Fulé, P.Z. Bottom-up control of a northern Arizona ponderosa pine forest fire regime in a fragmented landscape. *Landsc. Ecol.* **2012**, *27*, 983–997. [CrossRef]
25. Smith, H.Y.; Hood, S.; Brooks, M.; Matchett, J.R.; Deuser, C. Response of Fuelbed Characteristics to Restoration Treatments in Piñon-Juniper-Encroached Shrublands on the Shivwits Plateau, Arizona. In Proceedings of the Fuels Management–How to Measure Success Conference Proceedings, Portland, OR, USA, 28–30 March 2006.
26. Lindsay, B.A.; Strait, R.K.; Denny, D.W. *Soil Survey of Grand Canyon Area, Arizona, Parts of Coconino and Mohave Counties*; Natural Resources Conservation Service and National Park Service: Washington, DC, USA, 2003.
27. Huffman, D.W.; Roccaforte, J.P.; Springer, J.D.; Crouse, J.E. Restoration applications of resource objective wildfires in western US forests: A status of knowledge review. *Fire Ecol.* **2020**, *16*, 18. [CrossRef]
28. National Park Service. *Fire Monitoring Handbook*; National Interagency Fire Center: Boise, ID, USA, 2003.
29. The PLANTS Database. Available online: http://plants.usda.gov (accessed on 30 December 2024).
30. Soil Survey Staff. *Kellogg Soil Survey Laboratory Methods Manual*; Soil Survey Investigations Report No. 42; Natural Resources Conservation Service: Lincoln, NE, USA, 2022.
31. McGarigal, K.; Cushman, S.; Stafford, S. *Multivariate Statistics for Wildlife and Ecology Research*; Springer: New York, NY, USA, 2000.
32. Dufrêne, M.; Legendre, P. Species assemblages and indicator species: The need for a flexible asymmetrical approach. *Ecol. Monogr.* **1997**, *67*, 345–366. [CrossRef]
33. McCune, B.; Mefford, M.J. *PC-ORD: Multivariate Analysis of Ecological Data. User's Guide*; MjM Software Design: Gleneden Beach, OR, USA, 1999.
34. SAS Institute. *SAS/STAT User's Guide*; SAS Institute: Cary, NC, USA, 1999.
35. Bates, J.D.; Davies, K.W. Seasonal burning of juniper woodlands and spatial recovery of herbaceous vegetation. *For. Ecol. Manag.* **2016**, *361*, 117–130. [CrossRef]
36. Abella, S.R.; Crouse, J.E.; Covington, W.W.; Springer, J.D. Diverse responses across soil parent materials during ecological restoration. *Restor. Ecol.* **2015**, *23*, 113–121. [CrossRef]
37. Scasta, J.D.; Fuez, B. Post-wildfire shrub dynamics and ecological site controls in a sagebrush steppe: Successional shift or enhanced visibility? *Arid Land Res. Manag.* **2018**, *32*, 229–235. [CrossRef]
38. Abella, S.R.; Covington, W.W. Forest ecosystems of an Arizona *Pinus ponderosa* landscape: Multifactor classification and implications for ecological restoration. *J. Biogeog.* **2006**, *33*, 1368–1383. [CrossRef]
39. Duniway, M.C.; Palmquist, E.C. *Assessment of Rangeland Ecosystem Conditions in Grand Canyon-Parashant National Monument, Arizona*; Open-File Report 2020-1040; U.S. Geological Survey: Reston, VA, USA, 2020.
40. Abella, S.R.; Hurja, J.C.; Merkler, D.J.; Denton, C.W.; Brewer, D.G. Overstory-understory relationships along forest type and environmental gradients in the Spring Mountains of southern Nevada, USA. *Folia Geobot.* **2012**, *47*, 119–134. [CrossRef]
41. Minott, J.A.; Kolb, T.E. Regeneration patterns reveal contraction of ponderosa forests and little upward migration of pinyon-juniper woodlands. *For. Ecol. Manag.* **2020**, *458*, 117640. [CrossRef]
42. Puhlick, J.J.; Laughlin, D.C.; Moore, M.M.; Sieg, C.H.; Overby, S.T.; Shaw, J.D. Soil properties and climate drive ponderosa pine seedling presence in the southwestern USA. *For. Ecol. Manag.* **2021**, *486*, 118972. [CrossRef]
43. Gascho Landis, A.; Bailey, J.D. Reconstruction of age structure and spatial arrangement of piñon-juniper woodlands and savannas of Anderson Mesa, northern Arizona. *For. Ecol. Manag.* **2005**, *204*, 221–236. [CrossRef]
44. Huffman, D.W.; Crouse, J.E.; Chancellor, W.W.; Fulé, P.Z. Influence of time since fire on pinyon-juniper woodland structure. *For. Ecol. Manag.* **2012**, *274*, 29–37. [CrossRef]

45. Beatley, J.C. Ecological status of introduced brome grasses (*Bromus* spp.) in desert vegetation of southern Nevada. *Ecology* **1966**, *47*, 548–554. [CrossRef]
46. Jacobs, B.F.; Romme, W.H.; Allen, C.D. Mapping "old" vs. "young" piñon-juniper stands with a predictive topo-climatic model. *Ecol. Appl.* **2008**, *18*, 1627–1641. [CrossRef]
47. Tausch, R.J.; West, N.E. Differential establishment of pinyon and juniper following fire. *Amer. Midl. Nat.* **1988**, *119*, 174–184. [CrossRef]
48. Wangler, M.J.; Minnich, R.A. Fire and succession in pinyon-juniper woodlands of the San Bernardino Mountains, California. *Madroño* **1996**, *43*, 493–514.
49. Williams, R.E.; Roundy, B.A.; Hulet, A.; Miller, R.F.; Tausch, R.J.; Chambers, J.C.; Matthews, J.; Schooley, R.; Eggett, D. Pretreatment tree dominance and conifer removal treatments affect plant succession in sagebrush communities. *Rangel. Ecol. Manag.* **2017**, *70*, 759–773. [CrossRef]
50. Huffman, D.W.; Crouse, J.E.; Sánchez Meador, A.J.; Springer, J.D.; Stoddard, M.T. Restoration benefits of re-entry with resource objective wildfire on a ponderosa pine landscape in northern Arizona, USA. *For. Ecol. Manag.* **2018**, *408*, 16–24. [CrossRef]
51. Westlind, D.J.; Kerns, B.K. Repeated fall prescribed fire in previously thinned *Pinus ponderosa* increases growth and resistance to other disturbances. *For. Ecol. Manag.* **2021**, *480*, 118645. [CrossRef]
52. Clark, P.E.; Williams, C.J.; Pierson, F.B. Factors affecting efficacy of prescribed fire for western juniper control. *Rangel. Ecol. Manag.* **2018**, *71*, 345–355. [CrossRef]
53. Barney, M.A.; Frischknecht, N.C. Vegetation changes following fire in the pinyon-juniper type of west-central Utah. *J. Range Manag.* **1974**, *27*, 91–96. [CrossRef]
54. Koniak, S. Succession in pinyon-juniper woodlands following wildfire in the Great Basin. *Great Basin Nat.* **1985**, *45*, 556–566.
55. Humphrey, L.D. Patterns and mechanisms of plant succession after fire on *Artemisia*-grass sites in southeastern Idaho. *Vegetatio* **1984**, *57*, 91–101. [CrossRef]
56. Ralphs, M.H.; McDaniel, K.C. Broom snakeweed (*Gutierrezia sarothrae*): Toxicology, ecology, control, and management. *Invasive Plant Sci. Manag.* **2011**, *4*, 125–132. [CrossRef]
57. McCulloch, C.Y. Effects of wildfire on deer habitat in pinyon-juniper woodland. *J. Wildl. Manag.* **1969**, *33*, 778–784. [CrossRef]
58. Mueggler, W.F.; Blaisdell, J.P. Effects on associated species of burning, rotobeating, spraying, and railing sagebrush. *J. Range Manag.* **1958**, *11*, 61–66. [CrossRef]
59. Harniss, R.O.; Murray, R.B. 30 years of vegetal change following burning of sagebrush-grass range. *J. Range Manag.* **1973**, *26*, 322–325. [CrossRef]
60. West, N.E.; Hassan, M.A. Recovery of sagebrush-grass vegetation following wildfire. *J. Range Manag.* **1985**, *38*, 131–134. [CrossRef]
61. Power, S.C.; Davies, G.M.; Wainwright, C.E.; Marsh, M.; Bakker, J.D. Restoration temporarily supports the resilience of sagebrush-steppe ecosystems subjected to repeated fires. *J. Appl. Ecol.* **2023**, *60*, 1607–1621. [CrossRef]
62. Everett, R.L.; Ward, K. Early plant succession on pinyon-juniper controlled burns. *Northwest Sci.* **1984**, *58*, 57–68.
63. Donato, D.C.; Fontaine, J.B.; Robinson, W.D.; Kauffman, J.B.; Law, B.E. Vegetation response to a short interval between high-severity wildfires in a mixed-evergreen forest. *J. Ecol.* **2009**, *97*, 142–154. [CrossRef]
64. Abella, S.R.; Springer, J.D. Effects of tree cutting and fire on understory vegetation in mixed conifer forests. *For. Ecol. Manag.* **2015**, *335*, 281–299. [CrossRef]
65. Abella, S.R.; Schelz, C.D. Resilient plant communities and increasing native forbs after wildfire in a southwestern Oregon oak shrubland. *Northwest Sci.* **2024**, *97*, 151–166. [CrossRef]
66. Ott, J.E.; McArthur, E.D.; Sanderson, S.C. Plant Community Dynamics of Burned and Unburned Sagebrush and Pinyon-Juniper Vegetation in West-Central Utah. In Proceedings of the Shrubland Ecosystem Genetics and Biodiversity Proceedings, Provo, UT, USA, 13–15 June 2000.
67. Condon, L.; Weisberg, P.J.; Chambers, J.C. Abiotic and biotic influences on *Bromus tectorum* invasion and *Artemisia tridentata* recovery after fire. *Int. J. Wildland Fire* **2011**, *20*, 597–604. [CrossRef]
68. Barak, R.S.; Fant, J.B.; Kramer, A.T.; Skogen, K.A. Assessing the value of potential "native winners" for restoration of cheatgrass-invaded habitat. *West. N. Am. Nat.* **2015**, *75*, 58–69. [CrossRef]
69. Reed, C.C.; Hood, S.M.; Cluck, D.R.; Smith, S.L. Fuels change quickly after California drought and bark beetle outbreaks with implications for potential fire behavior and emissions. *Fire Ecol.* **2023**, *19*, 16. [CrossRef]
70. Ganey, J.L.; Vojta, S.C. Comparative trends in log populations in northern Arizona mixed-conifer and ponderosa pine forests following severe drought. *West. North Amer. Natl.* **2017**, *77*, 281–292. [CrossRef]
71. Welch, T.G.; Klemmedson, J.O. Influence of the biotic factor and parent material on distribution of nitrogen and carbon in ponderosa pine ecosystems. In *Forest Soils and Forest Land Management*; Bernier, B., Winget, C.H., Eds.; Les Presses de l'Université Laval: Laval, QC, Canada, 1975; pp. 159–178.
72. Renne, R.R.; Bradford, J.B.; Burke, I.C.; Lauenroth, W.K. Soil texture and precipitation seasonality influence plant community structure in North American temperate shrub steppe. *Ecology* **2019**, *100*, e02824. [CrossRef] [PubMed]

73. Heidmann, L.J.; King, R.M. *Effect of Prolonged Drought on Water Relations of Ponderosa Pine Seedlings Growing in Basalt and Sedimentary Soils*; Research Paper RM-301; U.S. Forest Service, Rocky Mountain Forest and Range Experiment Station: Fort Collins, CO, USA, 1992.
74. Moorhead, K.K. A pedogenic view of ecosystem restoration. *Ecol. Restor.* **2015**, *33*, 341–351. [CrossRef]
75. Hotta, W.; Morimoto, J.; Yanai, S.; Uchida, Y.; Nakamura, F. Environmental heterogeneity on landslide slopes affects the long-term recoveries of forest ecosystem components. *Catena* **2024**, *234*, 107578. [CrossRef]
76. Lecomte, N.; Bergeron, Y. Successional pathways on different surficial deposits in the coniferous boreal forest of the Quebec Clay Belt. *Can. J. For. Res.* **2005**, *35*, 1984–1995. [CrossRef]
77. Day, N.J.; White, A.L.; Johnstone, J.F.; Degré-Timmons, G.É.; Cumming, S.G.; Mack, M.C.; Turetsky, M.R.; Walker, X.J.; Baltzer, J.L. Fire characteristics and environmental conditions shape plant communities via regeneration strategy. *Ecography* **2020**, *43*, 1464–1474. [CrossRef]
78. Wonkka, C.L.; Twidwell, D.; West, J.B.; Rogers, W.E. Shrubland resilience varies across soil types: Implications for operationalizing resilience in ecological restoration. *Ecol. Appl.* **2016**, *26*, 128–145. [CrossRef] [PubMed]

diversity

MDPI

Article

Holocene Flora, Vegetation and Land-Use Changes on Dingle Peninsula, Ireland, as Reflected in Pollen Analytical, Archaeological and Historical Records

Michael O'Connell [1,*] and Steffen Wolters [2]

1 Palaeoenvironmental Research Unit, School of Geography, Archaeology and Irish Studies, University of Galway, H91 TK33 Galway, Ireland
2 Lower Saxony Institute for Historical Coastal Research, Viktoriastr. 26/28, D-26382 Wilhelmshaven, Germany; wolters@nihk.de
* Correspondence: michael.oconnell@universityofgalway.ie; Tel.: +353-86-3891-444

Abstract: Palaeoecological investigations connected with extensive pre-bog, stone walls, and field systems at Kilmore, Dingle peninsula, Ireland, are presented. The main pollen profile, KLM I, spans the last 4000 years. When the record opened, pine (*Pinus sylvestris*) was already a minor tree, oak (probably *Quercus petraea*) was the main tall-canopy tree, and birch and alder were dominant locally. Substantial farming is recorded between ca. 1530 and 600 BCE (Bronze Age) when the stone walls were likely constructed. From ca. 560 CE onwards, intensive farming was conducted for much of the time. A largely treeless landscape emerged in the late twelfth century CE. Fine-spatial reconstructions of landscape and vegetation dynamics, including the timing of blanket bog initiation, are made. Post-glacial change in the western Dingle peninsula, based on published Holocene lake profiles and drawing on the new information presented here, is discussed. Reported are (a) fossil spores of the filmy ferns *Hymenophyllum tunbrigense*, *H. wilsonii*, and *Trichomanes speciosum*; (b) the first fossil pollen record for *Arbutus unedo* (strawberry tree) in the Dingle peninsula (540 CE); and (c) the first published records for *Fagopyrum* fossil pollen in Ireland, indicating that buckwheat was grown at Kilmore in the late eighteenth/early nineteenth centuries.

Keywords: Holocene; pollen analysis; woodland dynamics; farming history; blanket bog; biodiversity; biogeography; Ireland

1. Introduction

South-western Ireland is defined geographically by three prominent peninsulas, namely Dingle (*Corca Dhuibhne* in Irish), Iveragh, and Beara (Figure 1, inset). Each peninsula has its own particular character, but they share many features, including bedrock dominated by Devonian sandstone [1], extensive uplands including the highest mountains in Ireland (Carrauntoohil in Iveragh (1038 m) and Mount Brandon in Dingle (952 m); elevations are given relative to average sea level (asl), as indicated in recent Ordnance Survey of Ireland (OSI) maps), and a particularly rich cultural heritage, including archaeological material from most periods, beginning in the Mesolithic and extending into the early medieval period and beyond [2–9]. Several interesting biota are also present, including species with predominantly Atlantic/Lusitanian distribution patterns. These include *Geomalacus maculosus* (Kerry slug), *Simethis mattiazzii* (Killarney lily), *Arbutus unedo* (strawberry tree), and filmy ferns (*Hymenophyllum tunbrigense*, *H. wilsonii*, and *Trichomanes speciosum*, i.e., the Killarney fern) ([10–12]; the online version of the BSBI Atlas at https://plantatlas2020.org

https://doi.org/10.3390/d17070456

(accessed on 1 March 2025); and Figures S6, S7 and S9). Note: The prefix S designates figures, tables, and text that are available in the Supplementary Information (SI).

The region has what is widely regarded as some of the best examples of temperate rainforest in Ireland and Britain, namely the *Quercus petraea*-dominated, fern- and bryophyte-rich woodlands in Killarney National Park, the premier and oldest national park in Ireland. On limestone terrain within this park, there is a *Taxus baccata*-dominated, species-rich woodland that is of particular importance, given the rarity of semi-natural yew woodlands in Ireland and Europe [13–15]. Another important recent development is the establishment of Kerry Seas National Park/Páirc Náisiúnta na Marra, Ciarraí, in 2023, which includes important coastal and marine habitats. The terrestrial component of this national park is centred on the Dingle peninsula between Mount Brandon and the Conor Pass, where our study area lies (https://www.nationalparks.ie/kerry-seas/; accessed on 3 July 2024; see also [8,16,17]).

Our study area lies in the western Dingle peninsula in Kilmore Townland (Td.), about 3 km south of Brandon Bay (Figures 1A and S9–S11). The area is sheltered by a horseshoe-shaped range of mountains that extends from Mount Brandon in the west to Sliabh a' Fhia (Slievanea; 627 m) and Binn na Naomh (670 m) to the south (core sites KLM I and II are on a blanket-bog covered plateau near the base of these mountains (see below), and Binn os Gaoith (Beenoskee; 826 m) and Stradbally Mountain (Cnoc an tSráidbhaile; 798 m) at the north-east end of the horseshoe.

Kilmore Td. and adjoining Ballyhoneen Td., both largely blanket-bog-covered, have particularly rich archaeology. This includes a wedge tomb (megalith assignable to the Late Neolithic/Early Bronze Age), two standing stones, eight *fulachta fiadh* (burnt mounds), rock art (cup and circle type) on three of the wedge tomb stones (all assignable to the Bronze Age), and several enclosures, hut sites, a mound, and pre-bog stone walls [18,19] (Figure S9). In what Ó Coileáin [19] refers to as the 'Central Complex', pre-bog stone walls, with a total wall length of 7.35 km, extend over approximately 80 ha and define at least 30 fields (the complexes are indicated in Figure S9B). In addition, 27 archaeological sites have been recorded here, and pollen profiles KLM I and II were obtained from this complex. A second complex of pre-bog stone walls has been recorded on the eastern side of Lough Adoon, i.e., the L. Adoon East Complex [19]. The complex extends over 69 ha. Twenty-three wall lengths and several archaeological sites have been recorded, including many hut sites that suggest habitation. However, only one large field (4.5 ha) and two smaller fields, <0.5 ha, have been identified [19]. The third, smaller complex was located on flat terrain to the east of the Scorid river and immediately south of L. Adoon. This complex, referred to by Ó Coileáin [19] as the Wedge Tomb Complex, includes the Ballyhoneen wedge tomb. There are also two standing stones and four *fulachta fiadh* (probably Bronze Age) that further emphasise the importance of the area during the Bronze Age. Ó Coileáin [19] regards this complex as possibly a place reserved for ritual purposes. The fourth complex is located on the steeply sloping western side of L. Adoon. Ó Coileáin [19] regarded these stone walls as related to the eighteenth or nineteenth centuries CE and associated with relatively recent sheep farming.

As regards palynological investigations in south-western Ireland, the main focus has been on the Killarney area on the Iveragh peninsula and also on Beara and Mizen peninsulas, where the overall emphasis has been on Holocene vegetation dynamics, particularly as mediated by human impact in the second half of the Holocene [20]. In the case of the Dingle peninsula, the investigations by Dodson [21,22] are from the same general area as those reported here. Dodson's investigations focused on the L. Adoon area (Loch a' Dúin; lake of the fort), i.e., the lowlands that lie to the north and in the shelter of the Brandon/Stradbally horse-shaped mountain range. From blanket-bog covered lowlands (<150 m asl) near L.

Adoon, there are two pollen profiles, namely LAH (L. Adoon Hill Bog) and LAV (L. Adoon Valley Bog). These are sited in the second and third archaeological complexes, respectively (see above; Figure S9B). Both relate to the later Holocene, the former spanning the last ca. 4000 years and extending into recent times, while the latter begins somewhat later (after 4 ka) and ends before the so-called secondary rise of *Pinus*, i.e., before the eighteenth century (Text S1). Both profiles showed substantial evidence of human impact, as expected, given their proximity to megalithic and other archaeological sites (Figures 1 and S9). The third pollen profile, LCN, relates to nearby Lough Camclaun (Loch Cham Calláin) and derives from lake sediment that was covered by a mat of floating vegetation at the southern end of this glacial lake when cored by Dodson [21]. This site is at ~250 m asl and abuts a steeply rising terrain (see Text S1).

The marginal sediments of a small lake (~1 ha but once probably ≥5 ha), situated in the lowlands (~70 m asl) west of Mount Brandon and ~4 km from the coastline to the west, have provided a 6-m-long pollen profile (BL) that spans the Lateglacial and Holocene (excluding recent centuries) (Text S1) [23]. Barnosky referred to this site as Ballinloghig Lake (Loch Chorr Áille in OSI, DS 70) [23]. This profile provides an interesting contrast to LCN (L. Camclaun), which lies only 10 km to the east. The significance of these pollen profiles by Barnosky and Dodson with regard to local and regional vegetation history is considered in the Discussion.

From Lispole, on the lowlands east of Dingle town, OCarroll [24] presented a short pollen profile (60 cm) from organic-rich deposits sampled above a *fulacht fiadh* that was one of several archaeological features recorded in preparation for local road-alignment works. The profile spans ca. 350 BCE to 1150 CE, i.e., from the mid-Iron Age to the end of the early medieval period. Tall woody vegetation during this time consisted mainly of hazel, alder, and birch towards the end of the period. Records of cereal-type pollen and *Plantago lanceolata* suggest local farming during much of the time. In addition, AMS ^{14}C dating of botanical remains associated with *fulachta fiadh* and charcoal pits provides additional evidence of trees, shrubs, and cereal cultivation. Wood and charcoal samples yielded ^{14}C dates as follows: 3617 ± 54 BP (holly) and 870 ± 37 BP (alder), and two charcoal samples (oak) dated to the late medieval period (ca. thirteenth century), while elm and yew charcoal provided post-medieval dates (295 ± 31 BP and 421 ± 31 BP, respectively). Grains of oat and barley were ^{14}C dated to the mid-/late medieval period, and a rye grain gave a late date (330 ± 35 BP) [24].

At the north-eastern end of the Dingle peninsula, immediately east of the medieval town of Tralee, archaeological investigations in the context of extensive roadworks have provided a wealth of evidence for human activity dating back to the early Neolithic [25,26]. The Bronze Age is particularly well represented, as is the Iron Age, which is unusual in the context of Irish archaeology [27,28]. Palaeoenvironmental investigations, consisting mainly of macrofossil, wood, and charcoal analyses, have greatly added to our knowledge of woodland composition in the area (especially plants that are poorly represented or silent in pollen records, such as blackthorn (*Prunus spinosa*) and bramble (*Rubus fruticosus* agg.)), woodland structure, trees/shrubs used for fires, and the farming economy, especially crops and associated weeds [28]. New insights into medieval (e.g., kilns for cereal drying and iron workings) and post-medieval developments (e.g., lime kilns, brick making, and land reclamation) have also been obtained from these excavations [28–30].

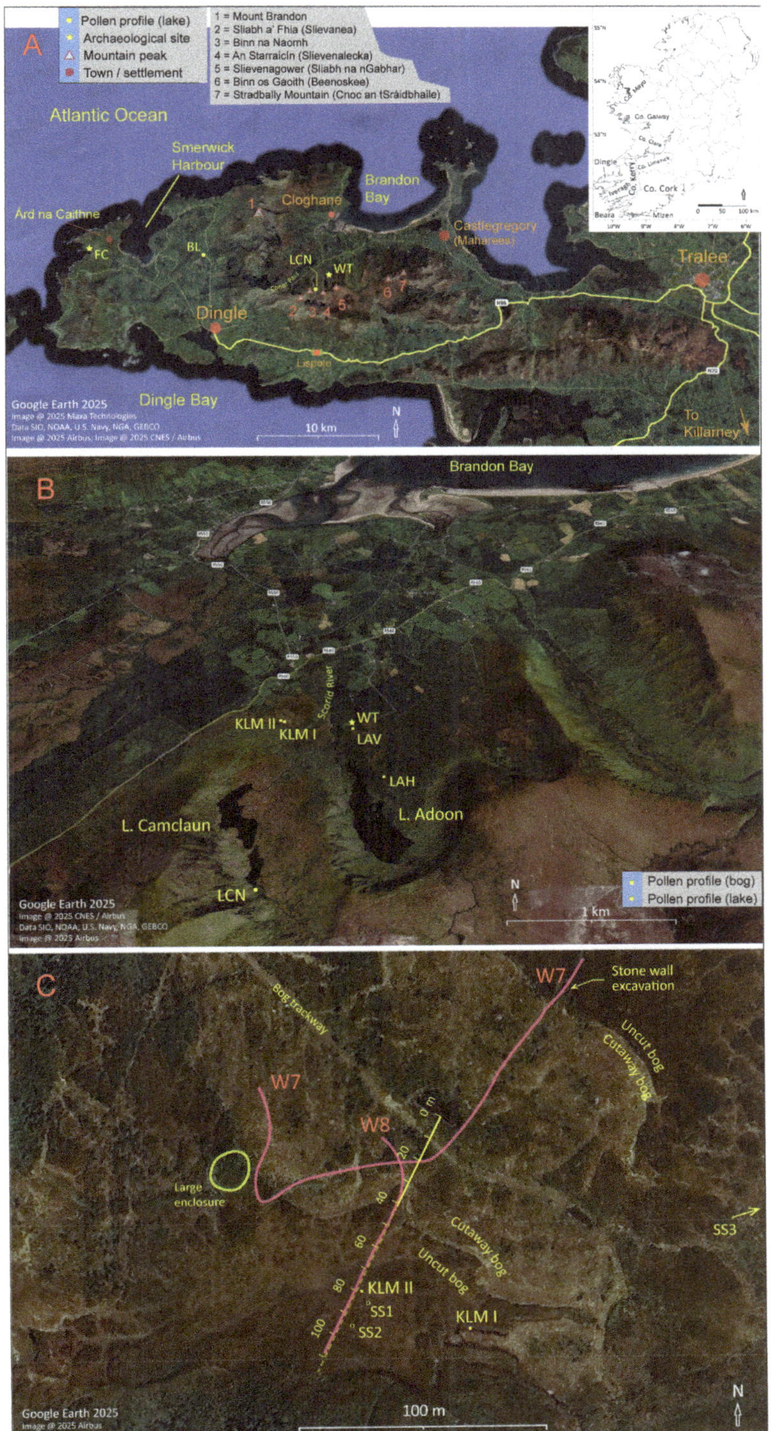

Figure 1. Maps showing the geographical details of the study region. The inset map of Ireland shows the county boundaries. The four main peninsulas of south-west Ireland, including Dingle, are

labelled. The main sites referred to in the text are shown in the images from Google Earth. (**A**) Dingle peninsula, the main peaks (numbered) in the Mount Brandon/Stradbally mountain range, pollen profiles BL (Ballinloghig) and LCN (L. Camclaun), archaeological sites at Ferriter's Cove (FC) and Ballyhoneen Wedge Tomb (WT), and main roads and settlements. (**B**) Tilted view towards the north. The main study area is shown in the image foreground. The locations of the Kilmore profiles KLM I and II, L. Adoon bog profiles LAV and LAH, and lake profile LCN are shown. (**C**) Detailed map centred on the main archaeological area ('Central Complex') in Kilmore. The locations of pollen profiles KLM I and II, surface pollen samples SS1 and SS2 (SS3 is off image), pre-bog stone walls W7 and W8, and the transect laid out along the line of W8 are shown in the image. The yellow dashed transect line indicates the transect is on uncut bog.

The investigations reported on in this paper aimed, in the first instance, at providing palaeoenvironmental contexts for the archaeological investigations that were being carried out at the time (the early 1990s) by Ó Coileáin [18,19]. Of particular interest to us, and also the archaeologist, was the woodland and land-use history, and dating of blanket bog initiation and spread, which provides a *terminus post quem* for stone-wall construction [31]. At the time of archaeological and palynological investigations, turf cutting for domestic fuel had removed peat from substantial areas within the Central Complex where KLM I and II were collected, but fortunately, considerable tracts of undisturbed peat remained and were available for investigation (Figure 1C; peat cutting has more or less ceased in the meantime; see photographs in Figures S10, S11, and S14). The KLM profiles lie close to those of Dodson [21], i.e., the peat profiles LAV and LAH, which lie within a kilometre to the east, and the lake-edge profile LCN, which is about 1.6 km to the south (Figure 1).

The nomenclature of higher plants in this paper follows Parnell and Curtis [32]. Bryophyte nomenclature generally follows Atherton et al. [33]. Placenames mainly follow those used by the OSI and, in the case of Dingle peninsula, the OSI Discovery Series (DS) map, no. 70, 6th edition. Cognizance, however, is taken of what is in common usage locally and in Ireland generally, as well as the placenames used in the EastWest 25series maps (www.eastwestmapping.ie; accessed on 1 December 2024). In some instances, both English and Irish versions of the placenames are given (especially where both are commonly used), the former often being merely an attempt by nineteenth-century cartographers at rendering in English spelling what the Irish version sounded like, though, in the process, the placenames invariably lost their significance as descriptors of landscape, vegetation, land use, etc., that were inherent in the Irish version. Dingle peninsula is often referred to as Corca Dhuibhne (in Irish; also in English, where other spellings such as Corkaguiny are not uncommon), which is the name of the barony that includes all but the easternmost part of the peninsula.

For details of the cultural periods referred to in the text, Waddell [34] is the main reference for prehistory, while for the historical period (beginning in Ireland at ca. 400 CE), various sources are used, including Aalen et al. [17] (see also Abbreviations). The early medieval period (CE) is regarded as extending from 400–1169 and includes the early Christian (ca. 400–800) and Viking periods (ca. 800–1169). The late medieval period, also referred to as the Anglo-Norman period, extends from 1169 to ca. 1550. Occasionally, reference is made to the classical pollen zones defined by Mitchell [35] for Ireland.

2. Materials and Methods

2.1. Sampling in the Field

In October 1993, a fresh face was exposed in a recently cut bank of turf, and a peat monolith, KLM I, was removed in three contiguous sections where the peat was deepest (2.23 m thick) (Figures S12 and S13). At the sampling site (52.20576 N, 10.16296 W; $\sim$138 m asl), there was a small localised depression of $\sim$10 cm in the mineral ground so that the base of the monolith was a few centimetres below the general level of the sub-peat mineral ground. A small pine stump that had been uprooted by peat cutters was observed nearby.

Prior to collecting KLM II, the uncut bog was cored along a transect that ran southwest from a recent substantial stone-wall constructed by peat cutters to facilitate drying of cut turf to a particularly large boulder that lay where the peat met steeply rising ground and ceased (Figures 1C and S14). This transect followed Wall 8 (wall descriptions and numbering after Ó Coileáin [18]). The surface along the transect was levelled with respect to the surface within the nearby large stone enclosure (see Figure 1C) using a theodolite. Peat thickness and stratigraphical features were investigated along the transect using a gouge corer. Core KLM II was collected using a Livingstone piston corer (core tube: 100 cm long, diameter: 5 cm), where the maximum thickness of peat was recorded, i.e., at 80 m from the start of the transect and about 40 m west of KLM I (co-ordinates: 52.20586 N, 10.16365 W; ~139 m asl).

Wall 8 intersects Wall 7, which runs across it (the stone walls in these parts were exposed by peat cutters; see Figure 1C). Wall 7 runs west towards the large stone enclosure referred to above, curves, and passes by the enclosure without touching it, i.e., it avoids it and presumably post-dates it (Figure 1C). This wall also ran east of Wall 8 and curved sharply northward to disappear under shallow peat. In an area where peat had not been cut, Wall 7 was excavated, and five ^{14}C dates were obtained by Ó Coileáin [18] (Figures 1C, S11 and S15; see also Section 4).

2.2. Preparation, Identification, and Counting of Fossil Pollen Samples

Monolith KLM I was sampled between 4 and 222 cm at intervals of 2 and 4 cm in the lower and upper parts of the monolith, respectively. The basal part only of core KLM II (255–285 cm) was sampled at 2-cm intervals. Samples were 1 cm thick, with the top being used to designate depth; e.g., sample 4 cm was from 4–5 cm from the bog surface at the sampling site.

Sample preparation for pollen analysis followed standard procedures. A *Lycopodium* spore suspension of known concentration was added at the start to enable determination of the pollen concentration. The samples were heated with 10% KOH for 10 min and sieved to remove large and unwanted debris. The sievings were later examined for macroremains (KLM I only). The mineral-rich samples were treated with 60% HF. All samples were acetolysed for 4 min, washed with glacial acetic acid, and then washed with water. Glycerin was used to store the pollen pellets and as a mounting medium.

Pollen identification and counting were carried out using a Leitz Laborlux D microscope (Leica, Wetzlar, Germany). A magnification of ×500 was used for routine counting, and ×1250 oil immersion with phase contrast was used for identifying critical pollen types. Where pollen density was low, a magnification of ×312 was used in conjunction with higher magnifications to check for difficult-to-identify pollen types.

Pollen identification was carried out to the highest taxonomic resolution wherever possible, i.e., genus or species level; however, in some cases, this was not possible. Taxa that present a particular challenge regarding identification and/or are important from the point of view of reconstruction of the palaeoenvironment include the following: *Corylus* and *Myrica* were distinguished according to the criteria of Mohr [36]. Monolete spores (-perine) consist of fern spores (probably derived from *Dryopteris* and related ferns, including *Thelypteris palustris*) that had lost their outer perine and so could not be further distinguished. Other fern spore determinations, excluding filmy ferns, follow Moore et al. [37]. *Rumex*-type is consistent with *Rumex* sect. *Acetosa* in Fægri and Iversen [38] and *Rumex acetosa*-type in Moore et al. [37]. *Rumex aquaticus*-type is as defined by Fægri and Iversen [38]). *Rumex*-type is expected to be derived from docks that are typical of grasslands (e.g., *R. acetosa*), while *R. aquaticus*-type probably derives from a dock of wet habitat, such as *R. hydrolapathum*.

Anagallis tenella-type (as in Moore et al. [37]) is most likely *A. tenella*, a common blanket-bog species. *Saxifraga hirsuta*-type, as described by Moore et al., potentially includes pollen of *S. spathularis* and *S. hirsuta*, both of which have been recorded in the locality [10]. Tricolporate, echinate Asteraceae (here referred to as Tubuliflorae, as distinct from Liguliflorae, e.g., *Taraxacum*) potentially includes pollen from several common composite daisy-like species. Only *Centaurea nigra*-type pollen (most likely derived from *C. nigra*) is plotted separately here (profile KLM I). Fabaceae (Leguminosae) potentially includes pollen from *Trifolium* and *Vicia* spp. (distinguished during counting but aggregated in the pollen plots). *Ulex* was distinguished and plotted separately (surface sample SS-3). Caryophyllaceae includes mainly *Cerastium*-type pollen (cf. Fægri and Iversen [38]).

Cereal-type pollen was defined according to the criteria described by Beug [39,40]. These pollen grains were further categorised based on their size (maximum length). Cereal-type pollen in the size range of 37 to 39 μm is, however, included in Poaceae and is regarded as most likely arising from non-cultivated grasses. *Secale* (rye) was distinguished from other cereal-type pollen based on the shape, size, and position of the pores, as described by Beug [39,40].

Vaccinium-type includes all ericoid pollen (tetrads) other than *Calluna vulgaris*, *Erica tetralix*, *E. cinerea*, *Empetrum nigrum*, and *Arbutus unedo*. *A. unedo* tetrads were distinguished with confidence from those of other tetrad-producing ericoid species based on large size (generally $\geq$50 μm), long colpi, and rather smooth surface (see Beug [39,40], Foss and Doyle [41]; Moore et al. [37]), and after comparison with modern reference samples (see below).

Non-pollen palynomorphs (NPPs) that were identified and counted include testate rhizopods such as *Amphitrema flavum*, *Assulina*, and *Hyalosphenia subflava*; crustaceans such as Copepoda; ascospores including *Gelasinospora*; and other fungal spores. *Mougeotia gracillima* and other Zygnemataceae taxa were also counted, as were fragments of the epidermis of *E. tetralix* seeds and charcoal particles $\geq$ 37 μm in size. Rhizopods were identified using Grospietsch [42], and other NPPs were identified using Van Geel [43]. Type numbers preceded by 'HdV-', in addition to the names that are in common use, are shown for NPPs in the pollen diagrams (cf. Shumilovskikh et al. [44]).

Type slides in the Palaeoenvironmental Research Unit (PRU), University of Galway, were consulted for all critical taxa, especially *A. unedo* and filmy fern spores. Regarding the filmy ferns, size statistics and other details derived from the study of modern materials were compiled and are presented in Table S1. These results show that *H. wilsonii* and *H. tunbrigense* can be distinguished with confidence based on size (88 μm, 74–111 μm vs. 48 μm, 36–65 μm; average sizes and size ranges are cited) and spore features, including shape, trilete marking, and size and distribution of projections on the spore surface. *T. speciosum* is similar to *H. tunbrigense* in that it is rather small (average size: 55 μm; size range, however, is large, i.e., 39–80 μm), but it can also be differentiated with confidence based on the shape of the spore, characteristics of the trilete mark, and shorter surface processes (Table S1).

High pollen counts ($\geq$1000) were striven for, but this was not always feasible in KLM I because of low pollen concentrations. In both KLM I and II, total terrestrial pollen (TTP) was used as the pollen sum (PS). The PSs in KLM I and KLM II averaged 689 and 932, respectively, which can be expected to provide an adequate basis for percentage calculations. Bog and aquatic taxa are excluded from the PS, as well as *Alnus* (excluded because of exceptionally high values at the base of the profile) and *Sphagnum*. The percentage values for the taxa outside the PS are based on TTP and the sum of the taxa in a particular category, e.g., bog/heath. In some instances, the particular category consisted of only a single taxon, e.g., *Alnus*, *Sphagnum*, *Equisetum*, and *E. tetralix* epidermal fragments. In these instances, and in the case of NPPs, percentages are expressed relative to TTP plus the particular taxon. Concentration (grains cm^{-3}) and influx (grains cm^{-2} y^{-1}) values were calculated in the usual way based on the number of *Lycopodium* spores (exotic) counted. Regarding the surface samples, *Alnus*, which has only moderate values, is included in the PS, as is the normal practice.

To summarise the pollen data, individual curves were combined into composite pollen curves, including AP (arboreal pollen) and NAP (non-arboreal pollen). NAP is subdivided into NAP1 (pollen indicative mainly of grasslands) and NAP2 (pollen of cultivated plants and weeds associated with arable/disturbed habitats). Pollen assemblage zones (PAZs), based mainly on the major movements of the percentage pollen curves, were distinguished by careful inspection of the percentage plots.

2.3. *Macrofossil Analyses*

The material retained on the 100-µm-mesh sieves during the initial stages of pollen preparation was examined using a Wild M8 binocular microscope (Wild Heerbrugg, Heerbrugg, Switzerland) and scored for abundance on a scale of 1–4, i.e., rare, occasional, frequent, and abundant. *E. tetralix* seeds were identified using type slides and photographs by Grosse-Brauckmann [45]. *Juncus effusus*-type seeds potentially include those of *J. effusus* and *J. conglomeratus* [46]. Mosses were identified using the descriptions provided by Smith [47]. The records mainly consist of *Sphagnum* leaves. *S. papillosum* was differentiated, as were other cucullate-leaved Sphagna that probably include *S. magellanicum* and *S. palustre* (referred to as *S.* sect. *Cymbifolia* (=Sect. *Sphagnum*). *S. austinii* (=*S. imbricatum* subsp. *austinii*) was not recorded. All other *Sphagnum* leaves (narrow-leaved, non-cucullate-leaved specimens) were referred to *Sphagnum* undiff. The abundance of mineral particles and charcoal fragments (referred to as macro-charcoal, i.e., fragments $\geq$ 100 µm) was also noted.

2.4. ^{14}*C Dating*

Six peat samples, each 1 cm thick and ~60 g wet weight, were taken from monolith KLM I and submitted for ^{14}C dating. ^{14}C dates were calibrated using OxCal ver. 4.4.4 [48] and the IntCal20 calibration curve [49]. Age/depth models for profile KLM I and other profiles were produced using Clam v. 2.3.8 [50].

2.5. *Determination of Ash Content and Tephra Detection*

Samples were taken for ashing from depths in monolith KLM I corresponding to the pollen-analysis samples. Core KLM II was sampled continuously every centimetre from 285 cm (base) to 255 cm (top). The samples were dried at 70 °C for 24 h, cooled in a desiccator, and the dry weight was ascertained (about 3 g and 1 g for KLM I and II, respectively). The samples were then ashed at 650 °C for 4 h. After cooling, the residue (ash) was weighed, and the ash content was calculated as a percentage of sample dry weight.

Later (in 2004/05), KLM I was sampled continuously for tephra by taking 4-cm-thick peat slices and ashing these following similar procedures to those described above [51]. Ashing took place, however, at a lower temperature (550 °C) to minimise damage to any tephra present, but for a longer time (5 h). After ashing, the mineral matter was washed with 10% HCl to remove the surface coating from the tephra shards and sieved using a 24-µm-mesh sieve to remove small particles. Glycerol was added to the material retained on the sieve, and the suspension was plated onto slides and examined microscopically at ×250 magnification. Tephra was identified based on viscularity with the aid of a polariser. Tephra shards do not rotate plane-polarised light (PPL); therefore, they become invisible upon rotation of the polariser [52]. In contrast, quartz grains remain visible as bright entities. Similar to tephra, phytoliths, i.e., silicic plant inclusions, do not rotate PPL, but they can be distinguished from tephra by their rough edges and lack of vesicles.

One or more tephra shards were detected in several of the 4-cm-thick samples but only in two contiguous samples, i.e., 49–53 cm and 53–57 cm, which had elevated numbers of shards (see Section 3). This 8-cm interval was subsampled by taking 1-cm-thick samples and ashing following the procedures outlined above. Tephra shards were distinguished and counted in the same way as described for the initial sampling.

2.6. Pollen Analysis of Surface Samples

Surface samples, consisting mainly of *Sphagnum* moss, were collected from within three relevés at a depth of ~2 cm, prepared for pollen analysis, and the pollen was counted (locations are indicated in Figure 1C). Sample preparation followed procedures similar to those used for preparing the peat samples, but an exotic was not added. Relevés were taken at the sampling sites as follows: Relevé 1 was taken in the vicinity of coring site KLM II. *Molinia caerulea* and sedges (*Schoenus nigricans*, *Eriophorum angustifolium*, and *Rhynchospora alba*) were abundant. Relevé 2, at about 95 m on the transect, was from the edge of a 1-m-wide, shallow, drainage channel on the bog surface. *M. caerulea* and aquatic species, especially *Potamogeton polygonifolius*, were abundant. Relevé 3 was from the east of the Scorid river in the immediate vicinity of a *fulacht fiadh*. Seven holly trees (*Ilex aquifolium*), five with berries, were recorded locally.

3. Results

3.1. Peat Stratigraphy, Ashing and Tephra Data

The results of the stratigraphical investigations along the bog transect are shown in Figure 2. The data relating to peat stratigraphy, ashing, and tephra investigations of monolith KLM I and core KLM II are shown in Figures 3 and 4, and details are given in Tables S2–S4.

The peat body along the transect was generally 2 to almost 3 m thick (maximum thickness recorded: 290 cm) (Figure 2 and Table S2). The lower peat was usually highly decomposed and had a fine-fibrous texture. Towards the surface, the peat was more fibrous, and decomposition was lower. Occasionally, *E. vaginatum* remains were recorded, mainly at mid-depth. Substantial pieces of *Betula* wood were recorded at the 90 m point. Charcoal was not conspicuous (see detailed stratigraphy for KLM II in Table S4). The sub-peat soil was silty in texture. Stones were encountered at the base of coring points 80 m, 90 m, and 100 m.

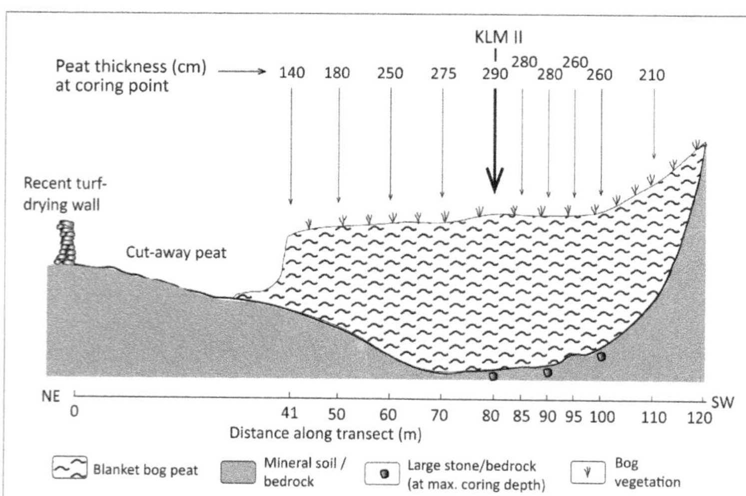

Figure 2. Schematic representation of the peat body and underlying mineral ground based on coring and levelling along a transect where core KLM II was collected (at 80 m). The vertical exaggeration is ×10 (approximately).

The stratigraphical features of monolith KLM I are as follows (Figures 3–5 and Table S3). The sub-peat mineral soil was silty and had a high stone content. In the peat column, changes in decomposition and texture were gradual. The basal part (192–223 cm), which contained substantial pieces of birch wood (*B. pubescens*), consisted of highly decomposed peat. The most pronounced feature is a distinct birch layer between 126 and 120 cm (Table S3). This layer, embedded in dark, well-decomposed peat, was traced laterally along the peat bank on either side of KLM I. A 3-cm-thick layer of dark, charcoal-enriched peat was present immediately beneath the birch layer. *E. vaginatum* remains were occasionally recorded but were not prominent.

Detailed stratigraphical information for core KLM II is available only for the lower part of the core (Table S4; also see Figure 6A_2). Substantial pieces of birch wood were recorded at 211 cm. Below 274 cm (to 281 cm), the peat was highly decomposed. A charcoal-rich layer was present between 278 and 280 cm. Peat with sand followed (281–284.5 cm). This rested on mineral ground consisting of light grey, silty soil (284.5–285 cm). During initial probing, 290 cm of peat was recorded in the vicinity of this coring point, which indicates a locally uneven mineral ground surface, as was the case at KLM I.

The basal peats in KLM I and II yielded elevated ash values (>20% starting at 210 cm and 283 cm in cores KLM I and II, respectively; Figure 3A,D, and Figure 6A_2). This suggests that as peat initially accumulated, it was under soligenous influence, and mineral ground was likely still exposed locally. The ashing results from KLM I, based on 1-cm- and 4-cm-thick slices, are broadly similar (Figure 3A,B). Average values of 2.1 $\pm$ 0.7% and 2.3 $\pm$ 0.8% (n = 48 and 50, respectively) were recorded (note: the lowermost highly minerogenic samples have been excluded from these statistics). Sample 153–157 cm gave 16% ash, which is regarded as an error and hence is excluded from the statistics. There was a general trend, though not pronounced, towards declining ash values until ~60 cm (Figure 3A,B). Above this depth, the values were generally elevated, especially near the top.

Elevated tephra-shard counts were recorded at 49–53 cm and 53–57 cm (16 and 10 shards, respectively; Figure 3C). In the 1-cm thick slices, sample 53 cm (i.e., 53–54 cm) yielded the greatest concentration of shards (110 shards; samples 52 and 54 cm yielded only 6 and 8 shards, respectively; Figure 3E), which suggests a distinct tephra layer at 53 cm. Most of the shards in sample 53 cm (93 out of 110) ranged in size from 20–90 μm (Figure 3F), and four shards > 100 μm were recorded. The shards were all colourless; most shards were vesicular (>80%), about 50% had air bubbles, and 13% had only elongated channels [51].

The tephra layer is bracketed by the uppermost two ^{14}C dates. These dates suggest that the tephra layer dates to the late thirteenth or early/mid-fourteenth centuries, i.e., at a time when there were several known Hekla eruptions [53,54]. $Hekla_{1300}$ lasted an entire year and was a major eruption with a Volcanic Explosivity Index (VEI) of 4. However, there are only a few records of the $Hekla_{1300}$ eruption from Ireland [53]. $Hekla_{1340}$ was of medium size (VEI-3), and $Hekla_{1222}$ and $Hekla_{1389}$ were smaller and thought to be of low magnitude (VEI-2 and VEI-3, respectively). These four eruptions are hence regarded as unlikely sources of the tephra layer recorded in KLM I. Hekla 1 (1104 CE), which is well-defined at several Irish sites [53,55], is clearly too old and is therefore excluded from consideration. This leaves $\text{Öræfajökull}_{1362}$ as the most likely candidate. $\text{Öræfajökull}_{1362}$ was a major eruption in south-east Iceland that began in the early part of 1362 CE, lasted a few months, had a substantial impact in Iceland, and produced widely dispersed tephra that has been recorded in Greenland ice cores and as far east as Germany and Poland [53,54,56]. There are also several $\text{Öræfajökull}_{1362}$ records from Ireland, including sites in western and south-western Ireland [52,57–59]. Thus, the available evidence points to $\text{Öræfajökull}_{1362}$ as the source of the KLM I tephra layer. Accepting this, the historical date of the eruption was used to provide a fix for the chronology in this part of KLM I. As there is no evidence in the stratigraphy or pollen for rapid changes in peat accumulation rates (which would follow if the tephra and the two uppermost ^{14}C dates were accepted as valid for the depths to which they pertain), the ^{14}C dates were not taken into account in the construction of the age/depth curve. Using the ^{14}C dates in preference to the tephra-derived dates results in somewhat older ages (but ≤100 y) for this part of the profile. Overall, however, the differences between accepting the tephra vs. the ^{14}C dates are small; therefore, the approach taken appears to be justified (see also ^{14}C dating, which follows).

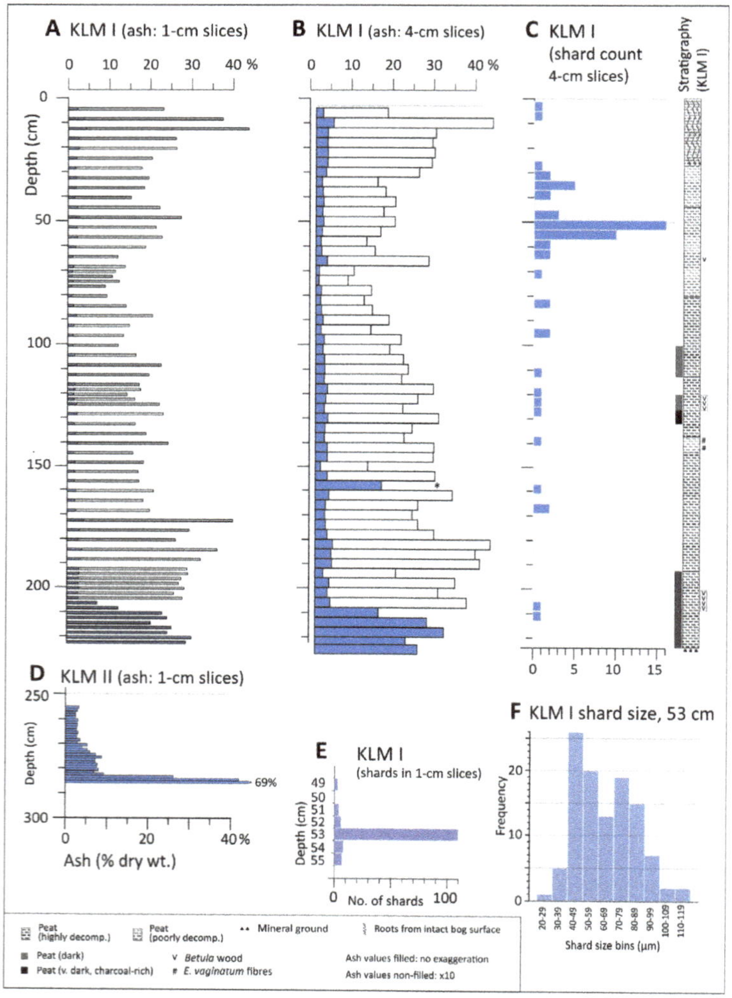

Figure 3. Plots showing the results of ashing (cores KLM I and II) and tephra-shard counts (KLM I). Ash values, cores KLM I and II (1-cm-thick slices) ((**A**,**D**), respectively); ash values, core KLM I (4-cm-thick slices) (**B**); shard counts, 4-cm-thick slices (**C**); shard counts in 1-cm-thick samples, samples 49–55 cm (**E**); shard-size distribution in sample 53 cm (**F**). In (**B**), sample 153–157 is marked with an asterisk. The ash value (16%) is unexpectedly high, due presumably to experimental error.

3.2. ^{14}C *Dating and Construction of an Age/Depth Model*

The six ^{14}C dates relating to KLM I appear to be generally acceptable (Figure 4; Table S6). The lower four ^{14}C dates were accepted as reliable evidence of age. The upper two dates, however, are regarded as somewhat too old, mainly on the basis of the presence of a micro-tephra layer at 53 cm that is attributed to Öræfajökull$_{1362}$ (see above) and so are not used in the age/depth model. The uppermost levels are constrained by two dates assigned on palynological grounds. At 20 cm, a curve for *Pinus* recommences, which is assumed to reflect an increase in pine pollen arising from the widespread planting of pine that had already commenced by the mid-eighteenth century in Ireland [60], including Co. Kerry [61–63]. In the uppermost pollen spectrum at 4 cm, *Pinus* representation did not greatly increase, which suggests that this spectrum predates the increased afforestation of the later twentieth century. Therefore, the top of the pollen profile is unlikely to extend beyond 1950 CE, i.e., the age assigned to depth 4 cm.

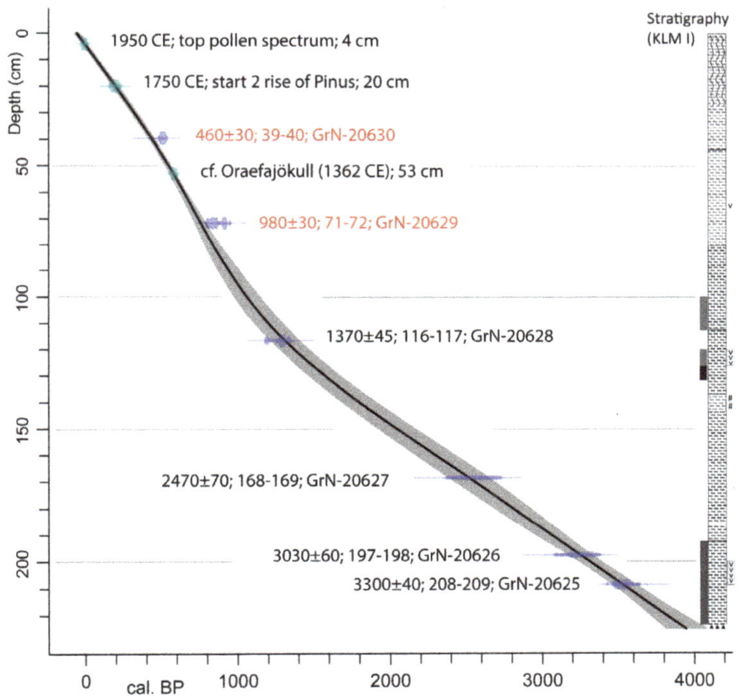

Figure 4. Plot of age/depth curve for profile KLM I, as produced by Clam. The spline curve (smoothing = 0.25) is based on ^{14}C dates (uppermost dates, in red, were excluded within Clam) and the tephra layer. The key to the stratigraphic symbols is shown in Figure 3.

Overall, the age/depth curve suggests more or less constant and low peat accumulation rates in the lower part of the profile (at/below 140 cm: average value: 0.4 mm y^{-1}), increasing rates between about 140 and 70 cm, and followed by a decline that stabilised towards the top of the profile (average 0.85 mm y^{-1}, from 52 cm to the top; this is probably artificially low as a result of drainage arising from peat cutting). Additional ^{14}C dates would reveal greater variation in peat accumulation rates. This is hinted at by changes that suggest slow rates, including dark and more decomposed peat, very high pollen concentration and influx values in zone KLM I-1 and briefly elevated values at about 130 cm (immediately below the woody birch layer) (Figure S1).

Profile KLM II does not have independent evidence of age; therefore, a precise chronology cannot be attached. Given its location and context, it is assumed that the base of KLM II is similar in age to KLM I (4000 cal. BP) or perhaps somewhat older (based on the *Pinus* curve). *P. lanceolata* begins to increase substantially at ca. 3300 cal. BP in KLM I.

A similar rise was recorded at 267 cm in KLM II (*P. lanceolata* attained >14% at 265 and 263 cm in mid-zone KLM II-3). As this development is probably of more than merely local significance, this feature in KLM II is assumed to be of a similar age as the rise in *P. lanceolata* in KLM I (3300 cal. BP). Accepting that this is broadly correct, the basal peat in KLM II has accumulated at 0.28 mm y^{-1}, which is within the same order of magnitude as that estimated for KLM I (0.4 mm y^{-1}). Assuming a steady accumulation rate in KLM II, this profile extends to ca. 2800 cal. BP. However, it may be older, as the rate of peat accumulation in the basal part most likely increased over time, as is the case in KLM I.

3.3. Pollen Analysis: General Considerations, Conventions in the Pollen Diagrams and Surface-Pollen Data

Plots of the pollen data, i.e., the surface samples and pollen profiles KLM I and II, are presented as percentage diagrams in Figures 5 and 6, and as concentration and influx diagrams in Figure S1 (KLM I only). A summary of the KLM profiles and the surface-sample pollen data is available in Table S8. Pollen profiles LAH and LAV by Dodson [21,22] are presented in Figure 7. An overview of the percentage pollen data derived from the peat cores and summary interpretations in graphic form are provided in Figure 8. The relevés from the sites where surface samples were collected are listed in Table S5.

The conventions followed in the construction of pollen diagrams are as follows. Curves that are based on PSs that deviate from what is usually employed have names that include an asterisk to alert the reader that there is a deviation from normal practice (see Section 2). A lighter shade of brown is used for bog/heath taxa indicative of wet mire conditions and similarly for NPP taxa.

The surface sample data (three spectra) demonstrated a major contribution from local pollen producers (Figure $6B_1$,B_2; Table S8). Poaceae, for instance, average 69%. This is attributable to *M. caerulea*, a grass that is important both locally and regionally on extensive intact and cutover bogs. The local presence of *P. polygonifolius* resulted in a strong representation of Eu-*Potamogeton* in SS2 (20%). The AP component in the surface samples was low (average: 15%, consisting mainly of *Pinus* and *Betula*). This undoubtedly derives from long-distance pollen transport, as there are few trees locally or in the wider region, apart from conifer plantations, which, at the time of sampling, were not yet extensive/mature. In SS3, *Ilex* achieved 5.9%, which, although high for a shrub with poor pollen dispersal, is not surprising given the presence of several mature holly specimens locally (though often female and hence not expected to produce pollen [64]). Despite the high Poaceae values that lower the percentage representation of other taxa, taxa indicative of grasslands on mineral ground are well represented. This suggests that these taxa, and especially *P. lanceolata*, produce substantial amounts of widely dispersed pollen.

A puzzling aspect of the surface-pollen spectra is the high cereal-type pollen values. This is especially true given that, during the later part of the twentieth century, arable farming greatly declined here, as in most parts of western Ireland ([65]; Figure $6B_1$; see Section 4.3). Not only cereal-type pollen of medium size (40–45 µm) but also large pollen, including pollen ≥ 50 µm (*Secale*, too; a single grain is recorded in SS2), are well represented. The most likely explanation is that several decades, rather than a few years of pollen deposition, as originally thought, are represented in the surface samples.

As regards bog/heath taxa, several are under-represented and, in particular, shrubby species, including ericoids such as *Calluna*, *E. tetralix* (not recorded at all), and *E. cinerea*. Bog myrtle (*Myrica*), which was present at the three sites (and plentiful at SS3), was poorly represented. This should not be attributed to misidentification, as *Corylus* is also poorly represented (*Myrica* and *Corylus* pollen are difficult to separate; see Section 2). Cyperaceae are strongly represented, which is not surprising given the important role of sedges (*Carex* spp., *Rhynchospora*, and *Eriophorum* (bog cotton), etc., especially at SS1 and SS2). Also interesting is the strong representation of *Amphitrema flavum*, *Assulina*, and Copepoda in SS1 and SS2, particularly in SS1, where humid rather than wet conditions prevailed. In SS3, where dry conditions prevailed, these taxa (also Eu-*Potamogeton*) were poorly or not at all represented.

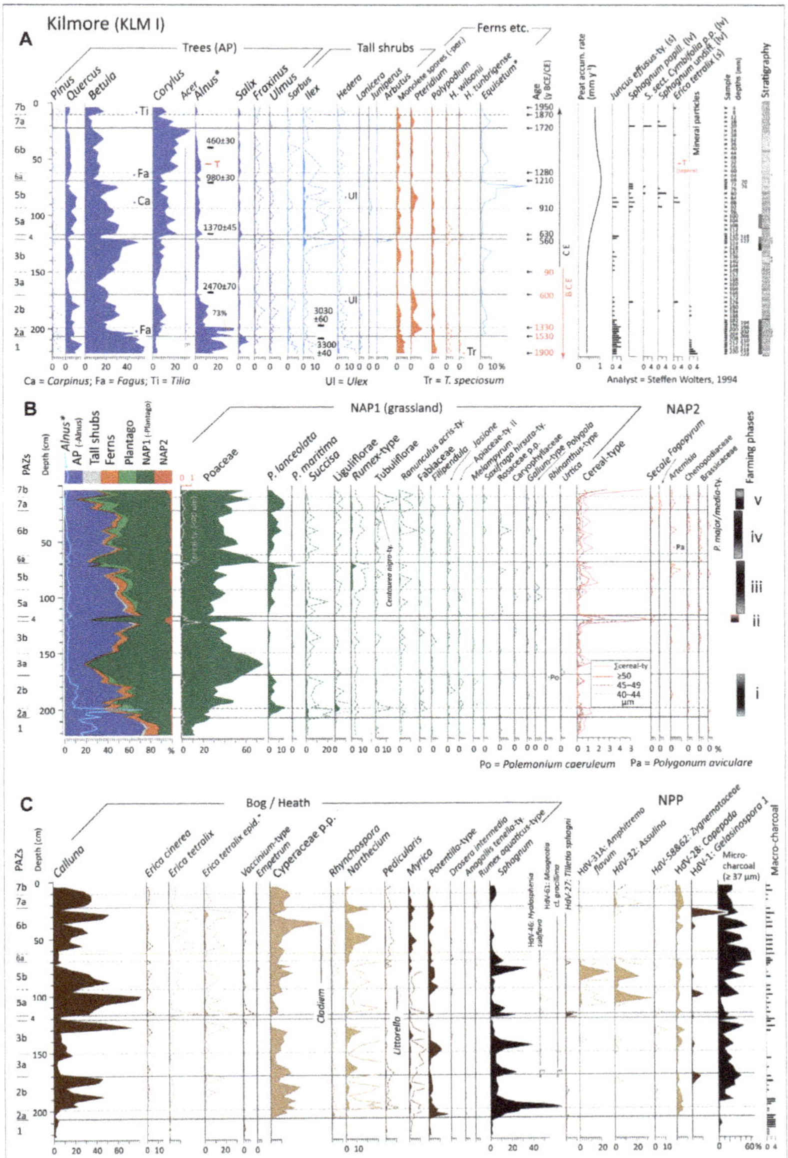

Figure 5. Percentage pollen diagram, profile KLM I. Curves with broken lines are exaggerated ×10. T = tephra layer (cf. Öræfajökull$_{1362}$). A taxon name with an asterisk indicates that the taxon in question is outside the PS and/or a PS is used that differs from that used in neighbouring taxa (see Section 2). Rare taxa are indicated by a dot, and taxon names are abbreviated. ^{14}C dates, dates for PAZ boundaries, a peat accumulation rate curve, macrofossil records, and stratigraphy are shown (see (**A**)). Intensive farming phases are shown in (**B**). Dark/light brown fill in curves for bog/heath taxa and NPPs (in (**C**)) is used to convey the idea of dry/wet conditions on the bog surface (also in Figure 8C$_1$).

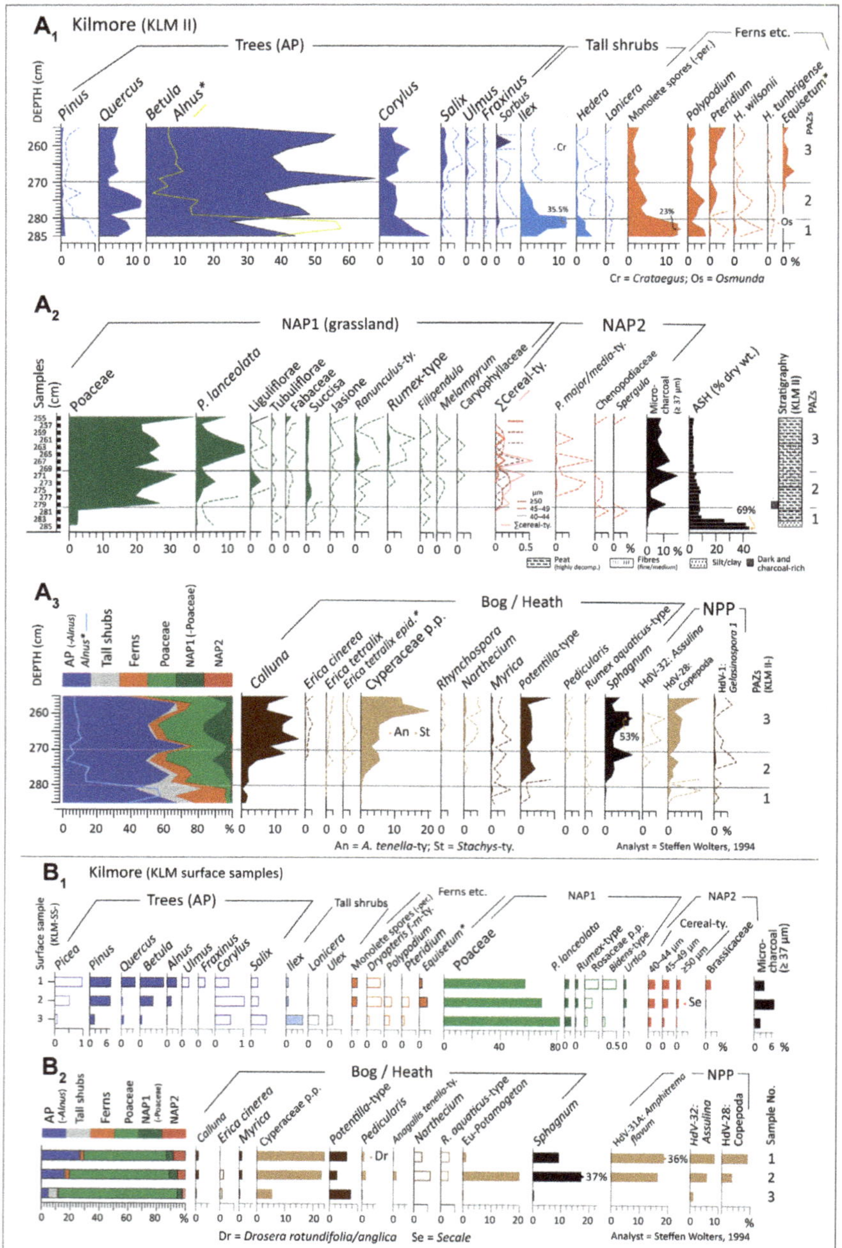

Figure 6. Percentage pollen diagrams, profile KLM II ($\mathbf{A_1}$–$\mathbf{A_3}$), and surface samples ($\mathbf{B_1}$,$\mathbf{B_2}$). The conventions followed are similar to those in Figure 5. In ($\mathbf{B_1}$,$\mathbf{B_2}$), the x-axis is magnified ×10 in the case of non-filled histograms; Poaceae is scaled ×0.5 in ($\mathbf{B_1}$). A taxon name with an asterisk indicates that the taxon in question is outside the PS and/or a PS is used that differs from that used in neighbouring taxa (see Section 2).

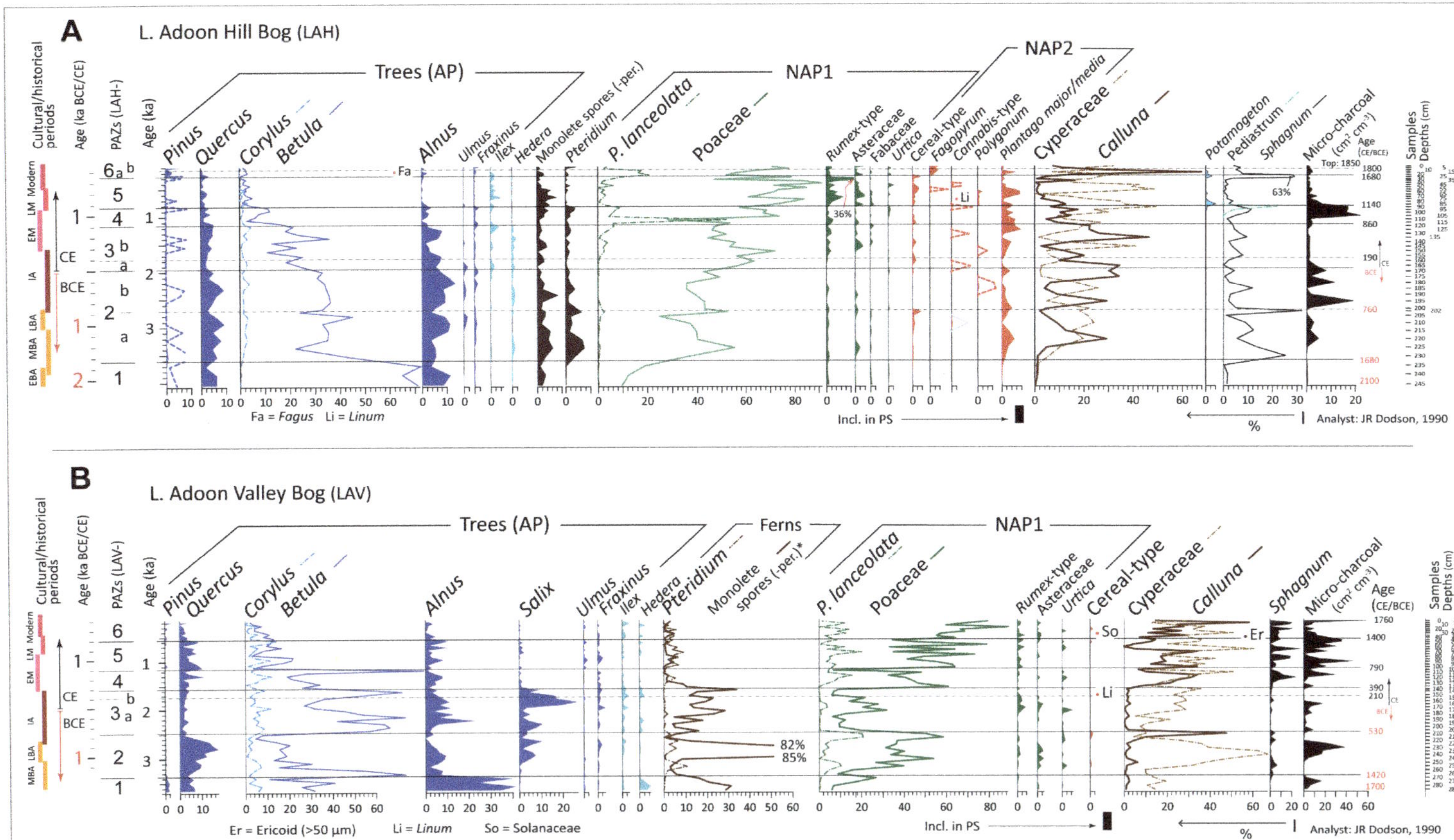

Figure 7. Percentage pollen profiles from blanket-bog peat in the L. Adoon area drawn to time scales (after [22]). The conventions followed are similar to those used in the KLM profiles. (**A**). Peat profile LAH from the eastern side of L. Adoon, 600 m SE of LAV. (**B**). Peat profile LAV from east of the Scorid river, near the Ballyhoneen wedge tomb. In (**B**) Monolete spores (-per.), i.e., Monolete fern spores without perine, an asterisk is added to the name to indicate that the taxon is outside the usual PS; its PS is TTP + Monolete spores (-per.) (see Section 2).

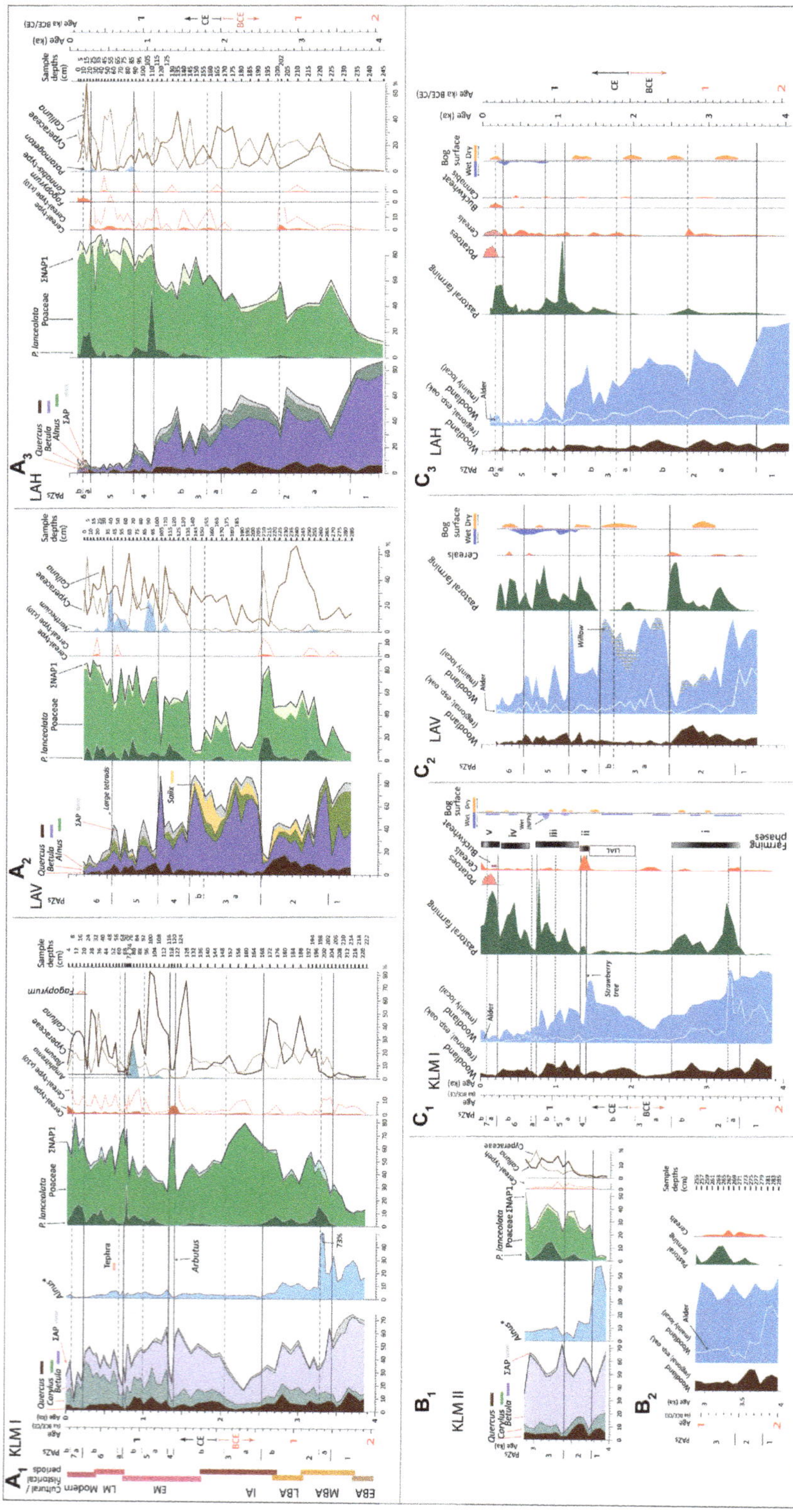

Figure 8. (**A$_{1-3}$**,**B$_1$**) Summary percentage peat pollen profiles from Kilmore/L. Adoon area drawn to time scales. (**B$_2$**,**C$_{1-3}$**) Schematic reconstruction of vegetation, land use, and local bog-surface wetness/dryness based on pollen profiles KLM I and II, LAV, and LAH. *Alnus* with an asterisk (in **A$_1$**,**B$_1$**) indicates this taxon is outside the usual PS; its PS is TTP + *Alnus* (see Section 2).

3.4. Pollen Profiles KLM I and KLM II: Woodland Dynamics and Human Impact

As regards the pollen source area for profiles KLM I and II, at the base of the profiles, it is undoubtedly mainly local. As peat accumulates and extends laterally, the source of the non-bog pollen takes on a more regional character, which, at least regarding farmland, relates to the lowlands that lie mainly to the north. Given the predominantly south-westerly and often strong winds, the vegetation of the uplands to the south and south-west, and especially that on the lower slopes near the sampling sites, can also be expected to be strongly expressed in the pollen records.

The Poaceae curve is assumed to include mainly pollen of *M. caerulea*, a species of wet grasslands and also blanket bog. *P. lanceolata*, on the other hand, is a species that is more or less confined to grassland and fallow and mineral ground generally [66], and so, in conjunction with other NAP such as Asteraceae (this includes Tubuliflorae (daisy, etc.) and Liguliflorae (dandelion, hawkweed, etc.)), provides a reliable indicator as to the origin of the Poaceae pollen. High Poaceae pollen representation, when *P. lanceolata*, etc., are poorly represented, can be regarded as mainly reflecting the dominance of *M. caerulea* at/near the sampling site, i.e., bog/heath vegetation rather than grasslands on mineral soils.

Profile KLM I, which spans from the Early Bronze Age (1900 BCE) to the near present time (mid-twentieth century), indicates that the local woodlands during most of this almost 4000-year-long period consisted mainly of birch, hazel, and alder (Figure 5; also Figure $8A_1$,C_1). Profile KLM II (Figures 6 and $8B_1$), which began somewhat earlier, probably did not extend beyond the Bronze Age (ca. 1000 BCE) (see Section 3.2). Both KLM I and KLM II indicate the local importance of woody vegetation as peat accumulated in the study area. As KLM I opened, tree cover and human impact appeared to be broadly comparable to that at KLM II, although alder was not as dominant, and the increase in human activity adversely affected birch more than alder. Both profiles suggest that pine is rare. However, a small pine stump, presumably from the lowermost peat in the general vicinity of KLM I (see Section 2), indicates that pine trees survived as peat accumulation commenced locally.

At the base of KLM II (Zone 1 and much of Zone 2), holly and oak, together with alder and birch, made important contributions. *Ilex* achieved 35.5% in Zone 1 but declined in Zone 2; *Quercus* achieved 12.5% in Zone 2, where *Alnus* declined and *Betula* expanded. The changing contribution of trees and tall shrubs at KLM II (and KLM I) in the context of local peat initiation may be the result of a relatively short period with low levels of human impact. The reduction in farming activity was probably the facilitator, rather than the result, of peat initiation and woodland regeneration. The subsequent increase in human activity (substantial rise in *P. lanceolata* in both KLM I and II; Figures 5B and $6A_1$) is probably largely responsible for the decline in alder and other associated changes. Records for filmy ferns, and especially *H. wilsonii* and *H. tunbrigense* in profile KLM II and a record for *T. speciosum*, as well as a curve for *H. wilsonii* and several records for *H. tunbrigense* in the basal part of KLM I, are noteworthy (Figures 5A and $6A_1$).

Regarding tree cover in the Iron Age and subsequently (from about 600 BCE; starting in zone KLM I-3), the main changes are as follows (Figures 5 and $8A_1$,C_2). In subzone KLM I-3a (600–90 BCE), AP values are very low, but this is probably attributable more to exceptionally high Poaceae values depressing percentage AP values rather than an actual decrease in AP pollen input; this finds some support in the concentration and influx pollen diagrams (Figure S1). Low values for *P. lanceolata* and other NAP taxa indicative of human impact support the conclusion that Poaceae pollen is largely of local (bog) origin. Subzone 3b to the top of Zone 5 (90 BCE to 1210 CE), with the exception of Zone 4 which is of short duration (see below), is characterised by elevated AP, particularly *Betula*, but *Quercus* and *Corylus* are also well represented. The role of human impact and climate change on vegetation development during this period, which spans the late Iron Age to the late medieval period, is considered below (Section 3.6).

Beginning at about 1380 CE, there was another shift that involved increased *Corylus* representation but substantially lower *Betula* and *Quercus* values. Profile KLM I ends with the lowest AP values overall in subzone 7a (1720–1870 CE), followed by an increase, especially in *Betula*, as the profile ends (subzone 7b; 1870–1950 CE). These changes in AP were probably mediated by changing levels of human impact resulting from particularly large demographic shifts (see Section 4.3).

3.5. First Fossil Records for *Arbutus* and Filmy Ferns from Dingle Peninsula

Of particular note are the records for *Arbutus* (almost certainly *A. unedo*) in sample 122 cm, profile KLM I (1.4%; nine *Arbutus* tetrads recorded), which dates to 540 CE (Figure 5A; Table S8). In this part of the profile, every second centimetre was sampled, and thus, a high time-resolution record is available. As each sample integrates about 20 years of pollen production, it is likely that there was sufficient time between the sample at 124 cm and the following sample at 122 cm for *Arbutus* to establish locally and produce the pollen recorded here (in cultivation *A. unedo* starts to flower when eight years old; under natural conditions at about 25 y [67]). Given that its pollen dispersal is poor [68,69], our records indicate the local presence of the strawberry tree. The precise circumstances under which this species grew and flowered at Kilmore are of particular interest. It should be borne in mind that its range seems to be largely static in Ireland since scientific recording of flora commenced [10]. In addition, *A. unedo* was once much more frequent in the Kerry/West Cork region [62,70], so past presence on Dingle peninsula, without any human intervention, cannot be ruled out. Until now, however, the evidence for presence has been limited to placenames, e.g., Árd na Caithne [71,72] (also rendered Ardnacaithne; in English 'high ground of the strawberry tree'; today usually referred to as Smerwick) near the tip of the peninsula (Figure 1A), but it is by no means certain that 'Caithne' in that placename translates to *Arbutus* [73].

Climate change may have been involved in the spread of *A. unedo*, given that there is evidence of climate fluctuations at this time [74,75]. Immediately preceding the *Arbutus* records in KLM I, *Betula* achieved exceptionally high pollen representation, which coincided with the local expansion of birch on the bog surface (Figures 5A and 8A_1,C_1). This development was preceded by a dry phase indicated by strong *Calluna* representation and a sharp decline in bog taxa, such as Cyperaceae and *Narthecium*, which are indicative of wet conditions (Figures 5C and 8A_1,C_1). The dry phase probably triggered colonisation of the bog surface by birch (for the phenomenon of birch colonising mainly shallow blanket-bog peat, see [76]; for recent examples of a birch on Irish bogs, see [77]). Although the evidence for climate change is rather strong, we favour introduction by people as the preferred explanation, especially given the static range of *Arbutus* in recent times (see above). Our fossil records coincide with an upsurge in human activity that is a feature of the early medieval (Christian) period in Ireland generally, and specifically on Dingle peninsula [6,27,78], including Kilmore. In profile KLM I, the *Arbutus* records coincide with the start of exceptionally high values for cereal-type pollen that signal the beginning of a short period (it lasted not more than about three human generations) centred on the seventh century CE (PAZ KLM I-4; farming Phase II; Figures 5 and 8C_1). During this period, arable farming assumed particular importance locally. Woody vegetation declined, presumably due to sustained human impact, which likely caused the local extinction of *A. unedo*. For recent further discussion of *A. unedo*, and specifically native versus archaeophyte status in Ireland and its biogeographic affinities, see [20,72,79].

The filmy fern record in KLM I is noteworthy. *H. wilsonii* has a continuous curve at the base of the profile (KLM I-1, 2a and the lower part of 2b; ca. 1900–1250 BCE; Figure 5A). It is also well represented at the top of KLM I-3b and in the lower spectra of KLM I-5a (early medieval period). In the upper part of the profile, there are occasional records. *H. tunbrigense* is not as strongly represented. It was recorded in only 1.5% of the spectra, mainly as single spores in the basal spectra. This contrasts with *H. wilsonii*, which was recorded in 34.3% of the spectra and attained a maximum of 6%, which suggests that it was locally common. Although poorly represented, *H. tunbrigense* was almost certainly present locally, especially in the early part of the record (prior to 1500 BCE). The only record of *T. speciosum* was a single spore in the basal spectrum. Local presence is possible, indeed likely, given its recent distribution in the Cork/Kerry region (Figure S6C,D)—but long-distance spore transport is also a possibility.

3.6. Farming Phases and First Fossil Record for Buckwheat Cultivation in Ireland

As regards human impact and farming, particularly intensive phases are identified (Figures 5 and 8).

Phase i. This corresponds mainly to zone KLM I-2 and spans at least 700 years, ending at ca. 600 BCE. This involved both pastoral and arable farming (Figures 5B and $8C_1$). In zone KLM I-1, cereal-type pollen is also well represented, so arable activity seems to have featured throughout the mid-and late Bronze Age. The impact on local woodland cover was substantial (high *Pteridium* values indicate increased openness), with alder and birch adversely affected. The filmy fern record is interrupted at the transition KLM I-2a/2b, which also points to a decline in woodland in the general vicinity of the coring location.

Phase ii. The main activity during this short-lived phase ($\leq$100 y; zone KLM I-4) is centred on 600 CE. It involved mainly arable farming (cereal growing). Maximum cereal-type pollen values for the profile as a whole are recorded. *P. lanceolata* is scarce, so the peak in Poaceae is probably best attributed to *M. caerulea*. The sharp decline in AP (mainly *Betula*) and the large increase in Poaceae suggest that woodlands were strongly and adversely affected. The first records of *Secale* pollen relate to this phase. However, only occasional grains of *Secale* have been recorded in the profile. Rye was not an important crop; indeed, it may have had only an accidental presence in other cereal crops (most likely barley, and perhaps oats and some wheat [24]; see Section 1).

Phase iii. This phase (ca. 750–1200 CE) broadly corresponds to KLM I-5. In the early years, there was little cereal cultivation. As farming intensified (cf. high *P. lanceolata*; also much cereal-type pollen indicative of a mixed farming economy), woody vegetation was impacted, including oak, as the phase ended (upper part of subzone 5b). At this point, *P. lanceolata* registered 30%, and *Rumex*-type also peaked. This indicates that typical grassland species, and especially ribwort plantain, were present in abundance but not locally, as presumably, the bog was, by now, extensive (at the sampling site, peat had achieved a thickness of almost 1.5 m, and peat accumulation was well initiated where Wall 7 was excavated).

Phase iv. This phase (ca. 1300–1700 CE) broadly corresponds with KLM I-6b and involved both pastoral and arable farming. Mid-fourteenth-century developments, such as the Black Death, adverse climate, and political upheavals (cf. [80]), are not obvious from the data. Pastoral farming was most important (cf. *P. lanceolata*) towards the end of the period, when oak (remnants of oak woodlands had probably survived locally up until then) was severely affected. Hazel, however, flourished, which suggested that farming, though important, was not so intensive in the region as to result in complete clearance of woody vegetation. A farming economy that relied on semi-open wood pastures rather than completely open pastures, as in recent times, is also a distinct possibility (cf. [81,82]).

Phase v. This phase corresponds mainly with KLM I-7a, where *P. lanceolata* averages 13.6%. This is the most intensive farming phase and spans ca. 1820 to 1950 CE. Potatoes, which do not appear in the pollen record [83], rather than cereals, were the main crops at this time. A detailed consideration of developments during this period is provided in Section 4.3.

Of particular interest are the fossil pollen records for *Fagopyrum* (presumed to be *F. esculentum*, i.e., buckwheat, the species that is usually sown as a crop in Europe; Figure S8B). Pollen was recorded in two spectra (depths 20 and 16 cm). The age/depth model suggests that these spectra date to ca. 1750 and 1800 CE, respectively.

Buckwheat has been a common crop in much of central Europe since the twelfth century CE [84]. In northern Germany, it was often grown in tandem with rye as the main crop in areas with low-fertility soils. Not surprisingly, many of the pollen diagrams reproduced in Feeser et al. [85] have a substantial *Fagopyrum* curve that frequently begins in the late medieval period, peaks in the seventeenth and eighteenth centuries, and continues into the early twentieth century [86,87]. On extensive bogs in the north German lowlands, buckwheat cultivation became associated with *Moorbrandkultur*, i.e., bog drainage, followed by firing of the surface vegetation and sowing buckwheat in the ash that resulted from the firing [88]. As regards Britain and Ireland, Pescott and Akeroyd in BSBI plantatlas2020.org (accessed on 12 March 2025) suggest that buckwheat was "a significant grain crop in Britain and parts of Ireland until the 19th century". However, Greig [89] is of the opinion that although fossil *Fagopyrum* pollen has been recorded at British sites, buckwheat was never a popular food crop in Britain. Data presented by Dark [90] also convey the impression that it was an infrequent crop in Britain during the medieval period. However, the Oxford English Dictionary (OED) has entries for buckwheat dating back to the sixteenth century [91]. As far as Ireland is concerned, there seems to be no published pollen evidence to date, although there is further palynological evidence for buckwheat cultivation from the study area (see Section 4.1.1) that supports the idea that it was at least trialled as a crop in Kilmore before the Great Famine.

Kelly [92], in his comprehensive account of early Irish farming that is based mainly on Old Irish law texts, regards it as unlikely that 'rúadán' (=the red one) refers to buckwheat, given that in Europe, buckwheat is a food-plant that post-dates the Irish law texts. Furthermore, Kelly ([92], p. 225) is of the opinion that it "has never been a regular crop in Britain or Ireland". Lucas [93] does not mention it in his account of Irish food, which also relies on early historical sources, and it is not referenced in a recently published tome on the history of Irish food [94].

From the above literature, it seems that buckwheat has never been cultivated widely in Ireland. However, more intensive searches—not made any easier on account of the various names by which it has been referred to; e.g., bockwheat, beechwheat, and French wheat [95]—indicate that it was rather widely known and used as a foodstuff. For example, Robert Payne (reprint edited by A. Smith [96], p. 10), advising on the feeding of rabbits (connies) in 1589, recommends to prospective English settlers in Ireland "you may fatte them with graines mixed with oates, brane or French wheate [i.e., buckwheat]". Many decades later, with reference to the Confederate Wars in Ireland, military stores, including "french wheat, rye, and wheat" etc., were "certified as delivered at Cork, Dublin and Carrickfergus" in September 1642 [97], p. 31. However, Professor Wade, according to the Dictionary of Irish Biography (DIB), "a noted Irish botanist", states in his 1814 memoir to the Bog Commissioners ([98], p. 24) that buckwheat "Is a crop not at all known in Ireland; but by good farmers in England is considered as very valuable for feeding horses, fattening hogs, and keeping poultry". On the other hand, Professor Mackey, in his landmark *Flora Hibernica* ([99], p. 224), gives the following entry for *Fagopyrum*/Buckwheat:

"Cultivated ground, but introduced by cultivation, being often sown as food for pheasants and other poultry". Given the contradictory statements by Wade and Mackey, it appears that although buckwheat as a crop was known in early nineteenth-century Ireland, it was not widely cultivated or appreciated as a food for human consumption (floristic records from Ireland are few and relate mainly to recent years; Figure S8A).

Mitchell ([100] p. 172; also in earlier and later editions of *Reading the Irish Landscape*) refers to an unnamed German traveller to Ireland in 1828 who recorded that buckwheat was frequently seen as a crop along with oats and potatoes. Based on this and Old Irish law texts, Mitchell assumed that buckwheat was grown as a crop in nineteenth-century Ireland.

We have identified the German traveller referred to by Mitchell [100] as the well-known German landscape designer Hermann von Pückler-Muskau, who travelled widely in Ireland between August and December 1828, and shortly afterwards published an extensive account in *Briefe eines Verstorbenen* in four volumes. Volume 1 contains Letter 31, in which Pückler-Muskau described his stay at Bermingham House, near Tuam, Co. Galway, in September 1828 [101]. He was particularly struck by the deep bogs (raised bogs) that were reclaimed and extensively farmed ("gedeiht die Bruchwirthschaft") and on which buckwheat—he refers to it as "Haidekorn", i.e., an old version of the more common German word 'Heidekorn', literally 'heath corn' [101], p. 218—and potatoes and oats were grown. These accounts—in the original German and also in translation—were very popular at the time and, indeed, subsequently. Recently, they have been republished in English in a large-format, illustrated, and annotated volume by Parshall [102], while the letters relating to Ireland have been translated in part and published with commentaries by Bourke [103]. In the latter publication, the passage relating to buckwheat in Letter 31 was omitted.

As regards buckwheat at Kilmore, there is the possibility that it was introduced to Dingle by the Palatines from the Palatinate/Pfalz region in south-western Germany. These farmers were settled in Co. Limerick in the early eighteenth century and, later that century, moved into north Kerry [104–106]. However, there is no mention of buckwheat in the accounts of these industrious German farmers. Arthur Young, the 'improving' English landlord who travelled extensively in Ireland in the 1770s, visited Newcastle West, Co. Limerick, to report specifically on the Palatine farmers. He does not mention buckwheat, although, in his accounts, he frequently reports on arable farming, including the increasing popularity of the potato as a crop [106]. Interestingly, however, he refers specifically to the Maharees, as "famous for their corn products... famous for the best wheat in Kerry. All under the plough... they have known two crops of barley gained from the same land in one year and the second better than the first" ([106], vol. II, pp. 127, 128). The area referred to is on the north coast of the Dingle peninsula, about 10 km north-east of Kilmore (Figure 1A). This serves to emphasise the major changes in land use that have taken place during the last two centuries, as a largely arable-based economy has been almost completely replaced by pastoral-based farming. It also provides context for the exceptionally high cereal-type pollen representation in KLM I, which is difficult to envisage from a modern perspective.

4. Discussion

4.1. Palaeoecological Records from Kilmore in the Context of Earlier Published Records

The geographically closest and most pertinent pollen profiles relating to the Kilmore area are those by Dodson [21,22] (LAV, LAH, and CLN) and Barnosky [23] (BL) (see Section 1; site locations in Figure 1A,B). To facilitate comparison with the KLM profiles, these profiles were replotted on time scales (Figures 7–9). Details of the profiles and the procedures followed in replotting the profiles, including age/depth models, are available in Text S1 and Figures S2–S5. The two peat profiles, LAH and LAV, from the L. Adoon area, are first considered and compared with the nearby KLM profiles from Kilmore.

4.1.1. Profile LAH (L. Adoon Hill Bog) (Figure 7A)

The pollen profile LAH is from near the north-eastern shore of L. Adoon, 620 m south-east of core LAV (see below), in an area rich in archaeology that relates mainly to later prehistory (Bronze Age and Iron Age) and also many pre-bog stone walls that have not been dated [19] (Figures 1B and S9B). In LAH, there are only occasional *Pinus* records until near the top of the profile (zone LAH-6), where *Pinus* reaches 5.7% (ca. 1850 CE or possibly younger). *Betula* dominated at the base of the profile (LAH-1; average: 70%). At the bottom of zone LAH-2, it fell to 43% and then averaged 33% over the rest of the zone. There were corresponding increases in Poaceae and bog/heath taxa, which suggests that local birch woodlands were largely replaced by bog/heath vegetation. Apart from a decline in *Quercus* and *Alnus* at the LAH-2a/2b boundary (there seems to have been local and regional woodland clearance at about 760 BCE, i.e., near the end of the Bronze Age, these AP taxa remained steady at about 6–10%. In LAH-2a, cereal growing was important locally, which is indicated by high cereal-type representation and exceptionally (indeed impossibly) high *Plantago major/media* values (averages 1.0% and 2.4%, respectively). The importance of grasslands is difficult to evaluate, as much of the Poaceae pollen may arise from bog/heath communities. However, the low *P. lanceolata* values suggest that bog/heath may be the main source of Poaceae pollen.

Zone LAH-3, which spanned most of the first millennium up to ca. 860 CE, shows an intensification of farming that starts gradually (first two centuries CE) and, based on the NAP taxa (especially *P. lanceolata*, *Rumex* type, and Asteraceae (includes Liguliflorae and Tubuliflorae pollen; these have comparable representation)), progressively increases. Substantial woodland clearance, probably at local and regional levels, did not occur until zone LAH-4, at the base of which *P. lanceolata* had an exceptional peak of 52% (ca. 900 CE), which suggests ribwort-plantain-rich vegetation close to the sampling site. Given that the Poaceae values are relatively low (28%), while indicators of arable farming are well represented (cf. cereal-type and *P. major/media*), the high *P. lanceolata* values are probably indicative of local arable activity rather than a grassland sward (cf. [66]). Micro-charcoal also attains the highest representation in zone LAH-4, which again supports the idea of much human activity locally and probably regionally.

Zone LAH-5 (1140–1680 CE) reflects a landscape that is more or less treeless, presumably due to further intensification of farming (exceptionally high values for both NAP1 and NAP2 taxa). Considerable arable and pastoral farming is taking place close to, or at least in, the general vicinity of the site. Cereal-type pollen, recorded in all spectra and averaging 1.5%, indicates substantial cultivation of cereals. Flax may have also been cultivated, at least within the region. A single *Linum* pollen grain was recorded at 75 cm, i.e., ca. 1260 CE. This probably arises from flax (*L. usitatissimum*), although the non-cultivated *L. bienne* as the pollen source cannot be excluded. *Cannabis*-type pollen was recorded in this zone (45 cm) and in earlier zones, which suggests hemp cultivation, although hop (*Humulus lupulus*), which produces a *Cannabis*-type pollen, cannot be excluded as the pollen source [107]. It is expected that hemp (*C. sativa*) was grown as a source of fibre rather than as an oil. Hemp was particularly important for the mercantile and navy ships in these times, and so its cultivation was actively encouraged, especially in the eighteenth and early nineteenth centuries ([108]; as regards south-western Ireland, see [109]).

Of particular interest is the single record of *Fagopyrum* pollen at 55 cm (ca. 1430 CE). The cultivation of buckwheat (*F. esculentum*; see also below) as a crop in central Europe had begun by this time [87], but it is doubtful whether it had reached Ireland or Britain. The isolated record in LAH-5 is therefore regarded as the result of long-distance pollen dispersal (see de Klerk et al. [110], who highlight records of *Fagopyrum* pollen from periods prior to widespread cultivation in Europe). Notably, there are peaks in *Pediastrum* (22%)

and Cyperaceae (49%) at the top of LAH-4 and considerable values for *Potamogeton* at the top and bottom of LAH-5. This indicates that the local bog surface was wet and probably waterlogged for most of the time.

Zone LAH-6 spans approximately 1700 to 1850 CE. The increase in *Pinus* at the base of the zone is assumed to reflect the beginning of pine plantations in the wider region. There are no records for cereal-type pollen, probably reflecting the increasing importance of potato cultivation (silent in the pollen record; see Section 3.6), especially on marginal lands at/close to the site. Of particular interest are the further records for *Fagopyrum* (recorded in the five spectra of Zone 6 and averaging 2.7%), which undoubtedly signals local cultivation of buckwheat. This supports the evidence for local buckwheat cultivation, as provided by profile KLM I (see Section 3.4).

Subzone 6b (uppermost two spectra) is distinct in that *P. lanceolata* declines from an average of 18.3% in subzone 6a (three spectra) to 3%, while Poaceae increases substantially and *Pteridium* peaks in the uppermost spectrum. These changes probably reflect the decline/abandonment of farming, especially in marginal areas, which followed the Great Famine (1845–1852 CE) in many parts of western Ireland, including Co. Kerry [17,65].

4.1.2. Profile LAV (L. Adoon Valley Bog) (Figure 7B)

This blanket-bog profile is from about 500 m east of KLM I and 600 m north of LAH (Figures 1B and S9B). Apart from the lowermost spectra (zone LAV-1), where *Pinus* approached 2%, there are only occasional *Pinus* records. Woody remains are plentiful in the lower part of LAV, so it is not surprising that AP values are high (average 60% in zones LAV-1 to 4), with the main contributors being *Alnus* at the base, *Betula* (especially LAV-3 and 4), and *Salix* (LAV-3, upper part). These trees/shrubs were probably present at or near the sampling site, i.e., growing on mineral soils and peat. AP values are distinctly lower in the upper part of both LAV and LAH, reflecting the completely open, more or less treeless landscape that characterised much of medieval and subsequent times, i.e., from about 1000 CE onwards. This feature is not as pronounced in KLM I, which suggests that complete openness may not have taken place until the early eighteenth century. *P. lanceolata* is well represented in profile LAV, apart from zones 1 and 3, where *P. lanceolata* averages only 0.7% compared with 7% in the other zones. This suggests that pastoral farming has been emphasised over long periods. In contrast, occasional cereal-type pollen suggests little or no arable farming in the vicinity of the site.

The records (six) of large ericoid tetrads in sample LAV 40 cm are noteworthy (Figure 7B). The sources of these pollen grains could be *A. unedo* and *Rhododendron ponticum*, both of which produce exceptionally large tetrads. If the age ascribed to the sample is correct (ca. 1420 CE), then *R. ponticum* is unlikely given that the shrub was first introduced to Britain in 1763, and probably several decades later to Killarney where, by the early nineteenth century, it was of importance in at least some parts of the Killarney woods [63,111]. Attributing the records to *A. unedo* is not justified based on the size of the tetrad alone. Furthermore, the *A. unedo* records in profile KLM I are much older (ca. 540 CE) (Figure 5). Given the demise of woody vegetation and relatively intensive farming, the local survival of *Arbutus* over the long interval between these records is unlikely. Therefore, the source of these large ericoid tetrads remains unresolved.

4.1.3. General Comparison of Peat Profiles from Kilmore and L. Adoon

The three peat-derived profiles from Kilmore (KLM I) and L. Adoon areas (LAV and LAH) enable fine-spatial reconstruction of vegetation and land-use change over approximately the last four millennia, to which the time-limited information provided by profile KLM II can be added.

To better facilitate the comparison of these four blanket-bog-derived pollen profiles, summary profiles are plotted on a time scale (Figure $8A_{1–3}$,B_1). To further simplify the data and facilitate the reconstruction of past vegetation and land-use cover, as well as local bog wetness at each site, curves are also shown that aim to give a general impression of vegetation and land-use changes, as well as local bog-surface wetness at the four sites (Figure $8B_2$,$C_{1–3}$). Curves showing the contribution of tall woody vegetation are presented under two categories: (a) regional woodland (mainly oak), which is based largely on *Quercus* values, and (b) local tall shrub/woodland, which corresponds mainly to AP excluding *Quercus* (oak). Birch (*Betula*) is the largest contributor to this latter category, but alder (*Alnus*) may also be important; therefore, the estimate of its contribution is shown by a curve within the composite AP curve. Willow (*Salix*) makes a substantial contribution to parts of profile LAV, as indicated schematically on that profile. Hazel (*Corylus*) is of some importance only in profile KLM I and so it is shown in Figure $8A_1$ (see composite curves on the left-hand side of pollen profile KLM I). The estimation of pastoral farming is based primarily on *P. lanceolata* rather than Poaceae values, the latter being regarded as largely indicative of grasses growing locally on bogs and other wet habitats. Cultivation of cereals, buckwheat, and hemp relies on relevant curves, that is, cereal-type, *Fagopyrum*, and *Cannabis*-type pollen. Potato cultivation relies on historical and statistical data. The reconstruction of local bog-surface wetness/dryness is based mainly on Cyperaceae, *Narthecium*, and *Calluna* as indicators of wet and dry conditions, respectively. In the case of profile KLM I, for which NPP data are available, *Assulina*, *Amphitrema flavum*, and Copepoda are regarded as indicators of local wet conditions [43]. The wet/dry signal is not always clear-cut, in large measure probably due to the small-scale variation in wetness/dryness that is frequent on blanket bog surfaces [112].

The main points arising from the comparison of the four blanket bog pollen profiles, as summarised in Figure 8, are as follows. Blanket bog initiation at the four sites was broadly contemporaneous, having started at about 2000 BCE. Peat accumulation rates in the three profiles for which there is independent evidence of age (KLM I, LAH, and LAV) are also broadly similar ($\sim$0.6 mm y^{-1}).

In the early years (until ca. 1500 BCE), the main tall canopy species was oak, presumably *Q. petraea*, which is the dominant species of Atlantic oak woodlands in the west Cork/Kerry region [113]. Birch (*B. pubescens*) is important and dominates locally at all sites. However, alder (*A. glutinosa*) is important at KLM I, KLM II (it dominates here), and LAV. As peat accumulation is initiated, human impact, as indicated by NAP (especially *P. lanceolata*), is largely imperceptible, which suggests that peat initiation is not directly connected with human activity; land abandonment is a more likely cause (cf. [114]). Other factors that were probably also at play include climate (lower temperature/increased precipitation) and pedogenesis involving natural and/or human-induced soil acidification.

Between about 1500 and 600 BCE, i.e., the mid-Bronze Age to the early Iron Age, varying but generally high levels of pastoral farming were recorded at KLM I (referred to as farming Phase i; Figures 5B and $8A_1$,C_1) and also at LAV (Figures 7B and $8C_2$). At KLM I, cereal growing appears to have been important early in the period (1500–1350 BCE). At KLM II, the situation is similar, but this profile is short and probably does not extend to the end of the interval under consideration (Figure $8B_{1,2}$). Pastoral activity is also recorded at LAH, but it is much weaker (Figures 7A and $8C_3$). The LAH profile at this time may merely reflect farming activity at a distance, e.g., near KLM I and LAV. During this time, peat surfaces at all sites supported heathy/blanket bog vegetation, with much ling (*Calluna*) and sedges (Cyperaceae). Overall, peat surfaces appeared to be dry based on the abundance of *Calluna* pollen.

From about 600 BCE to 500 CE at KLM I, pastoral farming was greatly reduced (Figure $8A_1$,C_1). Arable farming persists but at a reduced level. During the interval, starting shortly before the BCE/CE transition and continuing to about 500 CE, woody vegetation expanded, especially oak and birch (corresponding to the LIAL). It is at the end of this interval that *Arbutus* pollen is recorded, coinciding with the beginning of a sharp increase in arable farming (top of KLM I-3b; ca. 540 CE) and substantial woodland clearance, at least at a local level (KLM I-4; ca. 560–630 CE). In LAV, the record during this time is comparable (zone LAV-3), but scrub vegetation is much more important (as well as birch, willow is important, the high pollen representation suggesting that willow grew on the rather dry bog surface). A clearance feature corresponding to that in KLM I-4 was not recorded, so this appears to be of local significance only. In profile LAH (cf. LAH-2b), the contribution of woody species (especially oak and birch) remains steady until about the BCE/CE transition, when AP values decline. At this point, Poaceae expanded, and cereal-type pollen was recorded, but there was very little *P. lanceolata*. Overall, there are different vegetation dynamics at LAH compared with KLM I and LAV, and human activity at LAH seems to be lower during this time.

From about 750 CE onwards, pastoral farming became important once again at KLM I. The increasing trend continued until about 1200 CE. Although birch declines in importance, the role of oak is maintained or indeed increases until the end of the period (ca. 1100 CE) when exceptionally high *P. lanceolata* values indicate ribwort-plantain-rich grasslands in the vicinity of KLM I. Where precisely these grasslands were located is not easily envisaged. The ^{14}C dates from the pre-bog stone-wall excavation (Figure 9; Table S7) suggest that peat initiation began at this location as early as the end of the second century CE, i.e., in the late Iron Age, perhaps after the abandonment of the stone-wall field system. Therefore, it seems that stone-wall construction took place no later than the mid/late Iron Age, although earlier construction (Bronze Age) is more likely. This may have been connected with the rather intensive farming recorded at the transition KLM I-2a/b, i.e., ca. 1300 BCE (intensive farming was also recorded in KLM II and LAV).

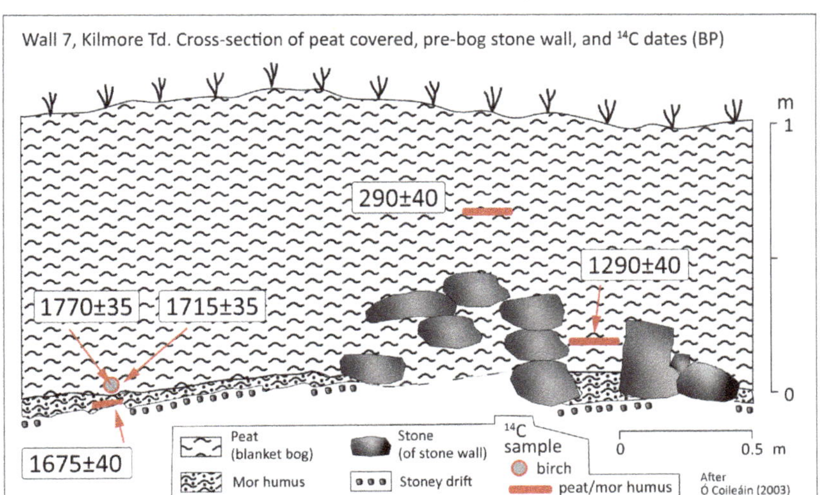

Figure 9. Diagrammatic cross-section of the excavated pre-bog stone wall (Wall 7), including ^{14}C dates (for details of the ^{14}C dates see Table S7). The sketch is after Ó Coileáin [19].

Regarding vegetation and land use, there was a step change in the upper part of all three profiles. This is characterised by a substantial decline in AP, including *Quercus*, and greatly elevated NAP1 and NAP2. These changes signal the creation of the open, largely treeless landscape of today in the context of substantial pastoral and arable farming practices. In KLM I, this is referred to as farming Phase v, which spans ca. 1720–1950 CE. It corresponds to zone KLM I-7, which is subdivided into 7a and 7b, with the boundary at ca. Beginning at ca. 1870 CE, the post-Great Famine population decline, bad weather, and general economic malaise resulted in rural decline and a switch away from arable farming (cereals and potatoes) in favour of cattle and sheep [65,115].

The correlation of the upper parts of profiles KLM I, LAV, and LAH, which is based largely on palynology, is not straightforward (Figure 8). However, the differences in the pollen assemblages appear to reflect small-scale and, hence, real differences in vegetation and land use rather than being an artefact of a weak chronology. The top of LAH (zone 6) has features similar to KLM I and is assumed to span more or less the same period, although probably not extending as far into recent times (the twentieth century may not be reached). Woody vegetation is scarce from ca. 900 CE (base of LAH-4), and particularly after 1100 CE (base of LAH-5). The LAV record is broadly similar in that pronounced woodland demise (birch rather than oak) started at 800 CE (base of LAV-5), but in this profile, intensive farming accompanied by birch clearance began earlier (base of LAV-4; ca. 400 CE). The overall evidence points to the late first millennium CE rather than later periods, such as after the Elizabethan plantation of Munster in the late sixteenth century [116], as a critical juncture in the creation of a landscape largely devoid of tall woody vegetation. In Ireland, the available historical and palynological evidence suggests considerable variation in the timing of final woodland clearances, with dates extending from the mid-/later first millennium CE to the early eighteenth century ([83,114,117–119]; see also Section 4.2.2).

The records of *Fagopyrum* pollen are particularly noteworthy. Fagopyrum was recorded in two spectra in KLM I-7a (dated to 1750 and 1800 CE, respectively) and in LAH, where it averaged 2.7% over five consecutive spectra at the top of the profile (LAH-6; 1700–1850 CE). *Fagopyrum* is not recorded in LAV, which is not surprising given that this profile, according to its age/depth model, ceased in the mid-eighteenth century. There can be no doubt that buckwheat was cultivated in this part of the Dingle peninsula in the eighteenth and probably into the nineteenth centuries. To the best of our knowledge, these are the only fossil records of buckwheat cultivation in Ireland to date.

The evidence for the wetness of the bog surface at the coring sites is also of interest in terms of possible climate change. Substantially increased bog-surface wetness was recorded in LAV, starting at about 900 CE and continuing until almost 1500 CE (Figure 8C_2). The record at LAH is rather similar (Figure 8C_3). At KLM I, wet conditions also began at about 900 CE but were only weakly expressed during the second millennium CE (Figure 8C_1). Given the high levels of human impact, the hydrology at this site may be affected by human activity, which is likely to have counteracted any tendencies towards wet bog surfaces mediated by precipitation/evaporation and/or temperature changes.

4.2. *Long-Term Holocene Environmental Change*

4.2.1. Pollen Profile LCN from Lough Camclaun [21]

Profile LCN, from the southern end of L. Camclaun (at 235 m asl; ~1.6 km to the south of KLM I), is regarded as reflecting mainly the Coumanara uplands, which include Binn na Naomh, which rises to 670 m asl at 560 m to the south-west of the site (Figure 1A,B). The profile spans most of the Holocene and is regarded as extending into the twentieth century (Figures 10A and S4). There are five ^{14}C dates available, but these have large errors ($\geq$110 y), and four of the dates relate to the lower part of the core. The age/depth model is based on a smooth spline curve (Figure S4) that uses three of the available ^{14}C dates and 0 BP assigned to the top of the profile (this is based on the *Pinus* curve, which fails to rise substantially as might be expected if the profile extended into the later twentieth century). While a more tightly constrained age/depth model is desirable, the curve is regarded as acceptable for the present purposes.

The selected percentage curves are plotted on a time scale in Figure 10A. Zonation is similar to that by Dodson [21], but here, subzones are used, so the divisions are finer. At the base of the profile (zone LCN-1), *Betula* is dominant, and the *Corylus* curve has been initiated. This is followed by the rise of *Pinus*, then *Quercus*, and later *Ulmus*, which seldom exceeded 2% in the profile. The sequence in the spread of tall trees, however, is not clear, which contrasts with the BL profile in which hazel dominates from an early stage, then oak and elm, and finally pine (see below), which may be regarded as more typical for south-western Ireland [20].

During the first half of the Holocene (zones LCN-2 to 4; ca. 10.4–5.4 ka), the local woodlands were dominated by pine and oak, and hazel was important until ca. zones LCN-3/4 transition (ca. 7.4 ka), after which alder began to expand. *Pinus* exhibits elevated values at ca. 8.4 ka, but this is probably as much an artefact of percentage calculations (monolete spores derived from ferns decline substantially at about this time) as the effect of a climate oscillation such as the 8.2 ka event [120]. Elevated *Pinus* values (highest in the profile) at and shortly after 6 ka are often seen in Irish pollen diagrams and probably reflect pine growing on expanding blanket bogs and, ultimately, climate change [20,121]. The classical Elm Decline follows (zones LCN-4/5 transition); however, as usual, where *Ulmus* values are low, the Elm Decline is poorly expressed.

The upper part of the profile (zones LCN-5 to 7) is characterised by a steep and inexorable decline in AP that starts in subzone LCN-5b and is largely complete at the base of zone LCN-6 (AP declines from 37% to 28% over the interval 4.2–3.8 ka, approximately). At about 3.8 ka (LCN-5/6 transition), there is a distinct shift in the limnic environment, probably more dystrophic as a result of the input of humic acids from the blanket bog, which is well initiated by this time, as reflected by distinct declines in *Isoetes* and *Pediastrum*.

Elevated *P. lanceolata*, Poaceae, and *Pteridium* values point to a substantial opening up of the landscape, presumably as a result of increased farming in the region relating to an upsurge in activity during Late Neolithic/Early Bronze Age times. The changes in the *Pinus* and *Quercus* curves are particularly noteworthy. At its highest, i.e., near the top of zone LCN-4, *Pinus* is at 39%; in LCN-5a, it averages 12% and finally drops to ~3% near the base of LCN-6. A small peak of 18% in LCN-5a dates to ca. 4.8 ka, probably reflecting the expansion of pine onto bog surfaces, i.e., the so-called 'pine flush' [114]. Expansion of bog first clearly manifests itself in LCN-5a (3450–2400 BCE; increase in *Calluna* and *Sphagnum*), during which *P. lanceolata* is consistently recorded. Human impact, as well as climate change (wetter and cooler, resulting in bogs too wet to support pine), were undoubtedly factors that favoured the expansion of bog and heath, starting in the Chalcolithic and continuing apace during the Bronze and Iron Ages.

Interestingly, while *Pinus* values declined, *Quercus* and *Betula* expanded, which suggests a distinct shift in woodland composition in favour of oak and birch at the expense of pine. By about 2.4 ka, pine was probably extinct locally and regionally, and other woody species, including oak, were scarce. *Quercus* values recovered somewhat in the interval 1.6–0.4 ka, i.e., in medieval times, which suggests a decline in human impact in this relatively remote upland region. This is supported by rather low *P. lanceolata* and cereal-type values; however, it should be noted that Poaceae (probably arising mainly from bogs), *Calluna*, and Cyperaceae are strongly represented (and included in the PS) so that the percentage values of all other taxa are artificially depressed.

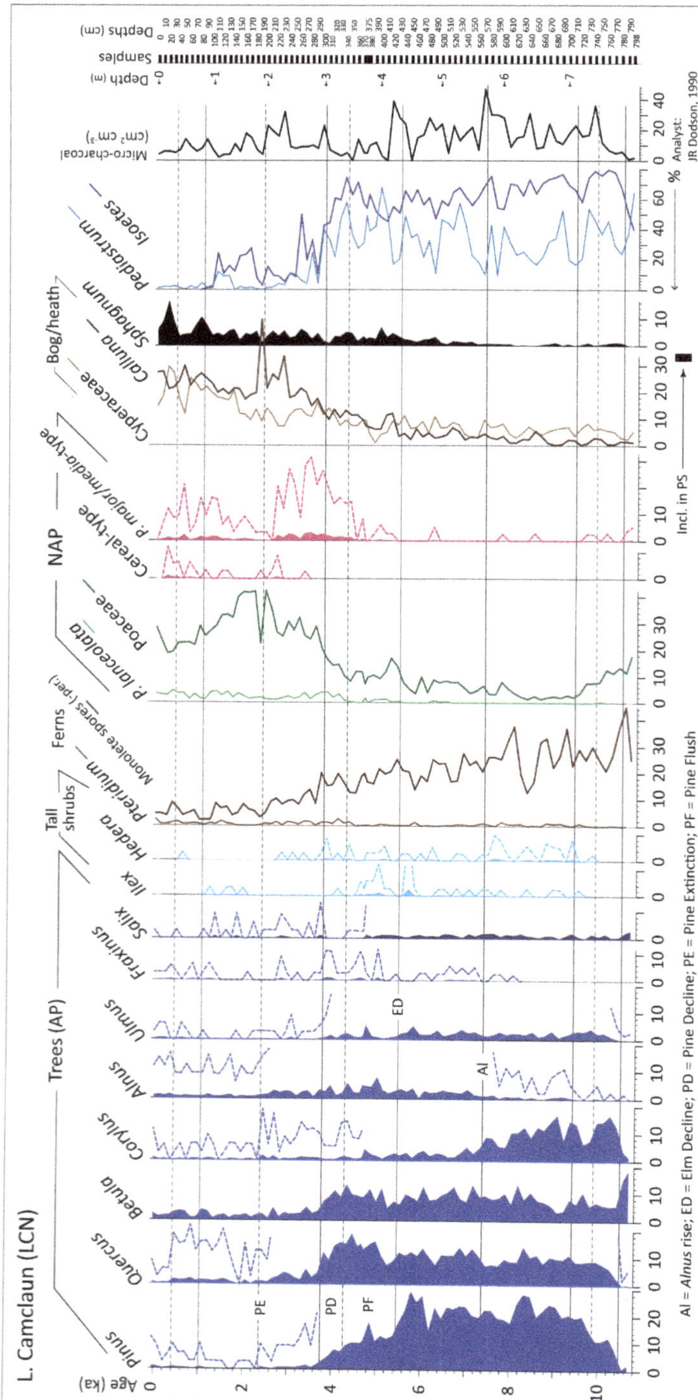

Figure 10. *Cont.*

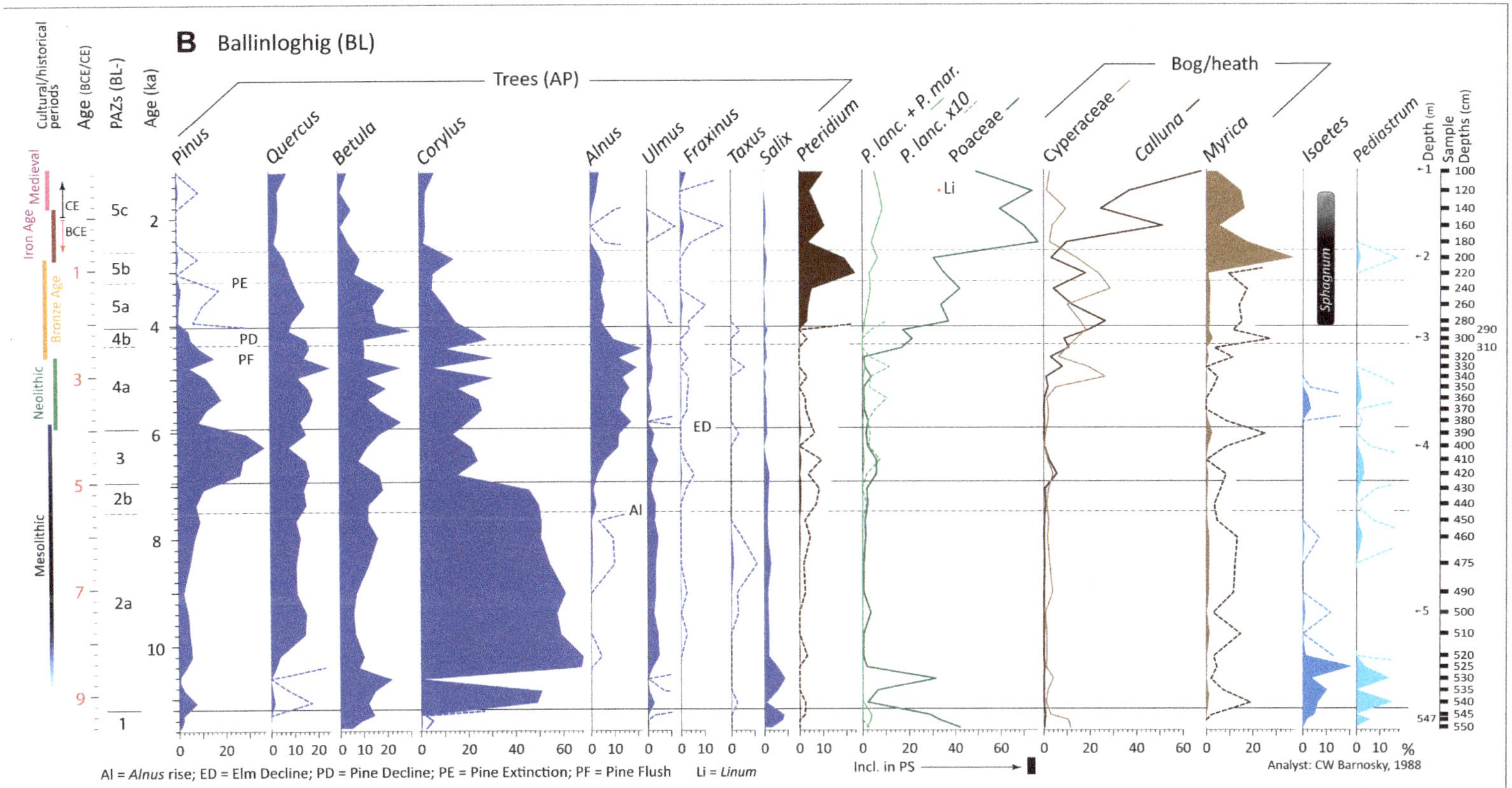

Figure 10. Percentage Holocene lake pollen profiles from Dingle peninsula plotted on a times cale. The conventions followed are similar to those used in the KLM profiles. (**A**) Profile LCN from L. Camclaun is after Dodson [21]. (**B**) Profile BL from Ballinloghig is after Barnosky [23]. In profile BL, bog taxa are excluded from the PS because of their dominance towards the top of the profile. Their dominance and also high *Sphagnum* values (shown schematically) are indicative of infilling and bog development at the coring location.

Regarding human impact, the evidence for early farming impact (Neolithic) is weak. In the mid/later Neolithic at ca. 5 ka, *P. lanceolata* values increased for the first time. *Pediastrum* values also increased substantially, which suggests increased mineral input to the lake, but charcoal values were low, indicating that fire was unimportant. There are consistently elevated charcoal values coinciding with or following the Elm Decline (ca. 5.4 ka) and also earlier (at ca. 7.4 ka (Boreal/Atlantic transition)), though it is unlikely that these latter are the result of human activity, given that population levels were probably low. In the Late Holocene, charcoal values are elevated between ca. 2.8 and 2.4 ka, i.e., in the late Bronze Age and early Iron Age; however, given the low *P. lanceolata* values, human activity may not be responsible.

The first major farming impact (pastoral) is registered at ca. 4 ka, which coincides with the Chalcolithic/Early Bronze Age transition. Cereal-type pollen is first consistently recorded (but at low values) in the early Iron Age (ca. 2.7 ka). The high *P. major/media* values (especially in subzones 5b and 6a; Chalcolithic/Early Bronze Age) are unexpected as this taxon of disturbed/trodden ground is usually associated with arable farming. The significance of this finding is unclear. In early medieval times (ca. 1.6–1.2 ka i.e., 350–750 CE), *Quercus* values are somewhat elevated (approaching 2%), but *P. lanceolata* and NAP2 taxa (indicative of arable farming) are well represented. This points to a farming economy, perhaps wood pasture, in which oak, the main surviving tall-canopy tree but now rather scarce, was selectively favoured.

PAZs LCN-6 and 7 correspond to, on palynological grounds (cf. especially *Pinus*), the time span represented in the peat profiles from the Kilmore and L. Adoon areas (see above). Human impact first manifested itself clearly in LCN-5b (substantial rise in *P. lanceolata*). At about 1800 BCE (LCN-6), i.e., at the same time or slightly later than the start of peat accumulation at KLM I and II, LAH, and LAV, AP values decline sharply, and NAP1 increases. These changes signal woodland clearance and the expansion of bog and grassland as a result of human impact, and probably a wetter and cooler climate. Interestingly, there is no hint in CLN of the important role that birch and alder continued to play elsewhere (sites KLM I, LAH, and LAV; all < 2 km apart). This serves to confirm that LCN reflects vegetation and land use at a broad regional scale (cf. [122]).

Changes at the transition LCN-6a/b (400 BCE) are rather small but significant in that they reflect the regional extinction of pine and greatly reduced oak and other tall woody vegetation. After this, changes continue to be small, although oak recovers somewhat, probably largely because of the regional character of the profile, smoothing of the curves due to mixing in the lake-sediment column, and the fact that the main land-use changes are taking place mostly at lower elevations on the relatively fertile lands to the north of LCN.

4.2.2. Holocene Pollen Profile from Ballinloghig (BL) [23]

The lake-sediment profile BL is from about 10 km west of the area of primary interest (Kilmore/L. Adoon), but the setting is very different in that the lake lies in the lowlands to the west of the Brandon mountain range and is hence exposed to the persistently strong Atlantic south-westerlies (the Atlantic is <5 km to the west) (Figure 1B). However, the site is largely surrounded by a moraine and is therefore sheltered from the prevailing winds. The 6.4-m-long core, which is from the western margin of the lake, includes more than a metre of Lateglacial sediment that is not considered here (see [23]).

Profile BL spans most of the Holocene (sediment from the top 100 cm was not analysed; the record probably stops prior to the end of the medieval period) (Figure 10B; Text S1). For most of the early and mid-Holocene (zone BL-2; ca. 11–7 ka; chronological control is weak here and in the profile as a whole; see Figure S5), *Corylus* dominates (at about 50–68%; mainly at ∼60%), *Quercus* is at ∼10% (after ca. 10 ka), *Pinus* has low values (often < 5%), and *Betula* has moderate values (generally 10%). *Ulmus* seldom exceeds 4%, and *Fraxinus* and *Taxus* are only sporadically recorded.

In the early Holocene, there was an anomalous pollen spectrum with no *Quercus* or *Pinus* and only 1.1% *Corylus*. The age/depth model suggests that the spectrum dates to ca. 10.6 ka, so it is very unlikely that the 8.2 ka climatic oscillation is represented. This view is supported by the behaviour of the pollen curves, especially *Pinus*, which is not typical of the 8.2 ka event [120].

The dominance of hazel and the relatively minor role played by pine contrast sharply with the record from L. Camclaun. A variety of factors are probably at play, especially the more montane character at L. Camclaun, differences in bedrock and drift geology giving more fertile soils at Ballinloghig, and local climatic conditions including greater exposure to Atlantic gales at Ballinloghig and less seasonal temperature variation on account of its distinctly maritime setting.

In the upper part of the profile (zones BL-3 to 5; 7 ka onwards), considerable changes are recorded. In the interval 7–6 ka (PAZ BL-3), *Pinus* achieved >30%, while *Corylus* drops to ~30%, from which it does not recover. A slow but steady rise in *Alnus* to ~12% also occurs during BL-3. Records for *P. lanceolata* (see exaggerated curve for this taxon in Figure 10B), as well as an increase in Poaceae, suggest a disturbance. Research over recent decades has shown that *P. lanceolata* records are a feature of the late Atlantic period (pre-Elm Decline, i.e., around 4 ka and earlier) in Irish Holocene pollen profiles, especially in profiles from western Ireland [20,123]. The adjustments in the contributions of pine, hazel, and alder, as well as the indications of the opening up of the woodland cover (increases in NAP and *Pteridium*) during BL-3, are probably the result of climate oscillations rather than human impact (Mesolithic presence at Ferriter's Cove, <10 km to the west, relates to the end of the Mesolithic, i.e., ca. 6.5 ka [9,124]).

The classical Elm Decline (here dating to ca. 5.8 ka), at the base of BL-4, is poorly defined, as is usual where *Ulmus* values are low. There is no evidence of early Neolithic Landnam, as is frequently recorded in Irish pollen diagrams [76]. Major changes, however, are recorded in BL-4b (4.4–4 ka), including a decline in *Pinus* to very low values. A rise in Poaceae at the end of BL-4a, as well as increases in Cyperaceae and *Calluna*, continue into BL-4b. This probably reflects lake infilling and wider landscape changes that involved the expansion of bog and heath. Given that there is little evidence for human impact up to this point, it is assumed that natural processes involving increasing soil acidification and increased wetness are the main drivers of these changes.

Zone BL-5 (4–1.2 ka) record deforestation and human impact. At the beginning of Zone 5, the first substantial increase in human impact is registered (cf. increase in *P. lanceolata* also Poaceae and *Pteridium*, and decline in *Corylus*). *Pinus* values are so low that local/regional presence of pine (*P. sylvestris*) cannot be assumed. However, pine probably persisted in the locality/region until the top of subzone 5a (ca. 3 ka). Final clearances followed (subzone 5b/c boundary; ca. 2.6 ka) that gave rise to the largely treeless present-day landscape. Given the weak chronology, especially in this part of the profile, the ages suggested by the age/depth model (Figure S5) for this part of the profile should, at best, be regarded as rough approximations. The overall pattern, however, is quite similar to that at L. Camclaun, but the treeless character of the landscape may have come about earlier at Ballinloghig, i.e., at about the Bronze Age/Iron Age transition (BL-5b/5c transition).

4.3. *Historical/Statistical Records with Reference to Recent Farming and Comparison with the Pollen Evidence*

Apart from historical records, including accounts by travellers [125] (see also Section 3.6), statistics relating to agriculture and population, systematically collected by central authorities since 1841, are an important source of information on population changes and the rural economy in Ireland. The collection of detailed records from all parts of the island commenced in 1847 [126–128]. Statistics were collected annually for the first 70 years and subsequently with less regularity and sometimes with considerable interruptions. Annual reports have also been published, providing a detailed breakdown of the statistics and general commentary, as well as useful overviews (e.g., [115,127]).

Rather than attempting to assess the detailed statistics that are available, the data compiled by the Central Statistics Office (CSO) for the publication 'Farming since the Famine' [127], and specifically those data relating to Co. Kerry, are now summarised. Statistics from smaller administrative units (e.g., barony and townland) are clearly more desirable for making comparisons with the pollen data, but these statistics have yet to be compiled; therefore, the data at the county level (Co. Kerry) are used and plotted in Figure 11. They are regarded as broadly indicative of the main trends in farming and population applicable in the study area, but it should be noted that the Dingle peninsula suffered greater population decline during the Great Famine and for decades subsequently compared with Co. Kerry as a whole [129].

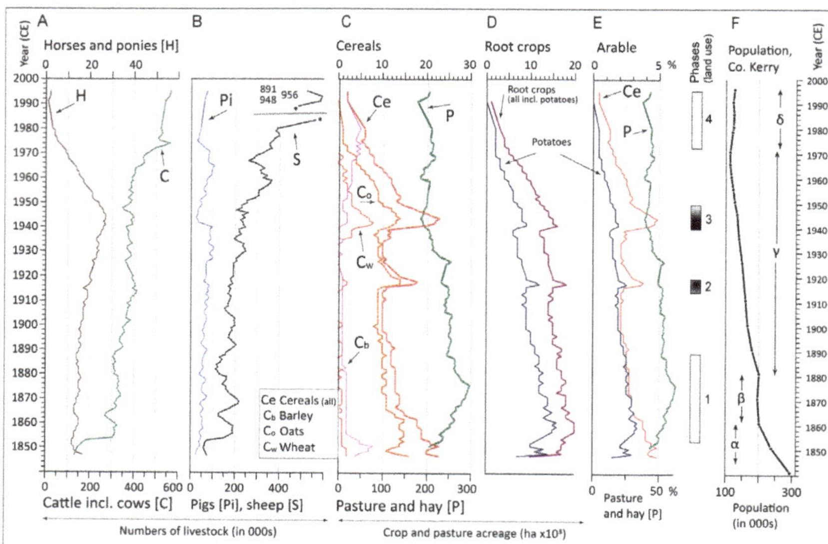

Figure 11. Agricultural and population statistics for Co. Kerry. The data in (**A–D**) are plotted as numbers or as acreage, as indicated on the plots. Annual records are available with some gaps. The main gaps relate to 1981–1990 and 1921–1924. Statistics are available for the latter period, but reliability is not guaranteed due to the unsettled political situation (source: CSO). In (**B**), sheep numbers for 1991, 1993, and 1995, which are exceptionally high, are plotted (numbers are also given) but to an x-axis that has a break with respect to the basal x-axis. Root crops mainly consist of turnips; potatoes are included in this curve and are also plotted separately (see (**D**)). In (**E**), acreage is expressed as a percentage of the land surface of Co. Kerry (4737 km^2). In this plot, cereals and potatoes are plotted separately, as well as pasture and hay combined, which reflect pastoral farming (this excludes rough grazing). The phases of agricultural activity (land use) are indicated, and infill is used to highlight periods with increased arable activity. In (**F**), population statistics are shown, and phases of population growth/decline are distinguished (indicated by Greek letters; details available in the text; source CSO, as given in Anon 2022 [65]).

Regarding the population, the main trends are as follows. The early decades of the nineteenth century leading to the Great Famine saw a rapid expansion of the population that was largely dependent on the potato as a food source. The maximum population at the base of plot F in Figure 11 relates to 1841; the population probably peaked a few years later before the severe famine conditions that commenced in 1845. After that, the demographic trends can be summarised in four phases. Phase α (1845–1860): a period of rapid decline connected with and following the Great Famine. This was caused by death from hunger and disease, and outwards migration. Phase β (1860–1880): the population stabilised for about two decades from the adverse effects of the Great Famine. Phase γ (1880–1970): a steady decline in population, mainly due to outwards migration. Phase δ (1970 and later): a gradual increase in the population, attributable to economic recovery connected with joining the European Economic Community (EEC; subsequently the EU). The population increase was largely confined to the main towns, such as Tralee and Killarney, while the rural population, especially those directly involved in farming, continued to decline [65].

The main trends in the rural economy are as follows: After the disaster of the Great Famine (which began in 1845, was the most severe in 1846 and 1847, but continued in Dingle for several further years [129]), agriculture stabilised over the course of about three decades (Phase 1, ca. 1853–1890), during which arable farming declined, while pasture (including meadow) increased and peaked at ca. 1876. Thus, the already dominant role of pastoral farming relative to arable farming was further emphasised (Figure 11C–E).

During Phase 2 (1914–1920) and especially Phase 3 (1940–1945), acreage under cereals increased due to factors associated with World War I and II (Figure 11C). Despite the government's emphasis on wheat growing during and immediately after World War II, oats, the traditional crop of western Ireland [92,105], continued to be the main cereal. At the end of the 1940s, arable farming, which was already in decline, rapidly fell to below 2% of the land area (Figure 11E), and horse and pony populations declined (Figure 11A) in response to changes in the farming economy and increased use of cars and tractors. At/before Ireland joined the EU in 1973 (beginning of Phase 4), a substantial increase in the bovine herd occurred. In the 1980s, there was a spectacular increase in sheep numbers (more than doubling; Figure 11B) as a result of incentives provided by the EU Common Agricultural Policy (CAP).

Only profiles KLM I and LAH extend into the period represented by the population and agricultural statistics but at a much lower temporal resolution. The intensive land use that followed the population explosion at the end of the eighteenth and early nineteenth centuries is well reflected in the substantial increase in *P. lanceolata* and NAP1 generally. Arable farming also increased with potatoes achieving great importance (*Solanum tuberosum* is silent in the pollen record; potato cultivation is shown in Figure 8C_1,C_3 based on historical records). The subsequent decline in farming and some resurgence of woodlands during the late nineteenth and early twentieth centuries are expressed at the top of both KLM I and LAH. Surface-pollen spectra derived from moss polsters clearly show a secondary rise in *Pinus* (reflecting pine plantations) and the planting of spruce (*Picea*) (Figure 6B_1). *Fagopyrum* pollen records in KLM I and LAH (Figures 5B, 7A and 8) indicate the cultivation of buckwheat in the Kilmore/L. Adoon area during the late eighteenth and possibly early nineteenth centuries. Whether it was cultivated as early as the fifteenth century is doubtful, given the scant evidence (a single pollen grain in LAH; Figures 7A and 8A_3).

5. Conclusions

Detailed palynological investigations in Kilmore Townland, Dingle peninsula, at a site with extensive pre-bog stone walls and other archaeological features, facilitated the reconstruction of vegetation history and land use during the last 4000 years. As the records opened (profiles KLM I and II), tall-canopy trees were rare and consisted mainly of oak. However, other trees and tall shrubs (mainly alder and birch) are important locally.

The main profile, KLM I, presents a dynamic picture of arboreal (AP) and non-arboreal pollen (NAP). The main factor determining the direction and magnitude of changes in woody vegetation and biodiversity is farming activity. This activity is reflected in the varying but generally high values for *P. lanceolata* (ribwort plantain) and cereal-type pollen.

Five distinct phases of active farming have been recognised (Figures 5B and 8C_1). Phase i spans 1300–600 BCE (mid/late Bronze Age, extending possibly into the Iron Age). This is followed by a long lull that includes a period of reduced activity, referred to as the Late Iron Age Lull, which spans ca. 90 BCE–550 CE. Phase ii is of short duration and is characterised mainly by arable farming, centred on ca. 600 CE and lasting for about a century. Phase iii is a period (ca. 750–1200 CE) of ever-increasing farming, mainly pastoral-based, but with cereal cultivation gaining importance towards the end. Farming is most intense at ca. 1200 CE, which coincided with early Norman times, i.e., the early years of the late medieval period. Phase iv bridges the late medieval and early modern periods (ca. 1300–1700 CE). Intensive pastoral and arable farming further reduces the already scarce woody vegetation, including oak. Phase v (ca. 1720 onwards), the most intensive farming period (especially the early part), coincides with the population explosion of the late eighteenth/early nineteenth century that was interrupted by the Great Famine (1845–1852), a catastrophic development that initiated a decline in population and farming that persisted well into the twentieth century.

Excavation by Ó Coileáin [18] of a pre-bog stone wall at Kilmore indicates that peat initiation at the excavation site began in the early first millennium CE (Figure 9). Palaeoecological investigations (profiles KLM I and KLM II; also LAH and LAV by Dodson [21,22]), however, show that peat initiation elsewhere in the study area was much earlier, i.e., at about the beginning of the Bronze Age or shortly thereafter (ca. 4 ka). As regards the extensive stone-wall network, the available evidence suggests construction in prehistory, i.e., prior to ca. 400 CE. As far as could be ascertained, wall construction predates peat initiation. Wall construction during farming phase i, and early in that phase (cf. KLM I-2a; KLM II-3; ca. 3.8 ka (Early Bronze Age)), is suggested as a distinct possibility based on the intensity of farming activity at that time, as revealed by the KLM profiles and profile LAV.

Additional noteworthy features of profile KLM I include a distinct tephra layer (most likely attributable to the Öræfajökull eruption of 1362 CE), and pollen records for the strawberry tree (*Arbutus*; 540 CE) and buckwheat (*Fagopyrum*; late eighteenth/early nineteenth centuries CE). The *Arbutus* records are the first independently dated Irish records from outside the main present-day distribution of the strawberry tree in Ireland, i.e., Killarney, the south-west Iveragh peninsula, and the Beara peninsula. *Fagopyrum*, which has seldom been regarded as a crop or food source in Ireland, is shown to have a long history on the island (from the late 1500s onwards), and its cultivation is demonstrated palynologically for the first time (fossil pollen was recorded by Dodson, but the records were not included in his publications [21,22]).

Records from the KLM profiles for filmy fern spores provide new insights into the history of this interesting Eu-Atlantic component of the European flora (see also [20,109]). Criteria are provided for distinguishing the spores of the three species recorded in Ireland, i.e., *Hymenophyllum tunbrigense*, *H. wilsonii*, and *Trichomanes speciosum*.

A lake-sediment pollen profile, LCN from L. Camclaun, by Dodson [21] provides information on Holocene woodland dynamics in the upland environment near the main study area. During most of the early Holocene, pine and oak were dominant. In the mid-Holocene (5.6 ka), pine began an inexorable decline, with the so-called 'pine decline' recorded at ca. 4 ka. After this, pine and woody vegetation generally had a greatly reduced role in the landscape. Pine likely became extinct in the region shortly before 2 ka.

A lake-sediment pollen profile, BL from Ballinloghig, by Barnosky [23] from the western lowlands abutting Mount Brandon suggests rather different woodland dynamics in the Atlantic-exposed end of the Dingle peninsula. Oak and hazel are much more important during the early Holocene. Pine, however, expands at the expense of hazel at about 7 ka, when alder also begins to expand. As at L. Camclaun, a distinct 'pine decline' is recorded at about 4 ka. Woody vegetation (especially oak and birch) seems to have persisted in greater quantities for a longer time at this far-western site (to ca. 2.6 ka). Occasional *Taxus* pollen is recorded in BL (until ca. 4 ka), but there are no records for *Taxus* in LCN or in the peat profiles from the Kilmore/L. Adoon areas. This suggests that yew never achieved any importance in these parts of Dingle peninsula, which contrasts with Killarney [68,130] and also with the Beara and Mizen peninsulas [20], where yew achieved moderate to substantial importance, especially in the mid-Holocene.

Supplementary Materials: The following supporting information, in a zip file, can be downloaded at https://www.mdpi.com/article/10.3390/d17070456/s1. Figure S1. Pollen profile KLM I. Pollen concentration (A) and influx (B) plotted to a depth axis. Figure S2. Age/depth curve, pollen profile LAH (L. Adoon Hill Bog). Figure S3. Age/depth curve, pollen profile LAV (L. Adoon Valley Bog). Figure S4. Age/depth curve, pollen profile LCN (L. Camclaun). Figure S5. Age/depth curve, pollen profile BL (Ballinloghig). Figure S6. Distribution maps of filmy ferns. Figure S7. Distribution maps of *Arbutus unedo*. Figure S8. Distribution map of *Fagopyrum esculentum* and plant illustration. Figure S9. Maps showing archaeological sites and other features of Dingle peninsula. Figure S10. Photo: Panoramic view to the north towards Cloghane/Brandon Bay. Figure S11. Photo: view to the north towards Cloghane/Brandon Bay (close-up of Figure S10). Figure S12. Photo: peat bank at KLM I. Figure S13. Photo: removal of monolith KLM I. Figure S14. Photo: view of Wall 8 in the 'Central Complex'. Figure S15. Photo: excavated Wall 7. Table S1. Spore characteristics of the filmy ferns, *Hymenophyllum tunbrigense*, *H. wilsonii*, and *Trichomanes speciosum*. Table S2. Results of investigations by gouge coring along the bog transect, Kilmore Td. Table S3. Stratigraphy, monolith KLM I. Table S4. Stratigraphy, core KLM II (from 80 m on bog transect). Table S5. Details of relevés recorded in Kilmore Townland. Table S6. Results of ^{14}C dating, monolith KLM I. Table S7. ^{14}C dates from the excavated pre-bog stone wall in Kilmore townland. Table S8. Overview of the surface-pollen spectra and pollen profiles KLM I and KLM II. Text S1. Summary details of pollen profiles from Dingle peninsula by Barnosky (1988) and Dodson (1990a, b) [21–23]. References [131–133] are cited in Supplementary Materials.

Author Contributions: Fieldwork was carried out by M.O'C. and S.W. S.W. carried out the pollen analytical investigations and wrote up the results towards an MSc degree (by research) at NUI Galway (now the University of Galway). The present manuscript was initially drafted by M.O'C.; subsequently, S.W. made inputs to all parts of the manuscript. All authors have read and agreed to the published version of the manuscript.

Funding: S.W. received financial support for his research stay at the University of Galway from the Friedrich-Naumann-Stiftung für die Freiheit within the framework of its student scholarship programme and from the Deutsche Bundesstiftung Umwelt (DBU).

Data Availability Statement: Pollen data have been submitted to Neotoma and Pangaea. Other data are available in (SI).

Acknowledgments: Mícheál Ó Coileáin assisted with fieldwork, provided invaluable guidance on local archaeology based on his detailed field surveys and archaeological excavation at Kilmore, and commented on the draft manuscript. Britta Schumacher, Anke Schelski, and Roland Meinecke assisted with fieldwork. The National Botanic Gardens of Ireland provided fertile fronds of *Trichomanes speciosum*. Advice and help in the laboratory were provided by Karen Molloy, Chung Chang Huang, Walter Dörfler, and Eneda Jennings. Tephra investigations were carried out by Laura Toner. John Dodson and Cathy Whitlock made available their pollen data from Dingle peninsula that was generated during research stays in the Department of Botany, Trinity College Dublin. Two anonymous referees provided helpful comments and suggestions. We thank all the above and others for their assistance with our research.

Conflicts of Interest: The authors declare no conflicts of interest.

Abbreviations

The following are the main abbreviations that have been used.

In connection with pollen data and diagrams

KLM, etc.	Pollen profiles. Main profiles: KLM I and KLM II (Kilmore), LAH and LAV (L. Adoon Hill and L. Adoon Valley), LCN (L. Camclaun) and BL (Ballinloghig)
AP	Arboreal pollen
NAP	Non-arboreal pollen (with subdivisions NAP1 and NAP2, i.e., grassland and arable/disturbed habitat indicators, respectively)
NPP	Non-pollen palynomorph
PS	Pollen sum (used as the basis for percentage calculations)
In connection with maps	
OSI	Ordnance Survey of Ireland
DS	Discovery Series maps (1:50,000) by OSI
asl	Average sea level. This is equivalent to the Malin Head Datum that is used as datum by OSI and also generally in all recent maps of Ireland. It replaces Ordnance Datum (OD) that is 2.71 m lower
Td.	Townland. In most instances the smallest administrative division of land in Ireland, mainly of ancient and Gaelic origin, and still widely used. The larger division Barony usually includes several townlands. Nowadays, the main local administrative division is County (Co.)
Other abbreviations	
BCE, CE	Before the Common Era, Common Era; equivalent to BC and AD, respectively
BSBI	Botanical Society of Britain and Ireland
L.	Lough (=lake)
LIAL	Late Iron Age lull, i.e., a period with low levels of human impact/farming that results in woodland regeneration; it dates mainly to the early centuries of the first millennium CE (more or less coinciding with the Roman period in Britain). It is a feature in many Irish pollen diagrams, especially from western Ireland
PPL	Plane-polarised light (in connection with tephra detection)
VEI	Volcanic Explosivity Index

Chronology of cultural periods

(as recognised here and in Ireland generally; abbreviations are given in some instances)
Mesolithic ((9000)-8000–3800 BCE); Neolithic (3800–2400 BCE); Chalcolithic (Chal; 2400–2000 BCE); Early Bronze Age (EBA; 2400–2000 BCE); Middle Bronze Age (MBA; 1800–700 BCE); Late Bronze Age (LBA; 1100–700 BCE); Iron Age (IA; 700 BCE–400 CE); Late Iron Age (LIA; ca. 1–400 CE); Early Medieval (EM; 400–1169 CE); Late Medieval (LM; 1169–1550 CE); Modern (1550–present)

References

1. Higgs, K.; Williams, B. *Geology of the Dingle Peninsula. A Field Guide*; McConnell, B., Ed.; Geological Survey of Ireland: Dublin, Ireland, 2018; p. 246.
2. Bartlett, T. Knockavrogeen West megalithic tomb: A unique discovery on the Dingle peninsula. *J. Kerry Archaeol. Hist. Soc.* **2024**, *24*, 69–89.
3. Bric, M.J. (Ed.) *Kerry. History and Society. Interdisciplinary Essays on the History of an Irish County*; Geography Publications: Dublin, Ireland, 2020.
4. Coyne, F.; Collins, T. *Mount Brandon, Co. Kerry—Dingle's Holy Mountain. Heritage Guide No. 29*; Archaeology Ireland (Wordwell): Bray, Ireland, 2005.
5. Crowley, J.; Sheehan, J. *The Iveragh Peninsula: A Cultural Atlas of the Ring of Kerry*; Cork University Press: Cork, Ireland, 2009.

6. Cuppage, J. *The Archaeological Survey of the Dingle Peninsula*; Oidhreacht Chorca Dhuibhne: Ballyferriter, Ireland, 1986.
7. Harte, A.; Ó Carragáin, T. Land tenure and farming in early medieval Kerry: A survey of field systems in the Lough Currane basin. In *Kerry. History and Society. Interdisciplinary Essays on the History of an Irish County*; Bric, M.J., Nolan, W., Eds.; Geography Publications: Dublin, Ireland, 2020; pp. 25–58.
8. Horner, A. *Mapping South Kerry: 450 Years of Maps and a Changing Landscape*; Wordwell: Dublin, Ireland, 2023; p. 464.
9. Woodman, P.C.; Anderson, E.; Finlay, N. *Excavations at Ferriter's Cove, 1983–1995: Last Foragers, First Farmers in the Dingle Peninsula*; Wordwell: Bray, Ireland, 1999; p. 219.
10. Stroh, P.A.; Walker, K.J.; Humphrey, T.A.; Pescott, O.L.; Burkmar, R.J. *Plant Atlas 2020: Mapping Changes in the Distribution of the British and Irish Flora*; Princeton University Press/Botanical Society of Britain and Ireland (BSBI): Princeton, NJ, USA, 2023.
11. Viney, M. *Ireland. A Smithsonian Natural History*; Smithsonian Books: Washington, DC, USA, 2003; p. 336.
12. Webb, D.A. The flora of Ireland in its European context (The Boyle Medal Discourse, 1982). *J. Life Sci. (R. Dublin Soc.)* **1983**, *4*, 143–160.
13. Benham, S.E.; Houston Durrant, T.; Caudullo, G.; de Rigo, D. Taxus baccata in Europe: Distribution, habitat, usage and threats. In *European Atlas of Forest Tree Species*; San-Miguel-Ayanz, J., de Rigo, D., Caudullo, G., Eds.; Publications Office of the European Union: Luxembourg, 2016; p. e015921+.
14. Kelly, D.L. The native forest vegetation of Killarney, south-west Ireland: An ecological account. *J. Ecol.* **1981**, *69*, 437–472. [CrossRef]
15. Kelly, D.L.; Kirby, E.N. Irish native woodlands over limestone. *J. Life Sci. (R. Dublin Soc.)* **1982**, *3*, 181–198.
16. Aalen, F.H.A.; Whelan, K.; Stout, M. (Eds.) *Atlas of the Irish Rural Landscape, 1st ed*; Cork University Press: Cork, Ireland, 1997.
17. Aalen, F.H.A.; Whelan, K.; Stout, M. (Eds.) *Atlas of the Irish Rural Landscape*, 2nd ed.; Cork University Press: Cork, Ireland, 2011.
18. Ó Coileáin, M. The Archaeology of the Loch a'Duin Valley, Cloghane, Co. Kerry. Master's Thesis, Department of Archaeology, NUI Galway [University of Galway], Galway, Ireland, 2003; p. 287.
19. Ó Coileáin, M. The archaeology of the Loch a'Dúin valley. *Kerry Mag.* **2012**, *5*, 44–45.
20. O'Connell, M.; Overland, A. Holocene vegetation and flora dynamics of the west Cork/Kerry region, south-western Ireland. *Veget. Hist. Archaeobot.* **2025**, *34*, 1–28. [CrossRef]
21. Dodson, J.R. The Holocene vegetation of a prehistorically inhabited valley, Dingle peninsula, Co. Kerry. *Proc. R. Ir. Acad.* **1990**, *90B*, 151–174.
22. Dodson, J.R. Fine resolution pollen analysis of vegetation history in the Lough Adoon Valley, Co. Kerry, western Ireland. *Rev. Palaeobot. Palynol.* **1990**, *64*, 235–245. [CrossRef]
23. Barnosky, C.W. A late-glacial and post-glacial pollen record from Dingle Peninsula, County Kerry. *Proc. R. Ir. Acad.* **1988**, *88B*, 23–37.
24. OCarroll, E. Landscape dynamics surrounding An Bhánóg Theas, Lispole, Co. Kerry. *J. Ir. Archaeol.* **2021**, *29*, 183–193.
25. Long, P. First settlers in the Lee valley. In *In the Vale of Tralee. The Archaeology of the N22 Tralee Bypass*; Long, P., O'Keeffe, P., Bennett, I., Eds.; Wordwell Books: Dublin, Ireland, 2020; pp. 83–100.
26. Long, P.; O'Keeffe, P.; Bennett, I. (Eds.) *In the Vale of Tralee. The Archaeology of the N22 Tralee Bypass*; Wordwell Books: Dublin, Ireland, 2020; p. 211.
27. Hession, J. The Iron Age: Grain, ore and death. In *In the Vale of Tralee. The Archaeology of the N22 Tralee Bypass*; Long, P., O'Keeffe, P., Bennett, I., Eds.; Wordwell Books: Dublin, Ireland, 2020; pp. 131–144.
28. Timpany, S. People and their environment over 6,000 years. In *In the Vale of Tralee. The Archaeology of the N22 Tralee Bypass*; Long, P., O'Keeffe, P., Bennett, I., Eds.; Wordwell Books: Dublin, Ireland, 2020; pp. 69–82.
29. Hession, J. The Lee valley in early medieval times. In *In the Vale of Tralee. The Archaeology of the N22 Tralee Bypass*; Long, P., O'Keeffe, P., Bennett, I., Eds.; Wordwell Books: Dublin, Ireland, 2020; pp. 145–160.
30. O'Flaherty, E.; Baker, L. Creating a modern landscape, 1540–1900. In *In the Vale of Tralee. The Archaeology of the N22 Tralee Bypass*; Long, P., O'Keeffe, P., Bennett, I., Eds.; Wordwell Books: Dublin, Ireland, 2020; pp. 161–176.
31. Wolters, S. Pollen Analytical Investigations Towards the Reconstruction of Late Holocene Environmental Change in Kilmore Td., Corca Dhuibhne, Co. Kerry. Master's Thesis, Department of Botany, NUI Galway [University of Galway], Galway, Ireland, 1996.
32. Parnell, J.; Curtis, T. *Webb's an Irish Flora*, 8th ed.; Cork University Press: Cork, Ireland, 2012; p. 504.
33. Atherton, I.; Bosanquet, S.; Lawley, M. (Eds.) *Mosses and Liverworts of Britain and Ireland. A Field Guide*; British Bryological Society: Plymouth, UK, 2010; p. 848.
34. Waddell, J. *The Prehistoric Archaeology of Ireland*, 4th ed.; Wordwell: Bray, Ireland, 2022; p. 396.
35. Mitchell, G.F. Post-Boreal pollen-diagrams from Irish raised-bogs. *Proc. R. Ir. Acad.* **1956**, *57B*, 185–251.
36. Mohr, R. Untersuchungen zur nacheiszeitlichen Vegetations- und Moorentwicklung im nordwestlichen Niedersachsen mit besonderer Berücksichtigung von *Myrica gale* L. *VAG Vechtaer Arb. Geogr. Regionalwiss.* **1990**, *12*, 1–144.
37. Moore, P.D.; Webb, J.A.; Collinson, M.E. *An Illustrated Guide to Pollen Analysis*, 2nd ed.; Blackwell Scientific: London, UK, 1991; p. 216.

38. Fægri, K.; Iversen, J. *Textbook of Pollen Analysis*, 4th ed.; Fægri, K., Kaland, P.E., Kryzwinski, K., Eds.; Wiley: Chichester, UK, 1989; p. 328.
39. Beug, H.-J. *Leitfaden der Pollenbestimmung für Mitteleuropa und Angrenzende Gebiete. Lieferung 1*; Gustav Fischer: Stuttgart, Germany, 1961; p. 63.
40. Beug, H.-J. *Leitfaden der Pollenbestimmung für Mitteleuropa und Angrenzende Gebiete*, 2nd ed.; Friedrich Pfeil: München, Germany, 2015; p. 542.
41. Foss, P.J.; Doyle, G.J. A palynological study of the Irish Ericaceae and *Empetrum*. *Pollen Spores* **1988**, *30*, 151–178.
42. Grospietsch, T. *Wechseltierchen (Rhizopoden)*, 3rd ed.; Kosmos (Franckh'sche Verlagshandlung): Stuttgart, Germany, 1972; p. 87.
43. Van Geel, B. A palaeoecological study of Holocene peat bog sections in Germany and the Netherlands. *Rev. Palaeobot. Palynol.* **1978**, *25*, 1–120. [CrossRef]
44. Shumilovskikh, L.S.; Shumilovskikh, E.S.; Schlütz, F.; Van Geel, B. NPP-ID: Non-Pollen Palynomorph Image Database as a research and educational platform. *Veget. Hist. Archaeobot.* **2022**, *31*, 323–328, Erratum in *Veget. Hist. Archaeobot.* **2023**, *32*, 109. [CrossRef]
45. Grosse-Brauckmann, G. Über pflanzliche Makrofossilien mitteleuropäischer Torfe. I. Gewebereste krautiger Pflanzen und ihre Merkmale. *Telma* **1972**, *2*, 19–55.
46. Körber-Grohne, U. Bestimmungsschlüssel für subfossile *Juncus*-Samen und *Gramineen*-Früchte. *Prob. Küstenforsch.* **1964**, *7*, 1–47.
47. Smith, A.J.E. *The Moss Flora of Britain and Ireland*, 1st ed.; Cambridge University Press: Cambridge, UK, 1978; p. 706.
48. Bronk Ramsey, C. Bayesian analysis of radiocarbon dates. *Radiocarbon* **2009**, *51*, 337–360. [CrossRef]
49. Reimer, P.J.; Austin, W.E.N.; Bard, E.; Bayliss, A.; Blackwell, P.G.; Bronk Ramsey, C.; Butzin, M.; Cheng, H.; Edwards, R.L.; Friedrich, M.; et al. The IntCal20 Northern Hemisphere radiocarbon age calibration curve (0–55 cal kBP). *Radiocarbon* **2020**, *62*, 725–757. [CrossRef]
50. Blaauw, M. Methods and code for 'classical' age-modelling of radiocarbon sequences. *Quat. Geochronol.* **2010**, *5*, 512–518. [CrossRef]
51. Toner, L. Tephra Investigations of a Peat Monolith from Kilmore Td., Corca Dhuibhne, Co. Kerry. Bachelor's Thesis, Department of Botany, NUI Galway (University of Galway), Galway, Ireland, 2005.
52. Chambers, F.M.; Daniell, J.R.G.; Hunt, J.B.; Molloy, K.; O'Connell, M. Tephrostratigraphy of An Loch Mór, Inis Oírr, W. Ireland: Implications for Holocene tephrochronology on the northeast Atlantic region. *Holocene* **2004**, *14*, 703–720. [CrossRef]
53. Plunkett, G.; Pilcher, J.R. Defining the potential source region of volcanic ash in northwest Europe during the Mid- to Late Holocene. *Earth-Sci. Rev.* **2018**, *179*, 20–37. [CrossRef]
54. Plunkett, G.; Sigl, M.; McConnell, J.R.; Pilcher, J.R.; Chellman, N.J. The significance of volcanic ash in Greenland ice cores during the Common Era. *Quat. Sci. Rev.* **2023**, *301*, 107936. [CrossRef]
55. Hall, V.A.; Pilcher, J.R. Late-Quaternary Icelandic tephras in Ireland and Great Britain: Detection, characterization and usefulness. *Holocene* **2002**, *12*, 223–230.
56. Sharma, K.; Self, S.; Blake, S.; Thordarson, T.; Larsen, G. The AD 1362 Öræfajökull eruption, S.E. Iceland: Physical volcanology and volatile release. *J. Volcanol. Geotherm. Res.* **2008**, *178*, 719–739. [CrossRef]
57. Hall, S.C. *Ireland: Its Scenery, Character, &c*; How and Parsons: London, UK, 1841.
58. Hall, V.A.; Bunting, L. Tephra-dated pollen studies of Medieval landscapes in the north of Ireland. In *Gaelic Ireland c.1250–c.1650. Land, Lordship and Settlement*; Duffy, P.J., Edwards, D., FitzPatrick, E., Eds.; Four Courts Press: Dublin, Ireland, 2001; pp. 207–222.
59. Pilcher, J.R.; Hall, V.A.; McCormac, F.G. An outline tephrochronology for the Holocene of the north of Ireland. *J. Quat. Sci.* **1996**, *11*, 485–494. [CrossRef]
60. Neeson, E. *A History of Irish Forestry*; Lilliput Press: Dublin, Ireland, 1991; p. 388.
61. Knightly, J. The evolution of a Kerry demesne, 1730–1830: Kilcoleman Abbey, Milltown. In *Kerry. History and Society. Interdisciplinary Essays on the History of an Irish County*; Bric, M.J., Nolan, W., Eds.; Geography Publications: Dublin, Ireland, 2020; pp. 277–298.
62. Smith, C. *The Ancient and Present State of the County of Kerry. Containing a Natural, Civil, Ecclesiastical, Historical and Topographical Description Thereof*; The Author: Dublin, Ireland, 1774; p. 268.
63. Watts, W.A. Contemporary accounts of the Killarney woods 1580–1870. *Ir. Geogr.* **1984**, *17*, 1–13. [CrossRef]
64. Peterken, G.F.; Lloyd, P.S. Biological flora of British Isles. *Ilex aquifolium L. J. Ecol.* **1967**, *55*, 841–858.
65. Anon. *County Kerry: A Socio-Economic Profile*; Kerry County Council: Tralee, Ireland, 2022; p. 162.
66. Behre, K.-E. The interpretation of anthropogenic indicators in pollen diagrams. *Pollen Spores* **1981**, *23*, 225–245.
67. Sealy, J.R.; Webb, D.A. *Arbutus unedo* L. [Biological flora of the British Isles]. *J. Ecol.* **1950**, *38*, 223–236. [CrossRef]
68. Mitchell, F.J.G. The vegetational history of the Killarney oakwoods, SW Ireland: Evidence from fine spatial resolution pollen analysis. *J. Ecol.* **1988**, *76*, 415–436. [CrossRef]

69. Mitchell, F.J.G. The biogeographical implications of the distribution and history of the strawberry tree, *Arbutus unedo*, in Ireland. In *Biogeography of Ireland: Past, Present and Future*; Costello, M.J., Kelly, K.S., Eds.; Irish Biogeographical Society (Occasional Publication No. 2): Dublin, Ireland, 1993; pp. 35–44.
70. Uí Chonchubhair, M. *Crainn Agus Toir [Trees and Shrubs]*; Dingle Publishing: Dingle, Co. Kerry, Ireland, 2024; p. 764.
71. Kelly, F. The old Irish tree-list. *Celtica* **1976**, *11*, 107–124.
72. Sheehy Skeffington, M.; Scott, N. Were the five rare heathers of the west of Ireland introduced through human activity? An ecological, genetic, biogeographical and historical assessment. *Brit. Ir. Bot.* **2023**, *5*, 221–251. [CrossRef]
73. Ó Cinnéide, S. Logainmneacha Duibhneacha [Dingle placenames]. *Feasta* **1983**, *38*, 13–14.
74. Charman, D.J. Centennial climate variability in the British Isles during the mid–late Holocene. *Quat. Sci. Rev.* **2010**, *29*, 1539–1554. [CrossRef]
75. McCormick, M.; Büntgen, U.; Cane, M.A.; Cook, E.R.; Harper, K.; Huybers, P.; Litt, T.; Manning, S.W.; Mayewski, P.A.; More, A.F.M.; et al. Climate change during and after the Roman Empire: Reconstructing the past from scientific and historical evidence. *J. Interdisc. Hist.* **2012**, *43*, 169–220. [CrossRef]
76. O'Connell, M.; Molloy, K.; Jennings, E. Long-term human impact and environmental change in mid-western Ireland, with particular reference to Céide Fields—An overview. *E&G Quat. Sci. J.* **2020**, *70*, 1–32.
77. Cross, J.R. Unusual stands of birch on bogs. *Ir. Nat. J.* **1987**, *22*, 305–310.
78. Connolly, M. (Ed.) Past Kingdoms. Recent Archaeological Research, Survey and Excavation in County Kerry. In *Proceedings of the 2005 Archaeological Lecture Series*; Kerry County Council: Tralee, Ireland, 2005; p. 96.
79. Kelleher, C.T. Phylogeographical studies indicate multiple origins of Ireland's tree flora. In *Origins of Ireland's Biodiversity*; Coughlan, N.E., Sleeman, D.P., Monaghan, N.T., Eds.; Irish Naturalist's Journal: Belfast, UK, 2024; pp. 17–25.
80. Yeloff, D.; Van Geel, B. Abandonment of farmland and vegetation succession following the Eurasian plague pandemic of AD 1347-52. *J. Biogeog.* **2007**, *34*, 575–582. [CrossRef]
81. Read, H.; Bengtsson, V. The management of trees in the wood pasture systems of south east England. In *Silvicultures. Management and Conservation*; Allende Álvarez, F., Gomez-Mediavilla, G., López-Estébanez, N., Eds.; IntechOpen: Rijeka, Croatia, 2019; Ch. 1; 25p.
82. Rackham, O. *Woodlands (New Naturalist, No. 100)*; Collins (HarperCollins): London, UK, 2006; p. 609.
83. O'Connell, M.; Molloy, K. Aran Islands, western Ireland: Farming history and environmental change reconstructed from field surveys, historical sources, and pollen analyses. *J. N. Atl.* **2019**, *38*, 1–27. [CrossRef]
84. Zemanek, A.; Zemanek, B.; Harmata, K.; Madeja, J.; Klepacki, P. Selected foreign plants in old Polish botanical literature, customs and art (*Acorus calamus*, *Aesculus hippocastanum*, *Cannabis sativa*, *Fagopyrum*, *Helianthus anuus*, *Iris*). In *Plants and Culture: Seeds of the Cultural Heritage of Europe*; Morel, J.-P., Mercuri, A.M., Eds.; EDIPUBLIA: Santo Spirito, Bari, Italy, 2009; pp. 179–193.
85. Feeser, I.; Dörfler, W.; Rösch, M.; Jahns, S.; Wolters, S.; Bittmann, F. (Eds.) *Vegetationsgeschichte der Landschaften in Deutschland*; Springer: Berlin/Heidelberg, Germany, 2025; p. 670.
86. O'Connell, M. Pollenanalytische Untersuchungen zur Vegetations- und Siedlungsgeschichte aus dem Lengener Moor, Friesland (Niedersachsen). *Prob. Küstenforsch.* **1986**, *16*, 171–193.
87. Wolters, S.; Bittmann, F.; Dörfler, W.; Behre, K.-E. Küstennahe Geestgebiete. In *Vegetationsgeschichte der Landschaften in Deutschland*; Feeser, I., Dörfler, W., Rösch, M., Jahns, S., Wolters, S., Bittmann, F., Eds.; Springer: Berlin/Heidelberg, Germany, 2025; pp. 477–489.
88. Behre, K.-E.; Bittmann, F. Moorbrand- und andere Moorkulturen. In *Vegetationsgeschichte der Landschaften in Deutschland*; Feeser, I., Dörfler, W., Rösch, M., Jahns, S., Wolters, S., Bittmann, F., Eds.; Springer: Berlin/Heidelberg, Germany, 2025; pp. 155–157.
89. Greig, J.R.A. The British Isles. In *Progress in Old World Palaeoethnobotany*; van Zeist, W., Wasylikowa, K., Behre, K.-E., Eds.; Balkema: Rotterdam, The Netherlands, 1991; pp. 299–333.
90. Dark, P. *The Environment of Britain in the First Millennium AD*; Duckworth: London, UK, 2000; p. 211.
91. OED. buckwheat, n. Oxford English Dictionary 2025. Available online: https://www.oed.com/dictionary/buckwheat_n?tl=true (accessed on 15 June 2025).
92. Kelly, F. *Early Irish Farming (Early Irish Law Series, Vol. IV)*; Dublin Institute of Advanced Studies (School of Celtic Studies): Dublin, Ireland, 1997; p. 751.
93. Lucas, A.T. Irish food before the potato. *Gwerin* **1960**, *3*, 8–43. [CrossRef]
94. Mac Con Iomaire, M.; Cashman, D. (Eds.) *Irish Food History. A Companion*; Royal Irish Academy: Dublin, Ireland, 2024; p. 800.
95. Britten, J.; Holland, R. *A Dictionary of English Plant-Names*; Trübner & Co. (for the English Dialect Society): London, UK, 1886.
96. Smith, A. (Ed.) *A Brief Description of Ireland: 1590 [Reprint of "A Brife Description of Ireland: Made in This Yeere 1589 by Robert Payne", Printed by Thomas Dawson, London, 1590]*; Irish Archaeological Society: Dublin, Ireland, 1841.
97. McNeill, C. (Ed.) Rawlinson Manuscripts (Class A). *Analect. Hib.* **1930**, *1*, 12–117.
98. Bog Commissioners. *The Fourth Report of the Commissioners Appointed to Inquire into Nature and Extent of the Several Bogs in Ireland and the Practicability of Draining and Cultivating Them*; House of Commons: London, UK, 1814; pp. 1–17 (report), 19–131 (appendices) and I–XVIII (plates).

99. Mackay, J.T. *Flora Hibernica*; Curry: Dublin, Ireland, 1836.
100. Mitchell, F. *The Shell Guide to Reading the Irish Landscape*, 2nd ed.; Country House: Dublin, Ireland, 1986; p. 228.
101. von Pückler-Muskau, H. *Briefe Eines Verstorbenen. Ein Fragmentarisches Tagebuch aus England, Wales, Irland, und Frankreich, Geschrieben in den Jahren 1828 und 1829*, 1st ed.; Franckh: München, Germany, 1830; Volume 1, p. 324.
102. Parshall, L.B. *Letters of a Dead Man [by] Prince Hermann von Pückler-Muskau*; Dumbarton Oaks Research Library and Collection: Washington, DC, USA, 2016; p. 753.
103. Bourke, E. *"Poor Green Erin" German Travel Writers' Narratives on Ireland from Before the 1798 Rising to After the Great Famine*, 2nd ed.; Peter Lang (Internationaler Verlag der Wissenschaften): Frankfurt am Main, Germany, 2013; p. 773.
104. Blume, H. Some geographical aspects of the Palatine settlement in Ireland. *Ir. Geogr.* **1952**, *2*, 172–179. [CrossRef]
105. Feehan, J. *Farming in Ireland. History, Heritage and Environment*; UCD, Faculty of Agriculture: Dublin, Ireland, 2003; p. 606.
106. Young, A. A Tour in Ireland: With General Observations on the Present State of That Kingdom. H. Goldney: Dublin, Ireland, 1780; Volume I and II.
107. Dörfler, W. Die Geschichte des Hanfanbaus in Mitteleuropa aufgrund palynologischer Untersuchungen und von Grosrestnachweisen. *Prähist. Z.* **1990**, *65*, 218–244. [CrossRef]
108. Gibson, K. Hemp in the British Isles. *J. Ind. Hemp* **2006**, *11*, 57–67. [CrossRef]
109. Overland, A.; O'Connell, M. Fine-spatial paleoecological investigations towards reconstructing late Holocene environmental change, landscape evolution and farming activity in Barrees, Beara Peninsula, southwestern Ireland. *J. N. Atl.* **2008**, *1*, 37–73. [CrossRef]
110. de Klerk, P.; Couwenberg, J.; Joosten, H. Pollen and macrofossils attributable to *Fagopyrum* in western Eurasia prior to the Late Medieval: An intercontinental mystery. *Palaeogeogr. Palaeoclimatol. Palaeoecol.* **2015**, *440*, 1–21. [CrossRef]
111. Cross, J.R. The establishment of *Rhododendron ponticum* in the Killarney oakwoods, S.W. Ireland. *J. Ecol.* **1981**, *69*, 807–824. [CrossRef]
112. Doyle, G.J. Phytosociology of Atlantic bog complexes in north-west Mayo. In *Ecology and Conservation of Irish Peatlands*; Doyle, G.J., Ed.; Royal Irish Academy: Dublin, Ireland, 1990; pp. 75–90.
113. Cross, J.R. The potential natural vegetation of Ireland. *Biol. Environ. (Proc. R. Ir. Acad. Ser. B)* **2006**, *106B*, 65–116. [CrossRef]
114. O'Connell, M.; Jennings, E.; Molloy, K. Holocene vegetation dynamics, landscape change and human impact in western Ireland as revealed by multidisciplinary, palaeoecological investigations of peat deposits and bog-pine in lowland Connemara. *Geographies* **2021**, *1*, 251–291. [CrossRef]
115. Coyne, W.P. (Ed.) *Ireland. Industrial and Agricultural*; Browne and Nolan: Dublin, Ireland, 1902; p. 532.
116. Pilz, A. Narratives of arboreal landscapes. In *A History of Irish Literature and the Environment*; Sen, M., Ed.; Cambridge University Press: Cambridge, UK, 2022; pp. 97–114.
117. Cole, E.E.; Mitchell, F.J.G. Human impact on the Irish landscape during the late Holocene inferred from palynological studies at three peatland sites. *Holocene* **2003**, *13*, 507–515. [CrossRef]
118. Everett, N. *The Woodlands of Ireland. A History, 700–1800*; Four Courts Press: Dublin, Ireland, 2014.
119. Hall, V.A. Vegetation history of mid- to western Ireland in the 2nd millennium A.D.; fresh evidence from tephra-dated palynological investigations. *Veget. Hist. Archaeobot.* **2003**, *12*, 7–17. [CrossRef]
120. Ghilardi, B.; O'Connell, M. Early Holocene vegetation and climate dynamics with particular reference to the 8.2 ka event: Pollen and macrofossil evidence from a small lake in western Ireland. *Veget. Hist. Archaeobot.* **2013**, *22*, 99–114. [CrossRef]
121. Plunkett, G.; Brown, D.M.; Swindles, G.T. *Siccitas magna ultra modum*: Examining the occurrence and societal impact of droughts in prehistoric Ireland. *Proc. R. Ir. Acad.* **2020**, *120C*, 83–104. [CrossRef]
122. Bunting, M.J.; Gaillard, M.J.; Sugita, S.; Middleton, R.; Brostrom, A. Vegetation structure and pollen source area. *Holocene* **2004**, *14*, 651–660. [CrossRef]
123. Kearney, K. Vegetation impacts and early Neolithic monumentality: A palaeoenvironmental case study from south-west Ireland. *J. Archaeol. Sci. Rep.* **2019**, *27*, 101940. [CrossRef]
124. Woodman, P.C. *Ireland's First Settlers. Time and the Mesolithic*; Oxbow Books: Oxford, UK, 2015; p. 366.
125. Durell, P.; Kelly, C. (Eds.) *The Grand Tour of Kerry*; Cailleach Books: Allihies, Co. Cork, Ireland, 2001; p. 230.
126. Chart, D.A. Two centuries of Irish agriculture. A statistical retrospect, 1672–1905. *J. Stat. Soc. Inq. Soc. Irel.* **1908**, *12*, 162–174.
127. CSO. *Farming Since the Famine. Irish Farm Statistics 1847–1996*; Central Statistics Office: Dublin, Ireland, 1997; p. 43.
128. de Bromhead, A.; Fernihough, A.; Ó Gráda, C. 175 years of mismeasuring Ireland?: Some traps set by historical data. *J. Stat. Soc. Inq. Soc. Irel.* **2023**, *51*, 128–155.
129. Foley, K. The Famine in the Dingle Peninsula. In *Atlas of the Great Irish Famine*; Crowley, J., Smyth, W., Murphy, M., Crowley, J., Smyth, W., Murphy, M., Eds.; Cork University Press: Cork, Ireland, 2012; pp. 398–405.
130. Mitchell, F.J.G.; Cooney, T. Vegetation history in the Killarney valley. In *Ross Island. Mining, Metal and Society in Early Ireland (Bronze Age Studies 6)*; O'Brien, W., Ed.; Department of Archaeology, NUI, Galway: Galway, Ireland, 2004; pp. 481–493.

131. Clark, R.L. Point count estimation of charcoal in pollen preparations and thin sections of sediments. *Pollen Spores* **1982**, *24*, 523–535.
132. Morbelli, M.A.; Piñeiro, M.R.; Giudice, G.E. Spore morphology and wall ultrastructure of Hymenophyllaceae Link (Pteridophyta) from north-west Argentina. *Grana* **2010**, *49*, 37–46. [CrossRef]
133. Page, C.N. *The Ferns of Britain and Ireland*, 2nd ed.; Cambridge University Press: Cambridge, UK, 1997; p. 540.

MDPI AG
Grosspeteranlage 5
4052 Basel
Switzerland
Tel.: +41 61 683 77 34

Diversity Editorial Office
E-mail: diversity@mdpi.com
www.mdpi.com/journal/diversity

www.ingramcontent.com/pod-product-compliance
Lightning Source LLC
Chambersburg PA
CBHW040817250926
51850CB00007BA/1189
9783725851232